THE ORIGINS OF MODERN BIOCHEMISTRY:
A Retrospect on Proteins

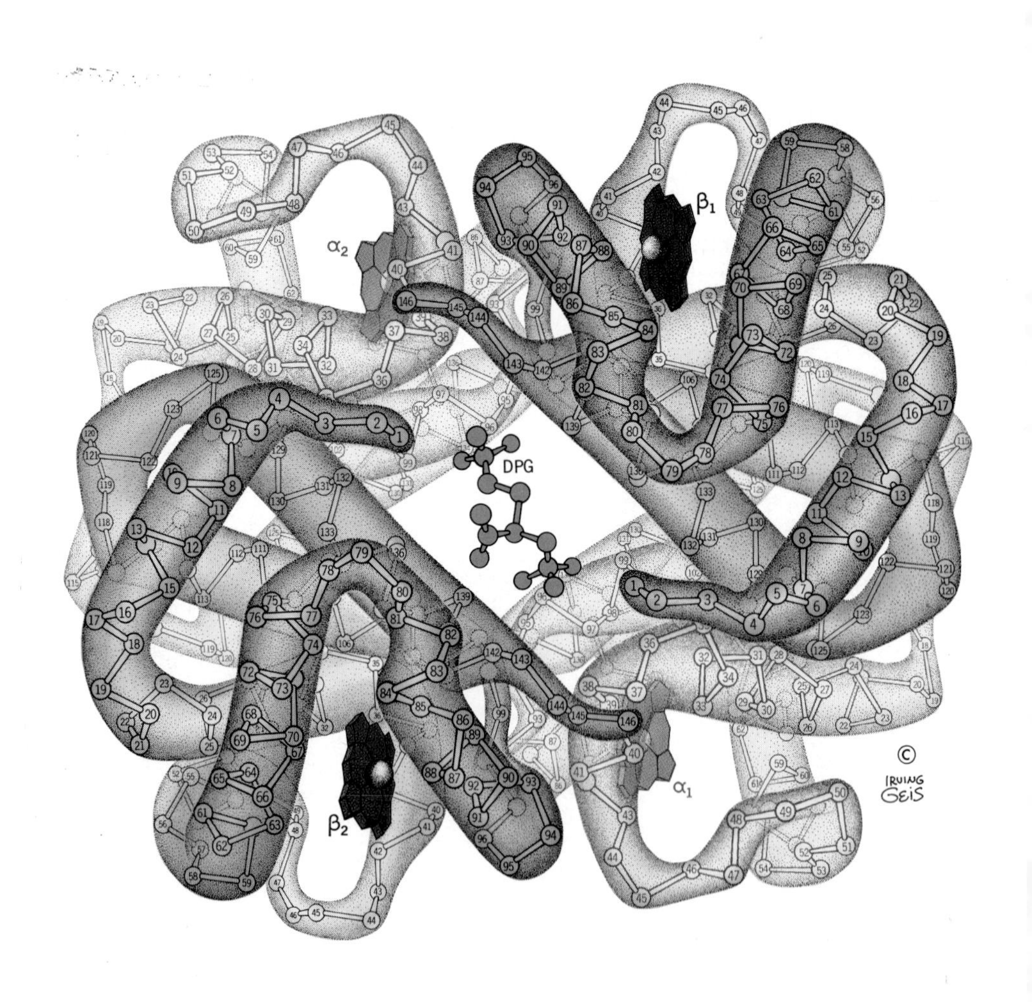

Alpha-carbon map of horse deoxyhemoglobin with bound DPG, from the hemoglobin coordinates of M. F. Perutz and G. Fermi, 1973. Illustration by Irving Geis.

ANNALS OF THE NEW YORK ACADEMY OF SCIENCES
VOLUME 325

THE ORIGINS OF MODERN BIOCHEMISTRY:
A Retrospect on Proteins

Edited by P. R. SRINIVASAN, JOSEPH S. FRUTON, AND JOHN T. EDSALL

In Memory
of
Dr. Turner Alfrey, Jr.

THE NEW YORK ACADEMY OF SCIENCES
NEW YORK, NEW YORK
1979

ACKNOWLEDGMENTS

Book design by Dick Lynch

Cover and frontispiece illustrations by Irving Geis, © 1979 by Irving Geis. The frontispiece illustration will appear in Proteins: Structure, Function, and Evolution, *by Richard E. Dickerson and Irving Geis, second edition, which will be published by Benjamin/Cummings, Menlo Park, California. It is printed here by the kind permission of the authors.*

Cover illustration: α-helix.

LIBRARY OF CONGRESS CATALOGING IN PUBLICATION DATA

Symposium on the Origins of Modern Biochemistry: a Retrospect on Proteins, New York Academy of Sciences, 1978.
The origins of modern biochemistry.
(Annals of the New York Academy of Sciences; v. 325)
1. Protein–Congresses. 2. Amino acids–Congresses. 3. Biological chemistry –History–Congresses.
I. Srinivasan, Parithychery R., 1927– II. Fruton, Joseph Stewart, 1912– III. Edsall, John Tileston, 1902– IV. Title. V. Series: New York Academy of Sciences. Annals; v. 325. Q11.N5 vol.325 [QP551] 508'.ls [574.1'9245] 79-12584

CCP

Printed in the United States of America

ISBN 0-89766-018-8

ANNALS OF THE NEW YORK ACADEMY OF SCIENCES

VOLUME 325

MAY 31, 1979

THE ORIGINS OF MODERN BIOCHEMISTRY: A RETROSPECT ON PROTEINS*

Editors
P. R. SRINIVASAN, JOSEPH S. FRUTON, AND JOHN T. EDSALL

Conference Chairmen
P. R. SRINIVASAN AND JOSEPH S. FRUTON

CONTENTS

* This volume is the result of a conference entitled *Symposium on the Origins of Modern Biochemistry: A Retrospect on Proteins,* held by The New York Academy of Sciences on May 31, June 1, and 2, 1978.

Financial assistance was received from:

- BECHMAN INSTRUMENTS, INC.
- CALBIOCHEM-BEHRING CORP.
- CIBA-GEIGY CORPORATION
- EXXON CORPORATION
- W. R. GRACE & CO.
- HOFFMANN-LA ROCHE INC.
- MERCK SHARP & DOHME RESEARCH LABORATORIES
- PHARMACIA FINE CHEMICALS
- SANDOZ, INC.
- SCHERING CORPORATION
- SEARLE LABORATORIES
- STAUFFER CHEMICAL COMPANY
- STUART PHARMACEUTICALS, DIVISION OF ICI AMERICAS INC.
- SYNTEX (U.S.A.), INC.
- USV PHARMACEUTICAL CORPORATION

J. W. Williams, John T. Edsall, Frederic L. Holmes
Felix Haurowitz, Emil L. Smith, Howard K. Schachman
P. R. Srinivasan, Philip Siekevitz, Heinz Fraenkel-Conrat

Dorothy Crowfoot Hodgkin, Joseph S. Fruton, A. H. Gordon
Charlotte Friend, Rollin D. Hotchkiss, N. H. Horowitz
David Shemin, Paul C. Zamecnik, Erwin Chargaff

SARAH RATNER, Z. DISCHE
F. LIPMANN, DAVID NACHMANSOHN, P. KARLSON
D. SPRINSON, HENRY GUERLAC

Introduction

P. R. SRINIVASAN
Department of Biochemistry
College of Physicians & Surgeons
Columbia University
New York, New York 10032

It was the best of times, it was the worst of times, it was the age of wisdom, it was the age of foolishness, it was the epoch of belief, it was the epoch of incredulity, it was the season of Light, it was the season of Darkness, it was the spring of hope, it was the winter of despair... .

CHARLES DICKENS
A Tale of Two Cities

THESE LINES were written in 1859 and they still ring true today. The present century, with two decades still ahead of us, has been remarkable in many ways. Many of us watched with wonder and awe at the landing of Man on the moon. Even in terms of human relationships, a greater awareness of the rights of individuals has taken place. I have not forgotten the darkness and despair. Let me stress the creative side of life this morning.

The present century will also be noted for its scientific advances. It has seen great developments in the field of physics: Einstein's theory of relativity, the elucidation of the structure of the atom, nuclear fusion and explosion in the field of particle physics, and so forth. The tools that grew out of these major advances—ultraviolet spectroscopy, infrared spectroscopy, electron spin resonance, nuclear magnetic resonance, x-ray crystallography, and mass spectrometry—have provided the means by which chemists could embark on the analyses and syntheses of complex structures and an understanding of chemical reactions on a more rational level. Needless to say, such marked achievements have influenced biology and resulted in the emergence of biochemistry as a discipline unto itself. Since the term biochemistry was first used by Hoppe-Seyler in 1877 we have seen great strides in this new discipline.

Modern biochemistry is primarily dynamic biochemistry. This comprises the various aspects of metabolism—synthesis, degradation, and conversion—and a study of the enzymes that catalyze these reactions. Since these unceasing changes take place in an orderly manner, regulation has been yet another area of investigation. Dynamic biochemistry is also concerned with all those chemical processes that take place on the structural elements, and a new field is slowly emerging: supramolecular biochemistry. Thus, one can say with enthusiasm that biochemistry is no longer a sideline of science but one of its major advancing frontiers.

A retrospect is therefore appropriate and timely. The idea for this conference originated two years ago at the Steering Committee of the Biochemistry Section of The New York Academy of Sciences. I undertook the task of looking into its prospects. I still recall vividly the faces of those present at that meeting: Drs. Mycek, Petrak, Sheppard, and P. Feigelson.

A few months later, I made two pilgrimages, one to New Haven to see Professor Fruton and another to Boston to see Professor Edsall. I was received with warmth and candor and out of these discussions the conference as you see here started taking its form and shape. I do want to thank Drs. Fruton and Edsall for their advice, timely help, and above all a willingness to listen to my problems when they arose.

Without minimizing the importance of other macromolecular substances of the cell it could be said that proteins do occupy a central part of biochemistry. The multiple nature of its functions is fascinating. Proteins could be enzymes or catalyze reactions. They could perform storage function, transport function. They are involved in contractile mechanisms. As antibodies they participate in protective functions. Some toxins are proteins. Hormones, regulators of metabolism, could also be proteins. Finally, proteins also play a structural role: keratin in skin and feathers; collagen in connective tissue and bone. More recently the discovery of the antifreeze protein in some fishes living in antartic waters is worthy of our attention. Whatever may be the logic, we are gathered here today to hear, learn, and discuss the topic, "A Retrospect on Proteins."

A symposium of this magnitude cannot be organized without the help of many persons. Permit me then to turn to a pleasant and important task. We convey our deep debt of gratitude to the outstanding speakers on the program from here and abroad and also to the illustrious scientists and historians of New York and its vicinity for kindly accepting our invitation to participate. We also appreciate the financial support which has been most generous but for which we could not have organized this symposium. Last, but not least, I want to express my thanks to the administrative staff of the Department of Biochemistry, Columbia University, and in particular to Stella Franco, Janette Loughlin, and Ruth DeBrunner for their devoted assistance. Since I am on sabbatical leave at the Department of Medical Genetics at the University of Toronto, I had to rely also on the help of that Department and I am grateful to M.

McConnell for her contribution. I will be failing in my duty if I forget to mention the cooperation of the staff of The New York Academy of Sciences. Let me mention the names of Ellen Marks, Renée Wilkerson, Ann Collins, Bill Boland, Richard Malloy, and India Trinley. Drs. Siekevitz and Frederick, members of the Conference Committee, have been wonderful to work with.

JOSEPH S. FRUTON took his Ph.D. at Columbia in 1934, and from 1934 to 1945 worked in Max Bergmann's group at the Rockefeller Institute in laying the basis for the understanding of the specificity of proteinases. He has been at Yale University since 1945 where he is presently the Eugene Higgins Professor of Biochemistry.

He has published over 200 articles on the chemistry of proteins, peptides, and amino acids and on the specificity and mechanisms of the action of enzymes. He and his wife, Sofia Simmonds, wrote a distinguished textbook on general biochemistry. More recently, he has written the book, *Molecules and Life,* a history of biochemistry.

His other contributions to biochemistry include service on a number of editorial boards, and on committees of the National Research Council. He has received the Eli Lilly Award and the Pfizer Award. Also, he is a member of the National Academy of Sciences, the American Philosophical Society, and the American Academy of Arts and Sciences.

Early Theories of Protein Structure

JOSEPH S. FRUTON
Kline Biology Tower
Yale University
New Haven, Connecticut 06520

THE MAIN PURPOSE of this paper is to provide some of the historical background for what others will say at this symposium about the origins of the present-day view of the primary structure of proteins. We all know that the decisive event in this development was the determination of the amino acid sequence of insulin by Frederick Sanger and his associates. It is appropriate to recall, therefore, that in a review published in 1952, Sanger began his discussion of the arrangement of amino acids in proteins as follows:

> As an initial working hypothesis it will be assumed that the peptide theory is valid, in other words, that a protein molecule is built up only of chains of α-amino (and α-amino) acids bound together by peptide bonds between their α-amino and α-carboxyl groups. While this peptide theory is almost certainly valid . . . it should be remembered that it is still a hypothesis and has not been definitely proved. Probably the best evidence in support of it is that since its enunciation in 1902 no facts have been found to contradict it.[1]

If the peptide theory was indeed only a hypothesis fifty years after Franz Hofmeister[2] and Emil Fischer[3] advanced it, there must have been reasons for doubting its validity and, as we shall see, it was repeatedly questioned during those years. I shall sketch some of the discussion of the problem, especially during the 1930s when I was involved in it, but before doing so, I wish to summarize the development of ideas about the structure of proteins during the latter half of the nineteenth century, before Fischer entered the field around 1900. In preparing this historical account, I have drawn heavily on my earlier writing on the subject.[4] I have also used the valuable history of protein chemistry recently prepared by Shamin.[5]

Throughout the nineteenth century, the substances we now call proteins occupied the center of the stage among the chemical materials obtained from biological sources. Their importance in animal nutrition was recognized by the second decade of the century, through the work of François Magendie, and by 1840 Gerardus Mulder had applied the new methods of elemental analysis to the so-called albuminoid substances,

0077-8923/79/0325-0001 $01.75/0

among them egg albumin, milk casein, and blood fibrin. By the 1860s, with the widespread acceptance of the protoplasmic theory, albumin acquired for some biologists the status of "the physical basis of life," and Thomas Graham had included it among the colloids. He defined the colloidal state as "a dynamical state of matter, the crystalloidal being the static condition. The colloid possesses ENERGIA. It may be looked upon as the probably primary source of the force appearing in the phenomena of vitality."[6]

Indeed, in the decades that followed, the energy-rich character of protoplasmic proteins was emphasized repeatedly. Thus, in 1875 Eduard Pflüger proposed that intracellular oxidation is effected by a labile energy-rich protoplasmic protein that contains cyano groups, which combine explosively with molecular oxygen to liberate CO_2.[7] The popularity of his idea encouraged others to produce variants, as in the case of the hypothesis advanced by Oscar Loew[8] in 1881 that the chemical group responsible for the dynamic properties of protoplasmic protein is the aldehyde group. A common feature of these hypotheses was the assumption that upon the death of the cell, the energy-rich living proteins are converted to dead proteins, which represent the well-known materials isolated from biological sources, and which were recognized to belong to the class of organic substances classified as amides. The ready acceptance of these ideas is indicated by their recurrence in biological textbooks of the time, as for example in Griffiths's treatise on the physiology of invertebrates.[9] Here he cites approvingly the proposal of P. W. Latham[10] that albumin has the "constitutional formula" shown in FIGURE 1. This is a physiologist's view of protein structure, and in Latham's words, albumin "is a compound of cyan-alcohols united to a benzene nucleus, these being derived from the various aldehydes, glycols and ketones, or that they may be formed in the living body by the dehydration of the amino-acids."

With the benefit of hindsight, we can now look at these forgotten and seemingly far-fetched ideas about protein structure with disdain, but it is well to remember that they were intended to explain the biological activity of proteins in terms of the chemical knowledge of the time. These speculations also attempted to define, in chemical terms, the difference between the so-called "living proteins" and the purified "dead" amide-like materials isolated from biological sources and analyzed by chemists like Mulder or Liebig. Thus, for Loew, this difference lay in the presence of aldehyde groups (derived from aspartic acid dialdehyde) in biologically active proteins, and in molecular rearrangements to produce amides upon death[8] (FIGURE 2). Although such ideas were roundly criticized by leading physiological chemists, notably Felix Hoppe-Seyler[11]

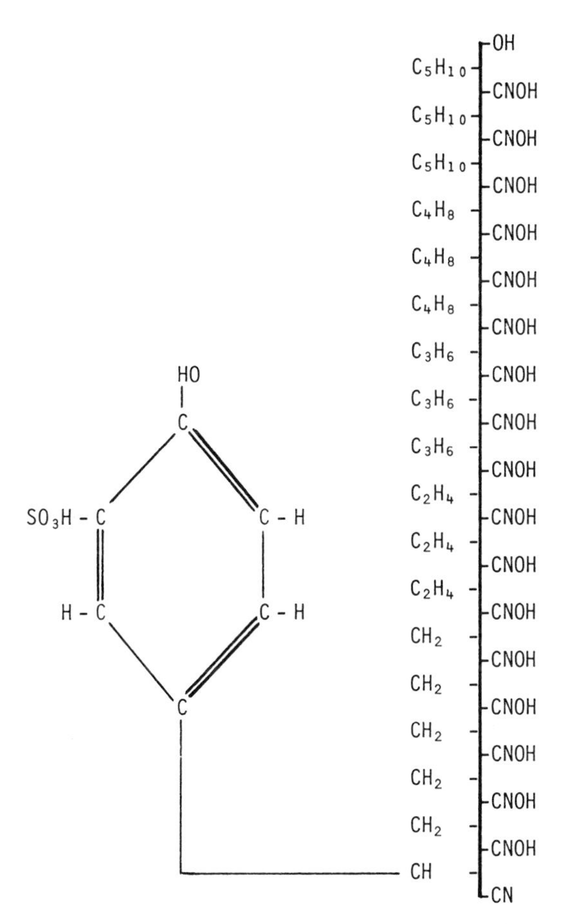

FIGURE 1. Latham's "constitutional formula" for protein. (After Latham.[10])

and Eugen Baumann,[12] they continued to have appeal for many biologists well into the first decade of the twentieth century.[13] Indeed, in 1909, Eduard Pflüger[14] felt it necessary to call attention to the limitations of Emil Fischer's contributions to protein chemistry, on the ground that the synthesis of polypeptides gave no information about the structure of biologically-active protens. As we shall see shortly, similar criticisms of the polypeptide theory of protein structure continued to be raised well into the 1930s.

During the last quarter of the nineteenth century, ideas about protein structure were also influenced by the results of studies on the nature of the products that arose from the action of pepsin and trypsin on proteins. Willy Kühne, together with his associates Russell Chittenden and Richard

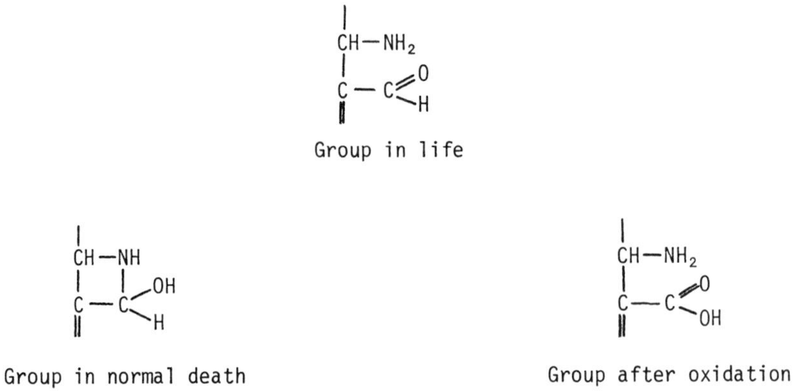

```
|
CH—NH2
|
C — C=O
‖    \H
```

Group in life

```
|
CH—NH
|    |  OH
C — C<
‖     H
```

Group in normal death

```
|
CH—NH2
|
C — C=O
‖    \OH
```

Group after oxidation

FIGURE 2. The aldehyde group in "living" protein. (After Loew and Bokorny.[8])

Neumeister, produced a complicated scheme in which various kinds of albumoses and peptones were thought to be successively produced. Again, this was a physiologist's view of protein structure because, before 1900, peptones were considered to represent the form in which the products of the digestion of dietary proteins were transported across the intestinal wall. This idea was developed most extensively by Franz Hofmeister, and was not laid to rest until after Otto Cohnheim[15] had shown in 1901 that the intestinal mucosa elaborates an enzyme he called erepsin, and which cleaves peptones to amino acids.

Indeed, one of the striking features of the nineteenth-century speculations about protein structure is their lack of emphasis on amino acids as structural units of proteins. Many amino acids had been isolated from natural sources by 1897, but only three—tyrosine, leucine, and aspartic acid—were considered by Neumeister[16] to be regular products of the decomposition of proteins.

Within only a few years, the situation had changed completely, largely as a consequence of the entry of Emil Fischer into the protein field. During the 1880s, Theodor Curtius had shown that amino acid esters could be distilled under reduced pressure without decomposition, and Fischer applied this to the separation of amino acids present in an acid hydrolysate of a protein. By 1906, about 15 amino acids were recognized to be protein constituents. Also, Fischer had isolated well-defined dipeptides (e.g., glycyl-alanine) from a partial acid hydrolysate of a protein. Most important of all was his systematic attack on the problem of synthesizing long-chain polypeptides.

In emphasizing the decisive role of Fischer, I do not wish to denigrate the importance in our story of two of his contemporaries—Franz Hofmeister and Albrecht Kossel. In 1902, Hofmeister concluded that the recurring unit of proteins is a structure he wrote as –CO–NH–CH=, largely on the basis of the facts that amino-nitrogen is released and the biuret reaction is lost upon protein hydrolysis.[2] Although he is usually credited with having proposed the peptide theory of protein structure, he did not use the word "peptide" in his famous 1902 review article; the word was introduced by Fischer. Kossel's contribution to the problem of protein structure arose from his work on the protamines, which appeared to have a simpler constitution than the well-known proteins, because upon hydrolysis they yielded large proportions of only a few amino acids, with a preponderance of arginine. It was perhaps Kossel who first emphasized, in 1900, the necessity for the quantitative accounting of a protein in terms of the amino acids produced on hydrolysis,[17] and initiated an effort that was to occupy protein chemists for several decades before the introduction of chromatographic methods by Martin and Synge and their refinement by Moore and Stein. Kossel followed the tradition of his time, and suggested that in more complex proteins, various amino acids are attached to a central nucleus similar to the protamines.[18] Although this idea reappeared a few years later in the so-called "kyrine" hypothesis of Max Siegfried, it was overshadowed by the peptide hypothesis and the massive evidence that Fischer was presenting in its favor. In relation to Kossel's contributions to protein chemistry, it might also be added that according to one of his biographers,[19] Fischer was led to work on proteins upon the urging of Kossel.

In a lecture he gave in 1906, Fischer[20] summarized his work on proteins:

> Six years ago, when I decided to devote myself to the study of the proteins, I began with the amino acids, in order to gain new viewpoints and methods for their complex derivatives. The success has not disappointed me. First it was possible to find a new separation method for the monoamino acids through the use of esters. . . . More significant seem to me to be the similarly found methods for the conversion of the amino acids into their amide-like anhydrides, for which I have chosen the collective name "polypeptides." The higher members of this group of synthetic materials are so similar to the natural peptones in their external properties, certain color reactions, and behavior toward acids, alkalis and ferments that one must consider them to be very close relatives, and that I may describe their preparation as the beginning of the synthesis of the natural peptones and albumoses.

By 1907, when he had reached the stage of the 14-amino acid compound composed of glycine and leucine units, and found that it gave a strong biuret reaction and was precipitated by protein reagents, he voiced the belief that at the 20-amino-acid stage he will have entered into the protein group. Later in the same year, however, he decided to stop at the octadecapeptide (molecular weight, 1213), and he wrote as follows:

> If one imagines the replacement of the many glycine residues by other amino acids, such as phenylalanine, tyrosine, cystine, glutamic acid, etc. one would soon attain 2–3 times the molecular weight, hence values assumed for some natural proteins. For other natural proteins the estimates are much higher, up to 12,000–15,000. But in my opinion these numbers are based on very insecure assumptions, since we do not have the slightest guarantee that the natural proteins are homogeneous substances.
>
> From the experience thus far, I do not doubt that the synthesis can be continued by means of the same methods beyond the octadecapeptide. I must however provisionally waive such experiments, which are not only very laborious but also very expensive.[21]

With the benefit of hindsight, two comments may be made about this quotation. First, the synthetic methods available to Fischer did not allow the replacement of glycine residues by most of the amino acids he listed. This possibility was only open after 1932, when Fischer's former associate Max Bergmann, together with Leonidas Zervas, invented the so-called carbobenzoxy method for peptide synthesis.[22] The elaboration of this method revolutionized protein chemistry during the past half-century,[23] and although problems still remain, any complex polypeptide can now be synthesized.[24] The other comment relates to Fischer's apparent conviction that the molecular weight of proteins was likely to be below 10,000. It reflects the reluctance of the organic chemists of his time to accept the possibility that large molecules could be stable structures. This prejudice was to play a large role during the 1920s in the negative response to Hermann Staudinger's insistence on macromolecules held together by covalent bonds.

If Fischer, with Hofmeister, is usually credited for having formulated clearly the peptide theory of protein structure, it must also be stressed that it was Fischer who provided the authority for questioning the idea that proteins are long-chain polypeptides. In 1906, he stated that

> simple amide formation is not the only possible mode of linkage in the protein molecule. On the contrary, I consider it to be quite probable that on the one hand it contains piperazine rings, whose facile cleavage by alkali and reformation from the dipeptides or their esters I have observed so frequently with the artificial products, and on the other hand the numerous hydroxyls of the oxyamino acids are by no means inert groups in the protein molecule. The latter could be transformed by anhydride

> formation to ester or ether groups, and the variety would increase further if poly-oxyamino acids are assumed to be probable protein constituents.[20]

The reference to poly-oxyamino acids arose from his colleague Emil Abderhalden's work that suggested the existence of a diaminotrioxydodecanic acid as a product of protein hydrolysis, but Fischer withdrew this claim in 1913.

The first part of this quotation, however, was frequently cited during the 1920s in support of the idea that the protein molecule is constructed of diketopiperazines held together by secondary valences. Its chief protagonist was Abderhalden,[25, 26] but the evidence he adduced was not impressive, and in the judgment of Vickery and Osborne, "it does not seem that the enormous labors of Abderhalden and his associates have really furthered his fundamental view of the structure of the protein molecule."[27] This negative opinion was not shared by others. For example, in 1929 Ross Gortner stated that "the evidence which has been presented by Abderhalden for the presence of diketopiperazine rings in proteins amounts almost to proof."[28] Indeed, in the same year (1924) that Abderhalden proposed his theory, Max Bergmann[29] offered a hypothesis that also assumed the association of diketopiperazine-like structures, with the difference that his presumed units were unsaturated diketopiperazine derivatives of the sort he had prepared from dipeptides containing serine or cysteine. At first, Bergmann thought that he had isolated such compounds in the form of double molecules, but his later work[30] showed this conclusion to be erroneous and after 1927 he no longer advocated the theory. The attraction of the idea that diketopiperazines formed structural units of proteins also drew Paul Karrer into the field, and he synthesized an extensive series of labile ring compounds of this sort.

A major factor in the upsurge in popularity of the diketopiperazine structure of proteins, aside from Fischer's authority, was the report by Rudolf Brill[31] in 1923 that silk fibroin gave an x-ray diffraction pattern that was consistent with either a long polypeptide chain consisting of glycine and alanine units, or the association of glycyl-alanine units. The latter alternative was preferred in keeping with the prejudice of the time against the concept of the macromolecule advocated by Staudinger.

Another reason for the scepticism of the 1920's about the polypeptide theory of protein structure was the fact that, in the words of Vickery and Osborne,[27] "the nature of the bonds attacked by pepsin is unknown. . . . pepsin has never been found to have an effect upon any synthetic model substance whether it contained a peptide bond or not." Also, Gortner[28] noted that "Fischer concluded that the chains of the polypeptides were not long enough for pepsin to act upon them, but it seems more probable that pepsin attacks some linkage other than the linkage in the peptide group."

In their memorable review article on the status of the problem of protein structure in 1928, Vickery and Osborne[27] emphasized the importance of studies with proteolytic enzymes. By that time, important progress had been made by Waldschmidt-Leitz in the partial purification of the enzymic components of the so-called erepsin of intestinal mucosa and of Kühne's pancreatic trypsin.[32] By means of the new adsorption techniques developed in Willstätter's laboratory, Waldschmidt-Leitz identified a dipeptidase and aminopolypeptidase, as well as a carboxypolypeptidase, in addition to the proteinase trypsin. He demonstrated the specificity of the peptidases by means of synthetic peptide substrates (Table 1), but these compounds were only those that could be made by the methods developed by Emil Fischer, which were not applicable to most amino acids with reactive side chains. Synthetic peptides were not available for the known proteinases, and the prevailing view around 1930 was that the specificity of the proteolytic enzymes is adapted to the molecular size of their substrates. The proteinases were supposed to act exclusively on substrates of high molecular weight but not on lower polypeptides and dipeptides, while the opposite was assumed for the peptidases. Also, it was widely thought that pepsin preferentially attacks protein cations, trypsin digests proteins anions, while papain acts on protein zwitterions, and that the sole function of these proteinases on proteins was "to disrupt the colloidal aggregation and to loosen the secondary valences that hold together the polymeric complexes."[33]

Table 1

Specific Splitting of Peptides[32]

Peptide	Dipeptidase	Amino-polypeptidase	Carboxypoly-peptidase
Dipeptides			
Glycyl-glycine	+	−	−
Leucyl-glycine	+	−	−
Glycyl-tyrosine	+	−	−
Histidyl-glycine	+	−	−
Phenylalanyl-arginine	+	−	+
Glutaminyl-tyrosine	+	−	+
Polypeptides			
Leucyl-tri-glycine	−	+	−
18-peptide	−	+	−
Leucyl-glycyl-tyrosine	−	+	+
Leucyl-di-glycyl-tyrosine	−	+	+
Leucyl-tri-glycyl-tyrosine	−	−	+

These views reflected the influence of the physical-chemical tradition of colloid chemistry, which had great appeal to many biochemists during the first three decades of this century. As a consequence of the popularity of this tradition, there was confusion not only about the structure of proteins, but also about the nature of enzymes. Enzymic catalysis was thought by many people to be a consequence of adsorption on colloidal surfaces[34] rather than stereochemically specific interaction with a region of a protein, as visualized by Emil Fischer in his famous lock-and-key analogy.[35] Sumner's isolation of urease in the form of a crystalline protein was dismissed as irrelevant to the problem, and even those, like Willstätter, who considered enzymes as discrete chemical substances, denied that they were proteins.[36] It was only after Northrop[37] had presented the massive evidence for the protein nature of the pepsin he had crystallized in 1930 that the tide began to turn.

I now wish to speak briefly of my personal activity during the 1930s in the study of the specificity of the proteinases, and the relation of the results to the problem of protein structure. Let me emphasize that it is impossible for an active participant in a scientific development to be historically objective in assessing his own work, and I am no exception to that rule. What I have to say should be taken as testimony to be checked, as in all historical research, against that of others.

I began work on proteolytic enzymes in 1934, when I joined Max Bergmann at the Rockefeller Institute for Medical Research. He had been Director of the Kaiser-Wilhelm Institute for Leather Research, and was forced to leave Germany after Hitler came to power.[38] Bergmann had been an associate of Emil Fischer, and his research strategy reflected the Fischer tradition. Where Fischer had used the synthetic glucosides he had prepared for the first time to study the action of invertase and emulsin, Bergmann capitalized on the invention of the carbobenzoxy method to embark on the study of the specificity of peptidases. My first assignment in Bergmann's American laboratory was to continue the work begun in Dresden on the peptidases of intestinal mucosa, but as I became familiar with the enzyme literature it was evident that the real challenge lay in the study of the proteinases pepsin, trypsin, and papain. In particular, it seemed that just because no one had yet found a synthetic peptide substrate for pepsin did not appear to be a valid reason for questioning the peptide theory of protein structure. The proper approach was rather to utilize the potential of the new carbobenzoxy method to make hitherto inaccessible peptides for test as substrates for pepsin and other proteinases. Needless to say, Bergmann encouraged me strongly to follow this line. Indeed, it proved to be relatively easy to find synthetic substrates for papain,[39] and by 1936 we had also found them for chymotrypsin[40] and trypsin,[41] which Kunitz, at the Rockefeller Institute in Princeton, had just crystallized. It turned out that trypsin acts preferentially at interior peptide bonds itvolving an arginyl or lysyl residue, and that chymotrypsin prefers a bond involving an aromatic amino acid, such as phenylalanine. It was evident, therefore, that this catalytic selectivity was a reflection of the specific interaction of the side-chain groups of the peptide substrate with the enzyme, along the lines of Fischer's lock-and-key hypothesis. Then in 1938 we reported that crystalline pepsin can hydrolyze simple peptides such as carbobenzoxy-glutamyl-tyrosine,[42] and it seemed that the last objections to the peptide theory of protein structure had been removed (FIGURE 3).

I should add that one of the factors in finding specific synthetic substrates for pepsin and trypsin was my difficulty in accepting the widely-held view that pepsin acts preferentially on positively-charged protein substrates and that trypsin cleaves negatively-charged protein substrates. This idea, which was based on the pH optima of the two enzymes, as well as the concept that protein-protein interaction is largely a matter of electrostatic attraction, was inconsistent with the available data on the amino acid composition of various proteins in relation to the extent to which they were cleaved by these two enzymes. On the assumption that proteins are long polypeptides, I concluded from the examination of such data that pepsin should act preferentially on acidic substrates, especially those containing tyrosine or phenylalanine units, and that trypsin should act best on basic substrates, with lysine or arginine as structural units. The reasoning was naive, and later work required revision of the idea, but the outcome of the experiments that followed from it was gratifying.

During the mid-1930s, however, there was considerable doubt among biochemists about the peptide theory of protein structure, and our claim that the finding of synthetic substrates for trypsin and pepsin had removed one of the important objections to that theory was challenged by several leading investigators. They were ready to grant that denatured proteins are linear polypeptides, but they insisted that, in order to explain the specific biological and physical properties that proteins lost upon denaturation, it was necessary to assume the presence, in native globular proteins, of covalent linkages, other than the peptide bond, which are decisive in holding the structure together. Nor were they ready to accept the idea of Mirsky and Pauling[43] that hydrogen bonding within a peptide chain or between peptide chains was sufficient to explain the structure of globular proteins and the phenomenon of denaturation. The concept of hydrophobic interaction as an important feature of protein structure did not receive wide acceptance until much later.

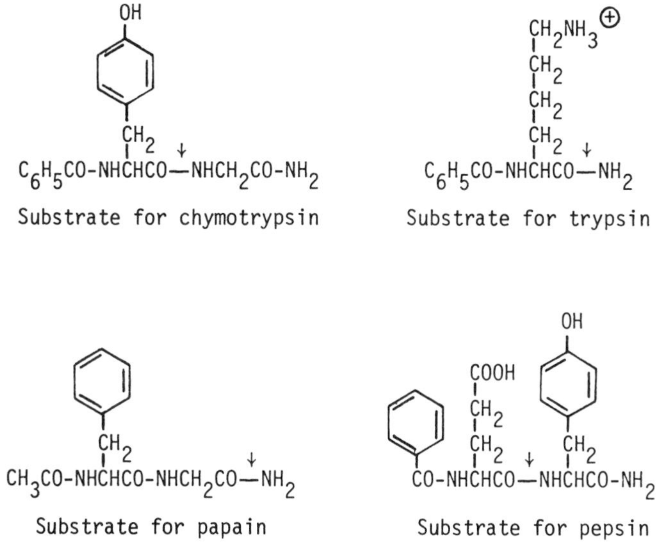

OH
CH_2
$C_6H_5CO\text{-}NHCHCO\text{—}NHCH_2CO\text{-}NH_2$
Substrate for chymotrypsin

$CH_2NH_3^{\oplus}$
CH_2
CH_2
CH_2
$C_6H_5CO\text{-}NHCHCO\text{—}NH_2$
Substrate for trypsin

CH_2
$CH_3CO\text{-}NHCHCO\text{-}NHCH_2CO\text{—}NH_2$
Substrate for papain

OH
COOH
CH_2
CH_2 CH_2
$CO\text{-}NHCHCO\text{—}NHCHCO\text{-}NH_2$
Substrate for pepsin

FIGURE 3. Early synthetic substrates for proteinases.

For example, in 1938, Linderstøm-Lang concluded from studies on protein denaturation that his data

> provide sufficient basis for giving warning against the conclusion that genuine proteins contain peptide bonds because they are split by proteinases like trypsin. They give a certain indication that peptide bonds are formed or "appear" (like SH-groups) upon denaturation, but they are not conclusive enough to decide whether or not some hydrolysable peptide bonds are pre-formed in the molecules of the genuine globular proteins.[44]

A sceptical note about the significance of the synthetic peptide substrates for pepsin was also struck by Northrop, who offered calculations indicating that they were hydrolyzed about 100 times more slowly than a protein substrate.[45]

For these reasons, considerable attention was given during the 1930s to hypotheses that assumed various ring structures involving covalent bonds. The idea of the 1920s that diketopiperazines were associated with each other by secondary valences had been largely abandoned, but the attraction of other possible ring structures was too great to be resisted. I shall pass over proposals such as those of Fodor[46] and of Troensegaard[47] because they did not receive much attention. The hypotheses that were taken more seriously stemmed in part from the x-ray data of William Astbury on keratin, from which he inferred in 1930 that this fibrous protein had hexagonal folds.[48] One of the suggested structures, developed by Dorothy Wrinch, involved a "cyclol" theory of protein structure.[49] In the Wrinch hypothesis, which was largely based on topological considerations, it was proposed that a native globular protein exists as a honeycomb-like structure constructed of six-membered rings formed by the covalent union of the NH of one amino acid unit with the CO group of another amino acid unit to give =N–C(OH)= bonds (FIGURE 4). One of the most influential protagonists of the Wrinch hypothesis was Irving Langmuir who stated in 1939:

> The original idea of native proteins as long chain polymers of amino-acid residues, while consistent with the facts relating to the chemical composition of proteins in general, was not a necessary deduction from these facts. Moreover it is incompatible with the facts of protein crystallography, both classical and modern, with the phenomena of denaturation, with Svedberg's results which show that native proteins have definite molecular weights, and with the high specificity of proteins discovered in immunochemistry and enzyme chemistry. All these factors seem to demand a highly organized structure for the native proteins, and the assumption that the residues function as two-armed units leading to long-chain structures must be discarded. The cyclol hypothesis introduced the single assumption that the residues function as four-armed units, and its development during the last few years has shown that this single postulate leads by straight mathematical deductions to the idea of a characteristic protein fabric which in itself explains the striking uniformities of skeleton and configuration of all the amino-acid molecules obtained by the degradation of proteins. The geometry of the cyclol fabric is such that it can fold round polyhedrally to form closed cage-like structures. These cage molecules explain in one single scheme the existence of megamolecules of definite molecular weights capable of highly specific reactions, of crystallizing, and of forming monolayers of very great insolubility.

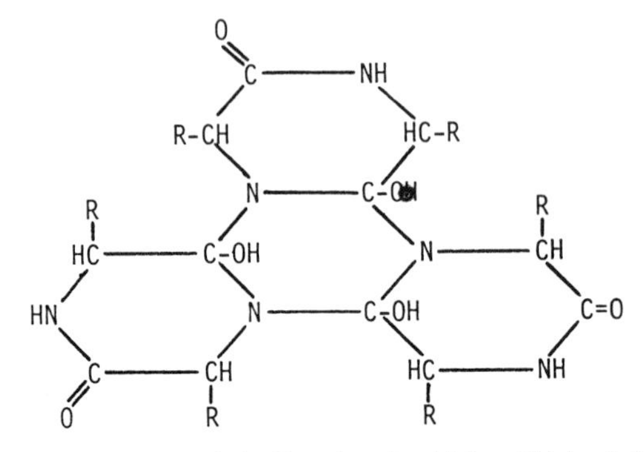

FIGURE 4. A "cyclol-6" molecule. (After Wrinch.[49])

> The agreement between the properties of the globular proteins and the cyclol structures proposed for them is indeed so striking that it gives an adequate justification for the cyclol theory. . . .[50]

In the same year, however, Pauling and Niemann[51] published a strong critique of the cyclol hypothesis, and its popularity waned rapidly thereafter. It should be noted that long after the cyclol hypothesis ceased to figure seriously in discussions of protein structure, the existence of the kind of bonding postulated by Wrinch was demonstrated for the ergot alkaloids.[52]

Along with the esthetic satisfaction gained from topological speculation about protein structure, the 1930s were also characterized by what Chibnall later termed "the hypnotic power of numerology."[53] An impetus to this development was provided by Svedberg, whose ultracentrifuge data led him to suggest that all proteins, regardless of apparent size, are aggregates of subunits having a molecular weight of about 17,500.[54] Indeed, one of the arguments offered in support of the cyclol theory was that it explained a Svedberg unit of 35,000 in terms of a cyclol containing 288 amino acid residues. This number also appeared in the proposal offered by Max Bergmann and Carl Niemann[55] during 1936–1938 that the number of each individual amino acid residue and the total number of all the amino acid residues contained in a protein molecule can be expressed as the powers of the integers two and three. The total number of amino acid residues per presumed molecule of protein appeared to fall into a series of multiples of 288, and a protein like fibrin was considered to have 576 (or $2^6 \times 3^2$) residues, corresponding to a molecular weight of 69,300. Furthermore, not only was the content of an amino acid denoted by the formula $2^n \times 3^m$, but every amino acid residue was thought to occur at a regularly periodic interval in the polypeptide chain of a protein. These remarkable conclusions were based on amino acid analyses performed by Carl Niemann on several proteins of doubtful homogeneity, and by means of analytical methods of uncertain reliability.

The Bergmann-Niemann hypothesis was welcomed by some geneticists, since it appeared to suggest the operation of a mathematical principle that might link Mendel's laws to the structure of proteins, which were considered at that time to be the material basis of heredity.[56] Among protein chemists who had worked for many years to improve the methods for the amino acid analysis of protein hydrolysates, the response, when expressed publicly, was one of scepticism tempered with the respect due to a chemist of Bergmann's distinction. As Chibnall put it in 1942: "Some of us . . . were not prepared to give this attractive generalization our immediate and unqualified support, for we questioned the reliability and completeness of the amino acid analyses on which it was based."[53] Indeed, opinions about the periodicity hypothesis were divided in Bergmann's laboratory, and by 1939 he was obliged to abandon the idea when his associate William Stein produced incontrovertible analytical data that could not be accommodated by the $2^n \times 3^m$ rule. Also, during the period 1939–1941 there were several critical articles by highly regarded protein chemists, among them Pirie[57] and Neuberger,[58] and it was clear that the Bergmann-Niemann hypothesis had suffered the fate of other mathematical theories of protein structure insecurely founded on experiment.

Perhaps the most important paper that emerged from the critical discussion of the Bergmann-Niemann hypothesis was the one by Gordon, Martin, and Synge,[59] for it marked the beginning of the development, by Martin and Synge, of chromatographic methods for the quantitative separation and analysis of the hydrolytic products derived from proteins. These methods made it possible for Sanger to attack the problem of the amino acid sequence of insulin. Sanger demonstrated that in the cleavage of the two insulin chains by trypsin and chymotrypsin, the peptide bonds that are attacked most readily are the ones indicated by the results with synthetic substrates. The situation with pepsin was less clear, but this was a consequence of the fact that more work needed to be done, with better synthetic substrates, to define the specificity of this enzyme.[60] The confusion of the 1930s was quickly forgotten, and chemists could get on with the business of determining the structure of proteins.

REFERENCES

1. SANGER, F. 1952. Advan. Protein Chem. 7: 1–67.
2. HOFMEISTER, F. 1902. Ergeb. Physiol. *1*: 759–802.
3. FISCHER, E. 1902. Chemiker-Zeitung *26*: 939–940
4. FRUTON, J. S. 1972. Molecules and Life. Wiley-Interscience. New York, N.Y.
5. SHAMIN, A. N. 1977. Istoria Khimii Belka. Nauka. Moscow.
6. GRAHAM, T. 1861. Phil. Trans. R. Soc. London *151*: 183–224.
7. PFLÜGER, E. 1875. Pflügers Arch. *10*: 251–369, 641–644.
8. LOEW, O. & T. BOKORNY. 1881. Pflügers Arch. *25*: 150–164.
9. GRIFFITHS, A. B. 1892. The Physiology of the Invertebrata. Reeve. London.
10. LATHAM, P. W. 1887. The Croonian Lectures on some Points in the Pathology of Rheumatism, Gout and Diabetes. Deighton Bell, Cambridge.
11. HOPPE-SEYLER, F. 1881. Physiologische Chemie. 982. Hirschwald. Berlin.
12. BAUMANN, E. 1882. Pflügers Arch. *29*: 400–421.
13. SCHRYVER, S. B. 1906. Chemistry of the Albumens. Longmans Green. London.
14. PFLÜGER, E. Pflügers Arch. *129*: 99–102.
15. COHNHEIM, O. 1901. Z. physiol. Chem. *33*: 451–465.
16. NEUMEISTER, R. 1897. Lehrbuch der Physiologischen Chemie. 30–34. Fischer. Jena.

17. Kossel, A. & F. Kutscher. 1900. Z. Physiol. Chem. *31*: 165–214.
18. Kossel, A. 1901. Ber. Chem. Ges. *34*: 3214–3245.
19. Mathews, A. P. 1927. Science *66*: 293.
20. Fischer, E. 1906. Ber. Chem. Ges. *39*: 530–610.
21. Fischer, E. 1907. Ber. Chem. Ges. *40*: 1754–1767.
22. Bergmann, M. & L. Zervas. 1932. Ber. Chem. Ges. *65*: 1192–1201.
23. Fruton, J. S. 1950. Advan. Protein Chem. *5*: 1–82.
24. Bodanszky, M., Y. S. Klausner & M. A. Ondetti. 1976. Peptide Synthesis. 2nd edit. Wiley. New York, N.Y.
25. Abderhalden, E. 1924. Naturwissenschaften *12*: 716–720.
26. Klarmann, E. 1927. Chem. Rev. *4*: 51–107.
27. Vickery, H. B. & T. B. Osborne. 1928. Physiol. Revs. *8*: 393–446.
28. Gortner, R. A. 1929. Outlines of Biochemistry. Wiley. New York, N.Y.
29. Bergmann, M. 1924. Naturwissenschaften *12*: 1155–1161.
30. Bergmann, M. & H. Ensslin. 1926. Ann. Chem. *448*: 38–48.
31. Brill, R. 1923. Ann. Chem. *434*: 204–217.
32. Waldschmidt-Leitz, E. 1932. Proc. R. Soc. London: *B111*: 286–291.
33. Oppenheimer, C. 1926. Die Fermente und ihre Wirkungen. 5th edit. Vol. 2: 812. Thieme. Leipzig.
34. Bayliss, W. M. 1925. The Nature of Enzyme Action. 5th edit. Longmans Green. London.
35. Fischer, E. 1894. Ber. Chem. Ges. *27*: 2985–2993.
36. Willstätter, R. 1927. Problems and Methods in Enzyme Research. Cornell University Press. Ithaca, N.Y.
37. Northrop, J. H. 1930. J. Gen. Physiol. *13*: 739–766.
38. Helferich, B. 1969. Chem. Ber. *102*: I–XXVI.
39. Bergmann, M., L. Zervas & J. S. Fruton. 1936. J. Biol. Chem. *115*: 593–611.
40. Bergmann, M. & J. S. Fruton. 1937. J. Biol. Chem. *118*: 405–415.
41. Bergmann, M., J. S. Fruton & H. Pollok. 1937. Science *85*: 410–411.
42. Fruton, J. S. & M. Bergmann. 1938. Science *87*: 557.
43. Mirsky, A. E. & L. Pauling. 1936. Proc. Nat. Acad. Sci. U.S.A. *22*: 439–447.
44. Linderstrøm-Lang, K., R. D. Hotchkiss & G. Johansen. 1938. Nature *142*: 996.
45. Northrop, J. H., M. Kunitz & R. M. Herriott. 1948. Crystalline Enzymes. 2nd edit. 73–74. Columbia University Press. New York, N.Y.
46. Fodor, A. & S. Kuk. 1933. Biochem. Z. *262*: 69–85.
47. Trønsegaard, N. 1944. On the Structure of the Protein Molecule. 2nd edit. Munksgaard. Copenhagen.
48. Astbury, W. T. & H. J. Woods. 1930. Nature *126*: 913–914.
49. Wrinch, D. M. 1937. Proc. R. Soc. London. *A160*: 59–86; *A161*: 505–524.
50. Langmuir, I. 1939. Proc. Phys. Soc. *51*: 592–612.
51. Pauling, L. & C. Niemann. 1939. J. Amer. Chem. Soc. *61*: 1860–1867.
52. Glenn, A. L. 1954. Q. Rev. (London) *8*: 192–218.
53. Chibnall, A. C. 1942. Proc. R. Soc. London *B131*: 136–160.
54. Svedbrg, T. 1937. Nature *139*: 1051–1062.
55. Bergmann, M. & C. Niemann. 1936. J. Biol. Chem. *115*: 77–85.
56. Goldschmidt, R. B. 1938. Sci. Monthly *46*: 268–273.
57. Pirie, N. W. 1939. Ann. Rep. Prog. Chem. *36*: 351–353.
58. Neuberger, A. 1939. Proc. R. Soc. London *B127*: 25–26.
59. Gordon, A. H., A. J. P. Martin & L. M. Synge. 1941. Biochem. J. *35*: 1369–1387.
60. Fruton, J. S. 1970. Advan. Enzymol. *33*: 401–443.

DISCUSSION OF THE PAPER

Herbert E. Carter (*University of Arizona, Tucson, Ariz.*): It seems to me that one could sum up most of what you said by saying that physical chemists confused the issue and that the organic chemists provided the data by which they confused themselves; but eventually they straighten the matter out.

David Nachmansohn (*Columbia University, New York, N.Y.*): Dr. Fruton stressed the importance of organic chemistry in the development of dynamic biochemistry. Nobody will question the importance of the elucidation of the structure of the most essential cell constituents as one crucial factor. An outstanding role was played by the school of Adolf von Baeyer, i.e., Emil and his cousin Otto Fischer, Richard Willstätter, Heinrich Wieland, and Eduard Buchner. But knowledge of structures alone would not have been sufficient for explaining the spectacular rise of biochemistry. The second and absolutely equally important factor was the rapid growth of physical chemistry at the turn of the century under the leadership of Wilhelm Ostwald, Jacobus Henricus van't Hoff, Svante Arrhenius, Fritz Haber, Walther Nernst, Sørensen, and others. Kinetics, bioenergetics, thermodynamics, electrochemistry, electrolyte behavior, and so forth, and the availability of new and highly sensitive physicochemical methods were just as necessary and indispensable.

Anybody who lived in that era in the 1920s in the Kaiser-Wilhelm-Institutes, where dynamic biochemistry really began its great era, knows how inseparable the two fields were for the work there, especially in Meyerhof's and Warburg's laboratories. As an illustration it may be mentioned that Emil Fischer, who got the Nobel Prize for his work on the structure of sugars and purines in 1902, but started his work in proteins in 1899, believed still in 1916 that the molecular weight of proteins did not surpass 5000. But shortly afterwards Sørensen using physicochemical methods arrived at values of about 30,000 to 40,000 and Adair, soon afterwards, at over 60,000. When Svedberg introduced the ultracentrifuge, combined with optical methods, about 1925, the weights he found were much higher and finally reached millions. Knowledge of subunit structure came of course only decades later. However, we also should not forget the new dimensions introduced into chemistry and later into biochemistry by the understanding of the electronic structure of atoms and molecules, an outgrowth of the Rutherford-Bohr atom model and of quantum mechanics. The pioneers in this field were G. N. Lewis, Langmuir, and Linus Pauling. All these developments decisively influenced biochemistry in general as well as protein chemistry.

Z. DISCHE (*Columbia University, New York, N.Y.*): I wanted to ask Prof. Fruton what is the largest protein for which the straight amino acid sequence is known.

FRUTON: I cannot answer that question in terms of numbers, perhaps either Emil Smith or John Edsall can help on that. About 80,000? Is that a reasonable number?

EMIL SMITH (*University of California, Los Angeles, Calif.*): For a single peptide chain to my knowledge the longest peptide sequence that has been established is that of β-galactosidase. This was done by Zabin and Fowler of UCLA, and there are something like eleven hundred residues in the peptide chain. There are four peptide chains in the active enzyme that are identical.

P. KARLSON (*University of Marburg, West Germany*): Just as a last point. There is another large molecule with a nearly equally long sequence. It is the collagen molecule and pro-collagen had I think around 1200 amino acid residues. But I think we should first thank Dr. Fruton for his excellent overview of the history of protein chemistry.

I would like to make a few comments if I may. First, let's go back to the time of Willy Kühne. On the basis of very sound experiments I should say, he divided protein molecules into things like albumins and antalbumins. That is, the molecule was believed to be composed of amino acids in peptide linkages and a core molecule which was believed at that time by Kühne and his followers to be of a different, unknown nature. This I think is quite important, it was a deduction from experiments which are valid in that sense.

The second point concerns Neils Troensegaard. His basic idea, as far as I know it, was that he wanted to study the protein molecule not only by hydrolysis as was done by Abderhalden, Fischer, Hofmeister, and others, but also by other degradative methods: oxidation, reduction, and so forth. By this means he came up with all kinds of artifacts which he believed to be a basic part of the protein molecule.

Another comment regarding the work of Wrinch: Felix Haurowitz published a paper in 1936 or 1938 disproving by chemical methods the cyclol model of Wrinch, because he could not detect hydroxyl groups by acetylation and other methods; namely the hydroxyl groups which should be present there if Wrinch's structure were correct. And a final comment on the ideas of Max Bergmann and Carl Niemann. The basic idea was at that time—I think we should acknowledge this—to find repeating units within large macromolecules. You know we have the repeating unit of maltose in glycogen and starch; we have the repeating unit of cellobiose in cellulose; and we have the repeating units of two different parts of uronic acids and other things like that. It was generally believed that the nucleic acids were made out of tetranucleotides in repeating units, and it was more or less tacitly assumed that a protein would also be built up as a form of repeating unit of amino acids. Therefore, the magic numbers of Bergmann and Niemann had something to do with this belief, I think, that there was a repeating unit in all macromolecules.

DAVID KAISER (*Columbia University, New York, N.Y.*): You were talking about the globular proteins and the phenomenon of denaturation *vis-à-vis* the type of bonds that were supposed to be operative. And I am wondering what these topological speculations on protein structure were able to disclose about these bonds? Are you really talking about the sulphur bridges or what?

FRUTON: It depends on exactly what time you are talking about. During the 1930s there was considerable discussion of what kinds of bonds might be broken in denaturation. The very persuasive idea was offered by Hsien Wu that denaturation corresponded to an unfolding of a folded molecule and then the question arose: what held the folds together? And there was the important paper of Mirsky and Pauling suggesting that hydrogen bonds held the fabric of the protein molecule together. But in addition to these ideas which have survived, there was a desire to find, or hope of finding, covalent bonds, stronger than hydrogen bonds, that would account for this specific globular structure, such bonds as for example in thiazoline rings. Experiments were conducted in model systems to see whether thiazoline rings might explain some of the properties of proteins in terms of the appearance of SH groups, which is what Linderstrøm-Lang was talking about in the passage I quoted. And then there were more speculative ideas based on no experiments whatever but an appeal to the asthetic sense such as in the Wrinch hypothesis, which postulated the structure that I showed. There were a lot of ideas to choose among and make things interesting.

N. W. Pirie went to Cambridge in 1925 and taught in the Biochemistry Department from 1931 to 1940. In the early 1930s his collaboration with F. C. Bawden started: first on Potato Virus X because Bawden worked in the Potato Research Institute, and then TMV when Bawden moved to Rothamsted. They found that TMV was a liquid crystalline nucleoprotein and that several other plant viruses were fully crystalline nucleoproteins. When Pirie moved to Rothamsted in 1940, their work became more physiological. Pirie is now working on the extraction of leaf protein as a human food.

Purification and Crystallization of Proteins

N. W. PIRIE
Rothamsted Experimental Station
Harpenden, Hertfordshire
AL5 2JQ United Kingdom

THE FIRST STEP towards unifying the category, to which the name protein was later applied, was taken by Beccari. *Mien chin* (literally, muscle of wheat) had been made in China for several thousand years by washing starch from wheat flour and collecting the cohesive mass of gluten. Beccari lectured on the process in 1728; he pointed out that a similar mass could be made from other seeds and that the stink produced when gluten was distilled, or allowed to putrefy, resembled the stink when animal matter was similarly treated. An account of the lecture was published in 1745; Bailey published a translation of it.[1] Rouelle, relying on the same criteria, called the material he coagulated from leaf extracts *matière végéto-animale;* I have published a translation of that paper.[2]

Others, working along similar lines soon after, did not claim that the substances studied were pure individuals. Then notional biochemistry started and has been with us ever since. Liebig deserves great credit for his analytical skill and his tireless advocacy of the outlook of Quesnay and Adam Smith—that the fundamental requirement for national prosperity is sound agriculture—but he confused the protein story. As Hopkins[3] remarked, he lacked "a biologist's instincts." Liebig[4] said it was "absolutely impossible to distinguish" the curds formed by heating egg white, serum, or vegetable extracts. He also said there were only four proteins. Though he clung to the idea that there were very few proteins, he soon had to admit more than four. The assumption that identical proteins could be isolated from different sources became improbable when Ritthausen observed differences in their amino acid compositions; he suggested that they would therefore differ in nutritive value.

The fault with the notional biochemistry of Liebig and his colleagues was excessive simplicity and rigidity. This was a mistake that Boyle had devoted most of the Fifth Part of the "Sceptical Chymist" to exposing. It consists in assuming that superficially similar attributes or functions must of necessity depend on the same structure or substance whenever they appear. But it persisted. Hartley dubbed it "Lavoisier's fallacy" because Lavoisier insisted that an acid must contain oxygen. It is still with us—even in biochemistry!

0077-8923/79/0325-0021 $01.75/0

Excessive simplicity was replaced by experimentally stultifying vagueness when Purkyně delved into his ecclesiastical background and introduced the notion of protoplasm. By the end of the 19th century this had become an obsession. Surprisingly, Huxley succumbed and wrote an article on protoplasm entitled "The Physical Basis of Life." Berzelius' sour comment on some of Liebig's flights of fancy is again apposite: "This easy kind of physiological chemistry is created at the writing desk, and is the more dangerous, the more genius goes into its execution. . . ." Protoplasm was given mystical or even magical properties, and was widely thought to lose its virtue and disintegrate into mundane proteins when extracted. Compared to protoplasm, Stahl's "phlogiston" and "sensitive soul" almost seem sensible. Other concepts thrown up by notional biochemistry, e.g., "biogen" and "inogen," had shorter lives.

Few chemically minded biologists doubted the inadequacy of the old concept of "parenchyma" (literally, what is poured in): structureless stuff filling the space between the visible and tangible parts of a tissue. Glanville[5] had said, "There is an inexhaustible variety of Treasure which Providence hath lodged in Things, that to the World's end will afford fresh Discoveries," and he advocated microscopic study of the "little Threds and Springs" in organisms. What Hopkins devoted so much effort to attacking was a purely verbal set of explanations of what went on between these Threds and Springs. He argued that the reactivity of metabolites should be explained in terms of comprehensible reactions catalyzed by enzymes. He detested "giant molecules." His objective was to "add to what the eye itself reveals an adequate mental picture of the invisible molecular events which underlie the visible."[3] He quoted with approval Boyle's interest in "motions and figures of the small parts of matter," and colourfully referred to enzymes as "events in progress."

It is difficult today to understand the attractiveness of the idea that cellular proteins, and even the proteins in fluids such as blood and milk, were often cleavage products from giant molecules. But ideas of this sort were supported at various times by such distinguished contributors to our knowledge of proteins as Hammarsten, Hardy, and Sørensen. Hardy thought at one time that globulins exerted no osmotic pressure and objected to the use of the word *molecule* for such large particles; to him they were like little cells and he stressed that such a particle has an inside with properties differing from the outside. Sørensen thought of the caseins as "reversibly dissociable component systems" held together by residual valencies. If "purification" involved the disruption of such complexes, the fragments need not be uniform in size and shape. A "pure" preparation would contain particles belonging to the same category, but not identical. Armstrong[6] had a similar outlook when he introduced the inappropriate metaphor "template" into biochemistry. "Matrix" would have been a more suitable term for the surfaces of such structures as enzymes, which, as he visualized things, were so large that only one side had to meet rigid specifications.

After purification, a substance of the type usually studied in conventional chemistry may still be contaminated with small amounts of other material. The possibility that the contaminant, rather than the main component, is responsible for a characteristic activity (e.g., pharmacological activity) can be eliminated by synthesis. This is not because the synthetic product is necessarily less contaminated, but because its contaminants are likely to differ from those in a natural product. What was envisaged thirty years ago[7-9] was the presence in purified preparations of comparable amounts of several proteins differing slightly in size and composition but all having the same activity. This outlook was reasonable because ideal homogeneity cannot be demonstrated. Heterogeneity may arise because material from different individuals (or even species) is pooled, because the specifications to which the protein-synthesizing mechanism works may change as a cell ages, because changes take place *in vivo* and during purification, or because the synthesizing mechanism does not work precisely. Although many enzymes occur in multiple forms other than genetically different isozymes, it is now clear either that the mechanism works with such precision that few aberrant molecules are made, or that there is an efficient scavenging mechanism[10] that removes them.

No thoughtful biochemist assumed that crystallinity, or even repeated recrystallization, had much bearing on the purity of a preparation. Mixed crystals were well known, and it was accepted that "the hospitality of the crystal lattice," to use Goldschmidt's phrase, could be extended to molecules with only superficial similarities.

The extent to which the concepts of conventional chemistry were thought to be applicable to proteins, depended on the dimensions assigned to them. The recentness of universal acceptance of the objective reality of atoms. and the consequent conclusion that molecular mass is a real mass and not just a ratio, is often not realized. Proteins were known to be colloids; without the acceptance of molecular reality the concept of purity was hardly applicable to them. It was perhaps for this reason that Fischer[11] disbelieved in proteins containing 100 amino acids; he thought 30 to 40 about the limit. Another reason for choosing such a small number was that a larger number seemed totally unnecessary. With help from Max Planck, he calculated that there could be more than 10^{27} permutations of a 30-unit chain made of only five amino acids, and an even larger number if all the amino acids were considered as well as modes of linkage other than the orthodox –CO–NH–CH= link. That seemed adequate to satisfy the idiosyncracies of half a million species of plants and microorganisms, and two or three million animals. By 1929, ultracentrifuge evidence made incontestable the existence of proteins containing several hundred amino acids. I increased Hopkins' disquiet at such large molecules by showing him that, if one accepted Eddington's numerological conclusions about the total number of protons in the Universe, there would not be enough matter in it to make simultaneously one molecule of each of the possible isomers of a protein as large as hemoglobin. This early realization that the potentialities of protein individuality so fantastically exceeded any conceivable biological requirement, was largely, though not wholly logically, responsible for my assumption that proteins would have to meet general rather than precise specifications.

During the 1930s attempts were made to constrain the possibilities of isomerism and introduce rules into protein structure. One hypothesis replaced the amide chain by a network of hexagons, another postulated that the number of appearances of each amino acid in a protein had to be divisible by two and three. Each had many metamorphoses and collected a surprising number of adherents. For ideological, objective, and esthetic reasons they were derided in the Cambridge Biochemistry Laboratory. Ideological, because we could see no biological advantage in such constraints. Objective, because there seemed to be no convincing evidence for them. And aesthetic because those who handled proteins regularly did not feel that they had that sort of rigidity. It is interesting to note that Kendrew *et al.*[12] reacted inversely. They were surprised at the absence of any apparent plan in the 3-dimensional structure of myoglobin; it seemed to me that theirs was the first protein structure to have a plausible appearance. Perhaps this illustrates the difference between a biological and a physical outlook.

In the period before immunology had become an important branch of science, interest in the nature of enzymes was an important reason for studying the separation and individuality of proteins. Traube, Pasteur. Buchner, Engels, Fischer, and Armstrong (father and son) assumed that enzymes were proteins. Mayer, Nenki, and Bayliss argued that they were not. Bayliss' conclusion is not surprising for he had a prejudice against proteins as is shown by his ridiculous nutritional dictum: "Look after the calories and the proteins will look after themselves." But perhaps he had not heard of people who depend on cassava, bananas, or yams! Willstätter gradually came to the conclusion that the enzymes he was studying were not proteins. Haldane[13] was undecided: "If, as many workers believe, the enzymes are all proteins, it is certainly remarkable that the majority of the successful attempts to purity them have led to the obtaining of substances which are at least predominantly nonproteins, although the original material from which they were derived consisted largely of protein." He went on to argue that the presence of protein could increase the stability of a non-protein enzyme so that protein, if "not part of the catalyst, is a part of the enzyme considered as a stable substance."

The influence of Hopkins and Haldane generated a sceptical and iconoclastic outlook in the students and junior staff of the Cambridge Biochemistry Laboratory. We understood Haldane's logic—but tended to think of enzymes as proteins. We even discussed whether every protein would be enzyme if only one could find the right substrate. The recent discovery[14] that serum albumin has a quasi-enzymic action on the bizarre substrate 1,1-dihydroxy-2,4,6-trinitrocyclohexadienate suggests that that discussion was not wholly absurd. Northrop,[15] writing ironically of the mental climate in the early 1930s said, "it was well known that enzymes were not proteins." That may have been the outlook in Princeton: it wasn't in Cambridge.

The general outlook in the Cambridge Biochemistry Laboratory was that, in biochemistry at any rate, experiments could prove that a supposition was wrong, but only rarely that it was right: this point of view, that science is essentially a technique for disproof, has become more widespread since its popularization by Popper. Recognition of the impossibility of demonstrating purity is an example of this outlook. The most that can be demonstrated is that less than some specified amount of a component, e.g., an impurity for which there is a trustworthy method of analysis, is present. Thus we did not doubt the presence of urease in the crystals that Sumner prepared: we wondered whether that was the only component. Early in 1936, while working with Landsteiner, I had met Northrop a few times, and in spite of doubts about crystals, I asked him later in that year for some crystalline trypsin because I wanted to use it to remove traces of protein from ribonucleic acid separated from tobacco mosaic virus (TMV). It turned out to be a potent ribonuclease. Dubos[16] had the same experience. At that date, the name given to a crystalline enzyme preparation depended more on the enzyme the experimenter was looking for, than on the dominant enzyme in the preparation. Our confidence in crystals, and in Sumner's critical faculties, was shaken when he gave 96,000 as the molecular mass of allegedly pure crystalline concanavaline A in which he found 0.023% manganese.[17]

Anomalies such as these have been largely removed by improvements in separatory techniques. They serve however to show the extent to which the beauty of a crystalline preparation can dull the critical faculties. Nevertheless, even today the search for contaminants in preparations made by conventional methods is often superficial. Thus TMV, unless incubated with trypsin or treated in other ways that radically alter its physical properties, contains enough normal tobacco protein to elicit an anaphylactic response in guinea pigs.[18] So far as I know, the preparations on which elaborate physical measurements are now being made are not tested for this type of contamination. Similarly, the treatments that are needed to remove leaf ribonuclease and seemingly extraneous nucleic acid from TMV, alter its precipitability by salts[19, 20] though they do not destroy its infectivity. This suggests, though it does not prove, that contamination with leaf ribosomes or their fragments[21] does not explain all these phenomena. But the existence of such problems should be remembered.

Viruses illustrate other aspects of purification and assessment of purity. Haldane[13] discussed the frequency of impact between enzyme and substrate, and concluded from the rates of some enzyme actions that some preparations might be nearly pure. If one virus particle could infect a plant, and if virus dimensions deduced from membrane filtration experiments were approximately correct, many liters of infective sap would contain only one or two milligrams of virus. Consequently, in early attempts to purify viruses such as TMV, most of the virus was probably discarded because it seemed incredible that bulky precipitates could be approximately pure. However, preparations made by precipitation with ammonium sulfate in the conventional manner[22, 23] differed in so many respects from preparations made when similar methods were applied to sap from uninfected plants, that it seemed reasonable to conclude that they had some connection with the virus. Bawden *et al.*[23] concluded cautiously: "These results have a certain intrinsic interest, but this would naturally be greatly enhanced could it be shown that these rods are in fact virus particles. This conclusion seems to us both reasonable and probable, but we feel that it is still not proved, nor is there any evidence that the particles we have observed exist as such in infected sap."

Preparations made from infective sap differed, not only from those made from uninfective sap, but also from one another. By 1938, Stanley had incorporated most features of our preparation[23, 24] into his description. This favored rapid progress in virus research, but it left unanswered the question "What did Stanley isolate in 1935?" The feature that excited most interest was the alleged crystallinity of the preparation. More nonsense was written, in both the popular and scientific press, on that than on any other scientific theme getting attention at the time. Although our preparations had very interesting liquid crystalline properties, we called the solid preparations, made by salting out or by acid precipitation, fibers. We chose that word partly because it suited the appearance of the structures once they were formed, and partly because of some observations on the process by which they were formed. For example: when dilute solutions containing TMV, glycine, and ethyl formate, initially at pH 5, are left undisturbed at room temperature for a few hours, they set to clear structureless jellies. The glycine acts as a buffer controlling the pH reached as the ethyl formate hydrolyzes. When the jelly is shaken, it breaks up instantly into a suspension of fibres with a characteristic sheen. Even more nonsense would probably have been written if the first virus we isolated had been one that is truly crystalline. Those who assumed an incompatibility between crystallinity and life, and who mistakenly thought that viruses were alive, were able to draw some comfort from the idea that, after all, TMV was "only a fiber."

Osborne & Wells used serology to demonstrate differences between seed proteins with similar amino acid compositions, and also to assess the degree of resemblance between proteins from related species. Sap from plants infected with TMV is an excellent antigen; sap from uninfected plants is feeble. The dilution at which our TMV solutions would precepitate with specific antisera, when compared with the precipitation end-points of other purified antigens, suggested that the dominant component in our preparations was a specific antigen, even if it was not the virus itself. We had in mind the possibility that what was being isolated was a product of the host's reaction to infection. Bence-Jones proteins—the light immunoglobulin chains synthesized in excess and excreted in urine by people with some plasma tumours—are an analogy. The character of an antigen:antibody precipitate, which can be loose and fluffy, or compact, depends on the geometrical form of the antigen. Changes in the appearance of the precipitate as TMV is being "purified" are *prima facie* evidence that, as presumed contaminants are removed, other changes take place. The processes are probably connected. Sometimes a "purified" virus may be the *écorché* rather than the body.

Our attitude towards the preparations of TMV that we made was unnecessarily cautious. Similar caution about the character of fully crystalline nucleoproteins from plants infected with several strains of necrosis virus[25, 26] was justified. The crystallizable nucleoprotein from one of them[27] was not infective, though the mother liquor was. This phenomenon, the failure of "satellite viruses" to multiply except in the presence of another virus, is now known to be common.

Popular attention concentrated on crystallinity: we paid most attention to the presence of nucleic acid in all our plant virus preparations. Nucleoproteins, that is to say, particles in which protein and nucleic acid are held together by more than electrovalencies so that they remain associated over a considerable pH range and in the presence of salts, were plausible biochemical postulates until viruses were studied. As a student, I had been particularly interested in nucleic acids[28] and knew that there was no experimental basis for Levene's "tetranucleotide hypothesis." Levene found only four nucleotides in a nucleic acid and had no reason to postulate anything more complex than a 1:1:1:1 ratio in a molecule containing one of each. It seemed to me that the physical properties of nucleic acid solutions were incompatible with the presence of molecules as small as tetranucleotides. That was an esthetic judgement. But it left us as ready to assume that the biological properties of TMV depended on its nucleic acid as on its protein component.[10] When we were isolating nucleoproteins from infected plants, few biochemists were interested in nucleic acids. Interest had greatly increased by 1950; but, even then, I was amazed to find how many people took the tetranucleotide hypothesis seriously.

The methods used at first to precipitate proteins, e.g., heat and the so-called "alkaloid reagents," were irreversible and therefore useless for fractionation. Even alcohol, later so successfully used for fractionating plasma proteins, denatures most proteins unless it is added very carefully with assiduous cooling. The technique of "salting out" had long been in use in soap-making, its transfer to protein chemistry in the middle of last century was an immense advance which culminated in the use of ammonium sulfate, which has dominated large-scale protein fractionation ever since. This dominance is mainly the result of its solubility. Thus, with some proteins, sodium sulfate is more effective at salting out, but limited solubility limits its use. Potassium phosphate was sometimes used, as were salts of divalent metals. The latter may however dissociate complex proteins. Thus strontium nitrate splits nucleic acid from TMV and denatures the protein,[29] and calcium chloride disrupts broad bean mottle and bromegrass mosaic viruses.[30] Differential precipitation or crystallization is one aspect of fractionation by salting out. Differential re-solution is another; contaminants often dissolve more incompletely or more slowly than the main product. For example, if bushy stunt virus is crystallized at room temperature, it redissolves on cooling whereas contaminants remain insoluble and can be removed by centrifuging at 0°.[31]

Fractionation with a volatile solvent, when possible, has the merit that salts do not have to be removed by tedious dialysis before the protein is analysed. The advantages of freeze-drying,[32, 24] to avoid the formation of a horny lump that dries slowly, were soon recognized. Partition between water and a solvent immiscible with it was little used, partly because Brønsted argued that, as molecular mass increases, movement from one phase to the other becomes more abrupt. It was feared that all the proteins would move together. Another reason was that protein-solvents immiscible with water, e.g., phenol, tend to denature proteins.

Fractionation between immiscible aqueous phases could be used with TMV[24] because the liquid crystalline layer excludes some contaminants. This type of fractionation, using immiscible aqueous solutions of polymers, has recently been thoroughly exploited by Albertson and his colleagues.

Ultrafiltration on membranes of graded porosity was used for small-scale fractionations. High-speed centrifuges, at first, air-driven spinning tops, and then electrically driven bucket centrifuges holding 20 to 40 ml, came into use in the early 1930s. These techniques were exploited most effectively by Bechold and by Schlesinger[33, 34] who made what, by hindsight, seems to have been a fairly pure preparation of bacteriophage.

Before the introduction of chromatography, gel filtration, and similar techniques, enzymes were often fractionated on calcium phosphate, hydrated alumina, and so forth. Considerable artistry was shown, notably by Willstätter and Kelin, in this technique. Its main defects are the difficulty of ensuring that different batches of adsorbent are comparable, and the large amount of adsorbent needed. The eluted enzyme was likely therefore to be dilute. This dilution probably explains Willstätter's sustained denial that the enzymes he worked with were proteins.

Before it was generally recognized that enzymes were proteins and could account for a considerable proportion of the total protein in an extract, precipitants (e.g., dyestuffs) were sometimes used in the hope that they would be fairly specific. This hope proved ill-founded and dissociation of the precipitant from the protein was often difficult. Attempts to use clupein[24] were also unsuccessful—mainly because of the inconvenient thixotropy of the precipitate. Antibodies were, however, purified by precipitating with the specific antigen, and dissociating the complex.

All these techniques were also used to assess purity. Obviously, it was prudent to use in the assessment a technique that had not been used in the later stages of purification. Chick & Martin had argued that egg albumin was a mixture because its solubility in ammonium sulfate solution depended on the concentration of the albumin. Sørensen offered the alternative explanation that the albumin abstracted water rather than ammonium sulfate from the solution, and so increased the effective concentration of the latter. In the elaborate Phase Rule studies of Northrop *et al.*[35] care was taken to ensure that the concentration of salt remained constant in spite of differences in protein concentration. Many of the proteins that were then prepared appeared to be homogeneous by the criteria of the Phase Rule, ultracentrifugation, and electrophoresis; more refined recent methods of study have, however, revealed inhomogeneity in some of them.

There are four interconnected themes in protein purification. Two are ideological and two technical. The extent to which it is legitimate to strip off parts of a particle and claim that thereby it is being purified must be decided. That problem arises particularly with lipoproteins and nucleoproteins, but some amino acids have been removed from enzymes and hormones without loss of activity. It is also uncertain whether every covalent change is relevant. Sometimes, a small change, e.g., the uncovering of a terminal amino group by deacetylation or an amino acid substitution at a critical point (as in sickle cell hemoglobin), has striking physical effects. But it is sometimes difficult to distinguish proteins from different species although sequence analysis shows that they differ in one amino acid. Were such a pair of proteins made in one species, the use of the term inhomogeneous would hardly seem to be justified. The great technical development in recent years is the introduction of micromethods for assay and analysis. This has stimulated a corresponding development of micro methods for fractionation which depend on techniques that would be difficult to use on a large scale. The interplay of factors such as these is not peculiar to protein purification.

Had a survey as brief as this been written by someone whose speciality was the purification of enzymes or antibodies, different examples of fractionation technique and changes in the concept of purity would have been used. An attempt to cover the whole subject would have made the article unwieldy. Furthermore, we might bear in mind Fruton's comment that he "relied largely on the published description and discussion of scientific work shortly after it was done. The limitations inherent in later recollections of an event, especially when its significance has become evident, are well known to historians."[36] It thus seemed likely that I would convey the outlook 30 to 40 years ago more accurately if I kept fairly close to published personal experience.

REFERENCES

1. Bailey, C. H. 1941. A translation of Beccari's lecture "Concerning grain" (1728). Cereal Chem. *18*: 555.
2. Pirie, N. W. 1978. Leaf Protein and Other Aspects of Fodder Fractionation. Cambridge University Press. London.
3. Hopkins, F. G. 1936. The influence of chemical thought on biology. Science *84*: 258.
4. Gregory, W. 1842. Animal Chemistry or Organic Chemistry in its Application to Physiology and Pathology by Justus Liebig. Taylor & Walton. London.
5. Glanville, J. 1668. Plus Ultra: Or the Advancement of Knowledge Since the Days of Aristotle. James Collins. London.
6. Armstrong. H. E. 1904. Enzyme action as bearing on the validity of the ionic-

dissociation hypothesis and on the phenomena of vital change. J. Chem. Soc. *73*: 537.
7. PIRIE, N. W. 1940. The criteria of purity used in the study of large molecules of biological origin. Biol. Rev. *15*: 377.
8. Amino acids and proteins. 1950. Cold Spring Harbor Symp. Quant. Biol. *14*: 198.
9. COLVIN, J. R., D. B. SMITH & W. H. COOK. 1954. The microheterogeneity of proteins. Chem. Rev. *54*: 687.
10. BAWDEN, F. C. & N. W. PIRIE. 1953. Virus multiplication considered as a form of protein synthesis. *In* The Nature of Virus Multiplication. P. Fildes & W. E. van Heyningen, Eds. 21. Cambridge University Press. London.
11. FISCHER, E. 1916. Isomerie der Polypeptide. Sitz. Kgl. Preuss. Akad. Wiss. 990.
12. KENDREW, J. C.. G. BODO, H. M. DINTZIS, R. G. PARRISH & H. WYCKOFF. 1958. A three-dimensional model of the myoglobin molecule obtained by x-ray analysis. Nature (London) *181*: 662.
13. HALDANE, J. B. S. 1930. Enzymes. Longmans, Green & Co. London.
14. TAYLOR, R. P. & A. SILVER. 1976. Bovine serum albumin as a catalyst. J. Amer. Chem. Soc. *98*: 4650.
15. NORTHROP, J. H. 1961. Biochemists, biologists and William of Occam. Ann. Rev. Biochem. *30*: 1.
16. DUBOS, R. J. 1937. The decomposition of yeast nucleic acid by a heat resistant enzyme. Science *85*: 549.
17. SUMNER, J. B., D. B. HAND & I-B ERIKSSON-QUENSEL. 1938. The molecular weights of canavalin, concanavalin A and concanavalin B. J. Biol. Chem. *76*: 45.
18. BAWDEN, F. C. & N. W. PIRIE. 1937. The relationships between liquid crystalline preparations of cucumber viruses 3 and 4 and strains of tobacco mosaic virus. Brit. J. Exp. Path. *18*: 275.
19. PIRIE, N. W. 1956. Some components of tobacco mosaic virus preparations made in different ways. Biochem. J. *63*: 316.
20. WHITFELD, P. R. & S. WILLIAMS. 1963. On the ribonuclease activity associated with tobacco mosaic virus preparations. Virology *21*: 156.
21. PIRIE, N. W. 1950. The isolation from normal tobacco leaves of nucleoprotein with some similarity to plant viruses. Biochem. J. *47*: 614.
22. STANLEY, W. M. 1935. Isolation of a crystalline protein possessing the properties of tobacco-mosaic virus. Science *81*: 644.
23. BAWDEN, F. C., N. W. PIRIE, J. D. BERNAL & I. FANKUCHEN. 1936. Liquid crystalline substances from virus-infected plants. Nature (London) *138*: 1051.
24. BAWDEN, F. C. & N. W. PIRIE. 1937. The isolation and some properties of liquid crystalline substances from solanaceous plants infected with three strains of tobacco mosaic virus. Proc. R. Soc. B *123*: 274.
25. PIRIE, N. W., K. M. SMITH, E. T. C. SPOONER & W. D. McCLEMENT. 1938. Purified prepartions of tobacco necrosis virus. (*Nicotiana* virus II). Parasitology *30*: 543.
26. BAWDEN, F. C. & N. W. PIRIE. 1942. A preliminary description of preparations of some of the viruses causing tobacco necrosis. Brit. J. Exp. Path. *23*: 314.
27. BAWDEN, F. C. & N. W. PIRIE. 1945. Further studies on the purification and properties of a virus causing tobacco necrosis. Brit. J. Exp. Path. *26*: 277.
28. PIRIE, N. W. 1970. Retrospect on the biochemistry of plant viruses. *In* British Biochemistry Past and Present. T. W. Goodwin, Ed. 43. Academic Press. New York.

29. Pirie, N. W. 1945. The fission of tobacco mosaic virus and some other nucleoproteins by strontium nitrate. Biochem. J. *56*: 83.
30. Yamazaki, H. & P. Kaesberg. 1963. Degradation of bromegrass mosaic virus with calcium chloride and the isolation of its protein and nucleic acid. J. Mol. Biol. 7: 760.
31. Bawden, F. C. & N. W. Pirie. 1943. Methods for the purification of tomato bushy stunt and tobacco mosaic viruses. Biochem. J. *37*: 66.
32. Pirie, N. W. 1931. The cuprous derivatives of some sulphydryl compounds. Biochem. J. *25*: 614.
33. Schlesinger, M. 1933. Reindarstellung eines Bakteriophagen in mit freiem Auge sichtbaren Mengen. Biochem. Z. *264*: 6.
34. Schlesinger, M. 1936. The Feulgen reaction of the bacteriophage substance. Nature, London *138*: 508.
35. Northrop, J. H., M. Kunitz & R. M. Herriott. 1948. Crystalline Enzymes. Columbia University Press. New York.
36. Fruton, J. S. 1972. Molecules and Life. Wiley-Interscience. New York.

DISCUSSION OF THE PAPER

Heinz Fraenkel-Conrat (*University of California, Berkeley, Calif.*): We will hear more about the history of protein chemists and TMV later. I will only make one comment. Dr. Pirie showed a slide of a fish floating around TMV. This was utilized at the Rockefeller Institute in Princeton by Stanley, who much later during World War II was studying hydrodynamics in reference to the construction of submarines. So you see that protein chemistry can have very useful consequences in unexpected ways.

Felix Haurowitz (*Indiana University, Bloomington, Ind.*): This morning both speakers repeatedly mentioned some fascinating controversies about the Wrinch theory, and since I was involved in it I want to say a few words. We should not underestimate the fascination of numerology as Dr. Fruton said. Whenever there is something fascinating in numbers, and often it appears to be simplifying, many people are inclined to believe in it. I remember an example of this. I was in Cambridge once to visit an International meeting. I was invited by Dr. Bernal to come with him to meet Dorothy Wrinch in the presence of Astbury and of Dorothy Hodgkin, who may or may not remember this meeting. So I was in the presence of a mathematician, Wrinch, and three crystallographers. I was the only biochemist. I wanted to prove that the cyclol theory couldn't be true but I could not convince the three crystallographers and the mathematician, because they found the numerology so fascinating that they could not abandon it.

Pirie: I used to say in the later 30s that I ought to get a grant from someone to keep Bernal sober whenever he was at a meeting where Wrinch was because Bernal, being an extremely helpful man, was always ready to get her out of the entanglement she got herself into because of her fertile imagination. Under the influence, the kindly influence of alcohol, he would always produce for her a new trick for getting out of some impossible tangle she had gotten herself into. This was very harmful.

DOROTHY HODGKIN (*Oxford University, Oxford, England*): I was involved at the very beginning with Dr. Wrinch's ideas and did, in fact, contribute to them. I think that Dr. Haurowitz is wrong about our attitude to her—the attitude of Bernal and myself. She was a friend of ours, and as a mathematician was very anxious to become acquainted with biology and biochemistry and to contribute mathematical ideas to the problems on which we were working. The date I first knew her was 1934-35 and the main events occurred after the first taking of the insulin photographs. The first idea she had about protein molecules was that in order that space should be filled by the amino acid residues, the residues could not only be joined in a linear sequence; another link besides the peptide link must be involved. This link, the cyclol link, was provided by Charles Frank (now Sir Charles Frank, and working in quite different fields) in the discussion on a lecture by W. T. Astbury; Bernal first brought it to the notice of Dorothy Wrinch. She showed that through this link amino acids could be joined to form planar fabrics and she first suggested that three fabric layers would fit into the unit cell I had found for insulin and constitute the insulin molecule. I ruled this out on the perhaps inadequate grounds that (003) of the insulin x-ray reflections was very weak, and I suggested, I am sorry to say, that she folded the fabrics over a surface. This had enormous strategic advantages because it closed the molecule and Dorothy Wrinch showed that the folding could lead to a variety of molecules of definite molecular weight. Bernal and I were always rather cautious about these ideas as we came to know more about proteins. We became tremendously worried when she received so much encouragement from such really eminent physical chemists as Langmuir. She found our attitude quite reprehensible. I think that John Edsall will confirm the difficulty of our relations in this period. We were friends of hers; we had helped her to develop her theories; but we did not believe in them, and that was our trouble.

COMMENT: Dr. Pirie showed a photograph of microsomes. As I understand it these microsomes are artifacts of certain procedures that are enacted on cells. I did not know if these microsomes were the result of, in this case, pathology induced by the tobacco mosaic virus or whether they were simply an artifact of any process enacted on the leaf in itself.

PIRIE: These ribosomes that I showed you were from an uninfected leaf. There is a paper that I wrote about the sort of structures that you got when you handled uninfected leaf by the same method used to handle infected ones to see what kind of material you would be likely to carry through. You are quite right that ribosomes are largely breakdown products of the rough membrane.

COMMENT: I am sorry. I did not know that you said that those were ribosomes. I thought you said they were microsomes.

PIRIE: I may have done so; we used to call them that.

ROLLIN D. HOTCHKISS (*Rockefeller University, New York, N.Y.*): While we are interested in burying ideas of numerology, or rather premature numerology, I would like to tell you something that I almost forgot. At an earlier New York Academy of Sciences meeting sometime in the 30s, Drs. Vickery and Cohn were reporting analytical data on purified crystalline proteins and attempting to work backward from the amino acid content to show that the data were quite compatible with the molecular weight, and at that particular time the molecular weights were alternatively the 35,000 of Svedberg or the 40,000 discovered by Pedersen. They were calculating from amino acid composition as if there were let's say 19 molecules in one case or 21 for the other molecular weight, and they were quite happy with the data. It struck me that almost any random data might give the same fit if you have allowable numbers, somewhere between 9 and 30 units, so I took *The New York Times* for that particular day (I have it still) and out of the bond quotations I picked a region where a variety of bonds prices were listed. I calculated the molecular weight of bonds from *The New York Times*. The fit of the protein data was something with an error of about 4.9, the fit of *The New York Times* bond quotation was about 2.3, less than half the error of the actual amino acid data.

CARTER: Those probably were not covalent bonds.

PIRIE: I did something similar (1940. Ann. Rep. Chem. Soc. *36*:351) and pointed out that the only long gaps in the earlier part of the sequence of numbers that can be expressed as $2^n \times 3^m$ are 19–23 and 37–47. The other gaps are small compared with the probable error in amino acid measurements. Therefore, the only numbers that would be excluded by the Bergmann-Niemann hypothesis are 22 and 42. With the 1939 level of analytical precision, the hypothesis seemed to me to have no observational basis.

Felix Haurowitz was born in Prague and went to medical school there. He received his M.D. degree in 1922 and his D.Sc. degree in 1923. He was then a member of the Department of Physiological Chemistry at the German University in Prague where he was a Docent and Associate Professor. He was carrying on work on hemoglobin and derivatives and then went into immunochemistry. In 1939 he went to Istanbul where he was head of the Department of Medical and Biological Chemistry until 1948, at which time he came to this country as Professor of Biochemistry at Indiana University. Since then he has done extensive work on immunological biochemistry and related areas.

He has published a large number of papers and has published books on the structure and function of proteins, on immunochemistry, and on the biosynthesis of antibodies.

He is a member of the National Academy of Sciences and the German Leopoldina Academy of Sciences.

Protein Heterogeneity: Its History, Its Bases, and Its Limits

FELIX HAUROWITZ

Department of Chemistry
Indiana University
Bloomington, Indiana 47401

AS CHEMISTS, we usually try to isolate homogeneous substances in order to be able then to describe their properties; this again allows us to verify their existence or to prepare them synthetically. For many years one of the best criteria for homogeneity was the crystallization of the investigated material. And for a long time nobody succeeded in crystallizing a protein. The problem was finally solved, for at least some of the proteins, in two steps. Its history is described in a book *Chemie der Proteine* published in 1900 by Otto Cohnheim,[1] professor of physiology at the University of Hamburg. Cohnheim dedicated his book to Wilhelm Kühne to whom he attributed the purification of proteins by high concentrations of sodium or ammonium sulfate.[2] He also emphasized the importance of the work of Hofmeister who described the first preparation of crystalline proteins.[3]

For a long time the crystalline proteins were considered as homogeneous. The first doubts on their homogeneity were raised in 1930 by Sörensen[4] who demonstrated that the solubility of crystalline ovalbumin and of other apparently pure proteins depends on their concentration. He concluded correctly that the dissolved protein consists of a mixture of reversibly dissociating particles. We would call them today polymers and monomers, or molecules and subunits. But there is no doubt that we deal in solutions with a heterogeneous mixture of particles of different size.

Another type of heterogeneity deals with contamination of proteins by slightly different metabolic precursors or derivatives. We have to distinguish these from polymorphism, i.e., the genetically determined production of two or more very similar forms of a protein. In the following sections of this communication I will discuss different types of heterogeneity. Since my own research dealt preferentially with hemoglobin

* This work was supported by the National Institutes of Health (Grant No. AI09752) and the National Science Foundation (Grant No. PCM76-01886).

0077-8923/79/0325-0037 $01.75/0 © 1979, NYAS

and with the immunoglobulins, I will frequently, but not exclusively, refer to examples from these two groups of proteins.

HETEROGENEITY IN PROTEINS CONTAINING SUBUNITS

All of us who worked during the years 1920–1925 on hemoglobin considered it as a homogeneous protein containing one iron atom per molecule; the calculated molecular weight was approximately 16,000. Indeed Hüfner and Gansser[5] found for horse hemoglobin, by means of osmometry, a molecular weight of about 15,000. In 1928, while I worked as Docent in the Department of Physiological Chemistry at the German University in Prague, Adair in Cambridge, England published a paper[6] in which he refined the osmometric method and found for hemoglobin a molecular weight close to 64,000. This was quite a shock for Professor von Zeynek in whose department I worked. Von Zeynek had been a student of Hüfner and knew that Hüfner was a very careful and precise worker; he asked me to prepare crystalline horse oxyhemoglobin and to redetermine its molecular weight by osmometry. I found the same value as Adair. At about the same time Svedberg[7] had developed his ultracentrifuge and confirmed the value obtained by Adair for the molecular weight of hemoglobin. The much lower value found by Hüfner was possibly caused by contamination of his hemoglobin solutions with NaCl or other salts. Nothing was known at the time of Hüfner's work on the influence of salts and pH on the osmotic pressure.

The precise measurements of Adair[6] and Svedberg[7] left no doubt that the hemoglobin molecule consisted of four subunits. It was considered as obvious that these four subunits were identical. This view was supported by the finding of four subunits of equal molecular weight in the hemoglobins of the horse and other animals. The belief in four subunits of equal composition was further supported by the finding that hemoglobin on slight acidification dissociated in two half-molecules, each of these containing two of the four subunits. It took about 30 years to demonstrate by improved methods of chromatography that the two half-molecules obtained by treatment with dilute acids were different.[8]

For a few years it seemed that the tetramer structure of hemoglobin was an unusual phenomenon. However, we know at present that most of the proteins of high molecular weight are macromolecules formed by the combination of several smaller subunits. Darnell and Klotz[9] published recently a list of approximately 500 proteins, each of them consisting of two or more subunits; in some of them all subunits were identical; others contained two or more different kinds of subunits. Changes in the ratio of subunits obviously caused the changes in the solubility discovered by Sörensen.[4] The extent of the heterogeneity in solutions can be demonstrated also by ultracentrifugation.[10] Even if there are only two types of subunits, A and B, they may occur in different ratios such as A_3B, A_2B_2, AB_3, and also as A_4 and B_4.[11] The term "molecule" becomes here ambiguous. We can designate the tetramers as molecules or we may call them complexes of four molecules. Nature evidently does not care too much about our classifications.

In some of the macromolecular proteins the subunits are held together by hydrogen bonds, by electrostatic attraction of positively and negatively charged amino acid side chains or by other noncovalent interaction. However, in many of the macromolecules the subunits are linked to each other by disulfide bonds. Chemists designate these bonds usually as covalent. However, the disulfide bond is easily and reversibly split by mild reducing agents such as mercaptoethanol, dithiothreitol (Cleland's reagent) or by other thiols. Moreover, disulfide bonds are easily formed from thiols in the presence of molecular O_2, oxidized glutathione, or other hydrogen acceptors. This process is catalyzed by a glutathione transhydrogenase and also by traces of heavy metal ions. Evidently, the equilibrium in the reaction

$$\mathrm{RSSR} + 2\,\mathrm{R'SH} \rightleftharpoons 2\,\mathrm{RSH} + \mathrm{R'SSR'}$$

can be shifted easily to the right by an excess of R′SH, or to the left by an excess of R′SSR′. In view of this lability of the disulfide bond the structure of disulfides can be considered as an equilibrium between two isomeric sulfenium sulfides:

$$\overset{+}{\mathrm{R}}\overset{-}{\mathrm{S}}\mathrm{SR} \rightleftharpoons \mathrm{RS}\overset{-}{\mathrm{S}}\overset{+}{\mathrm{R}}.$$

The disulfide bonds in macromolecules are usually supported by other noncovalent interactions. Thus, it has been shown by Nisonoff and Palmer[12] that the half-molecules of the IgG immunoglobulins combine easily with each other by a disulfide bond, but do not combine so easily with half-molecules from IgG molecules of other specificity. It is therefore clear that mutual close fit of the two half-molecules is brought about not only by the disulfide bond but also by other noncovalent interaction between closely fitting parts of the two half-molecules. We find here a definitely higher tendency for the formation of homogeneous rather than heterogeneous polymers.

HETEROGENEITY PRODUCED DURING DIFFERENTIATION AND MATURATION

Differences between the hemoglobins of different animal species had been claimed more than 100 years ago by E. Körber[13] in Dorpat, the capital of Estonia, which at that time was still part of the Russian empire.

Körber's work was continued by F. von Krüger. After the Frst World War Estonia became an independent republic and dismissed professors who were not of Estonian origin. Professor von Krüger returned to the University of Rostok in Germany. Von Krüger had discovered that the hemoglobins of different animal species can be distinguished from each other by their resistance to 0.05 N NaOH. The time to convert the bright red solution of oxygenated hemoglobin (HbO_2) into the dark brown product of denaturation was designated by von Krüger as decomposition time. The decomposition times found by means of this very simple method were 1 minute for human adult hemoglobin, 3 minutes for canine Hb, 21 minutes for rabbit Hb, 80 minutes for horse Hb, and more than a day for the hemoglobins of sheep or cattle.[14] When I investigated the rate of denaturation spectrophotometrically I found that it proceeded in six different species as a first-order reaction suggesting the presence of a homogeneous hemoglobin in each of the different species.[15]

In 1910 Wakulenko, one of the co-workers of von Krüger in Dorpat, who had returned to Tomsk in Siberia, described a surprising reaction. He found that the hemoglobin of new-born children was about 100 times more resistant to denaturation by NaOH than the blood of the mothers.[16] When I applied the kinetic spectrophotometric method to the umbilical blood of newborn children I found a break in the curve indicating the presence of two different hemoglobins in the newborn child.[17] I succeeded in separating the two hemoglobins from each other and in crystallizing the fetal hemoglobin.[18] The discovery of the fetal Hb was the first definite proof for the presence of more than one HB in an individual.

My finding was at first not accepted because Barcroft, who at that time was investigating the hemoglobin of fetal sheep and their mothers, never found any difference between fetal and adult hemoglobin. I met Barcroft for the first time in 1935 at the International Congress for Physiology in Leningrad. He and I were invited to act as co-chairmen of a symposium on hemoglobin. When I introduced myself to the famous British physiologist he told me: "I never believed in your finding of different fetal hemoglobins, but we repeated your experiments and by God you were right!" Today we know more than 150 different human hemoglobins. Most of them contain two different globins in their tetramers.

We also know that there is a period in which some erythrocytes produce and contain both hemoglobin A (adult) and F (fetal). There is a slow switch from the production of Hb F to that of HB A.[19] It is apparently dictated by the genome. It is remarkable that a similar switch occurs in the production of the human immunoglobulins. Shortly after immunization the first formed antibody molecules are macromolecules which contain heavy chains of the μ type. Both IgM and IgG molecules contain the same light chains. At the time of the switch from IgM to IgG some cells contain both IgM and IgG antibody molecules.[20] Evidently, we deal here with a temporary polymorphism of hemoglobins and immunoglobulins. The genetic background of this phenomenon will be discussed later.

HETEROGENEITY PRODUCED BY POSTTRANSLATIONAL CHANGES

Cohnhein[1] attributes to Kühne the classification of proteins as (I) Proteins, (II) Proteids or conjugated proteins, and (III) Albuminoids. This classification was also customary in the British and American literature. We need not discuss here the albuminoids because they were artifacts obtained by the enzymatic degradation of proteins. The term conjugated proteins has been used for many years, but has lost its importance because most of the so-called true proteins contain covalently bound carbohydrate derivatives. Indeed, it seems that all plasma proteins except serum albumin contain carbohydrate groups and thus are glycoproteins. Another large class of conjugated proteins are the so-called lipoproteins. They are essential components of the biological membranes. However, their lipid content and also the ratio of their lipid components vary within a wide range. Evidently, these lipoproteins are heterogeneous complexes formed by a possibly homogeneous protein and a great variety of lipids.

While it is not yet clear whether the lipids in all lipoproteins are bound covalently, covalent combination of proteins with carbohydrates has been found in the glycoproteins in which the glycosidic residues are usually bound covalently to asparagine, serine, or threonine residues. Other posttranslational changes in proteins are N-terminal acetylation, N-methylation of lysine residues, hydroxylation of lysine and proline (particularly in collagen) and cyclization of N-terminal glutamine, converting it into N-terminal pyroglutamate. If these reactions take place in only some of the proteins produced by the translation of the corresponding m-RNA, a heterogeneous mixture of substituted and unsubstituted proteins will be produced. Indeed, microheterogeneity of the myelin basic protein has been attributed to phosphorylation of some of the serine and threonine residues and to the metabolic conversion of glutamic acid to glutamine.[21]

While the reactions discussed in the preceding paragraphs are anabolic, catabolic reactions can also lead to heterogeneity. The most important of these catabolic reactions are those that result in the production of the final protein from its precursors. Reactions of this type lead to the formation of fibrin from fibrinogen, pepsin from pepsinogen, insulin from proinsulin, and other proteins from their precursors. Even serum albumin, which was mentioned earlier as an apparently homogeneous, crystalline protein is the product of two subsequent hydrolytic cleavages. It is formed by the hydrolytic removal of a strongly basic N-terminal hexapeptide from a substance called pro-albumin.[23] The hexapeptide owes its strongly basic properties to the content of three arginine residues. Pro-albumin, in turn, is formed from a much larger pre-pro-albumin by the hydrolytic removal of its N-terminal 18 amino acid residues.[23] Eight of these 18 amino acid residues have hydrophobic, nonpolar side chains. We do not yet know whether the pre-pro-albumin has another precursor.

THE GENETIC BASES OF HOMOGENEITY AND HETEROGENEITY

The primary structure of a protein, its amino acid sequence, is determined by the DNA sequence in the genome. We know that the process of coding is not infallible and that on rare occasions errors in the duplication of DNA or in the processes of transcription and translation may occur. Such deviations from the "Central Dogma of Genetics" affect usually only one of the three nucleotides of a trinucleotide. For this reason these one-step mutations frequently result in the replacement of the original amino acid by a related amino acid residue. A typical example is the occurrence in the constant portion of human IgG light kappa chains of either leucine or valine in position 191. This small difference has been discovered by producing in rabbits antibodies directed against human kappa chains. It was found that the occurrence of leucine or valine is genetically determined. The two different kappa chains were thus recognized as allotypes; they were designated as Inv^+ (with leucine) and Inv^- (with valine). This example is mentioned here because it demonstrates the extreme sensitivity of the immunochemical methods which made it possible to detect this small difference in one out of approximately 200 amino acids of a peptide chain whose amino acid sequence was not yet known. The origin of this difference is evidently a mutation that must have occurred during the evolution of the human kappa chain several thousands of years ago.

The occurrence of different allotypes must be distinguished from the phenomenon of polymorphism, which also has been observed in the constant regions of the human light chains of the lambda class. These chains occur in two variants designated as Oz^+ and Oz^-; the former contain lysine in position 190; the latter contain arginine in the same position. Both types of lambda chains are found in each human individual, usually in approximately equal concentration. Consequently they cannot be alleles of each other. They are controlled by two different, non-allelic genes and therefore are designated as isotypes.

The antibodies produced in vertebrates after the administration of different antigens are a highly heterogeneous mixture of immunoglobulins differing from each other chiefly in the variable V_H and V_L regions of the heavy and light chains. It is not yet clear whether lymphoid cells that produce immunoglobulins are unipotent, each cell producing only one type of immunoglobulin, or whether they are pluripotent and thus produce antibody molecules of different specificity. Most of the recent investigations indicate pluripotency of antibody-producing cells; the frequent observation of unipotency is attributed to competition between two or more antigens; in this competition the antigen which penetrates into the cell first might initiate the production of the homologous antibodies from their precursors, the heavy and light chains, and thus might prevent other antigens to interfere with this process. In the absence of extrinsic antigens the organism produces a highly heterogeneous mixture of different immunoglobulins.

The formation of allotypes or isotypes, discussed in the preceding paragraph, may be considered as a minor deviation from the Central Dogma of Genetics according to which the amino acid sequence of proteins is determined strictly by the sequence of ribonucleotides in the messenger-RNA whose nucleotide sequence in turn is determined by the deoxyribonucleotide sequence of the DNA molecules. Before 1977 it was extremely difficult to test this dogma by comparing the respective RNA and DNA sequences with the amino acid sequence. Owing to the brilliant methods developed by Sanger and his co-workers[24] and by W. Gilbert at Harvard University[25] it is at present possible to determine within a short period of time the deoxynucleotide sequence of long chains of DNA. Comparison of such sequences with the final amino acid sequence revealed that only certain parts of the DNA and RNA chains are expressed in the protein formed. Since both transcription and translation require the presence of several enzymes and also of soluble factors, these enzymes and factors may initiate or stop transcription and translation. We do not yet know how these enzymes and factors are formed. However, it is clear that the genetic mechanism is much more complicated than originally assumed.

In a recent note Gilbert[26] designated those regions of DNA which are expressed in the final protein as "exons" and those which are lost in transcription or translation as "introns." He suspects that the introns might serve as signals for enzymes such as those that excise tRNA from its precursors, and thus cut out a section of introns. Changes of this type would be quite different from the typical mutations. Tonegawa *et al.*[27] find in immunoglobulins transfer of parts of the original DNA sequence to other parts of the genome; they designate such translocations as changes in the "context" of the genome in contrast to typical mutations, which are changes in the trinucleotide "context" of the gene. These fascinating observations and hypothses make us wonder how any protein preparation can ever remain homogeneous in the presence of all the deviations from the original Central Dogma of Genetics.

HETEROGENEITY IN DISSOLVED PROTEIN MOLECULES

The term heterogeneity has been used in the preceding paragraphs for differences in the primary structure of the proteins, which is determined by their amino acid sequence. While this may result in a single conformation in the solid state of dry proteins or in hydrated protein crystals, it is well known that the amino acid side chains of dissolved proteins continually undergo perturbations, fluctuations, flip-over, and other motions. The intensity of these motions increases at higher temperatures and may then lead to the cleavage of noncovalent bonds or disulfide bridges between amino acid side chains, resulting finally in denaturation. The over-all extent of these changes in conformation can be measured in some instances by means of spectrophotometry, fluorometry, determinations of the relaxation time, and by other methods. In the last few years much more information has been obtained by means of the proton magnetic resonance of the amino acids, and particularly by determinations of the magnetic resonance of small amounts of natural ^{13}C-amino acids, or by labeling the side chains of amino acid residues with ^{13}C-containing or other paramagnetic substituents.[28, 29] The picture that emerges from these investigations of protein solutions shows the dissolved protein as a globule whose hydrated side chains continually fluctuate, collide with each other and thus are quite different from the rigid structure of the dehydrated protein molecules.[30]

Owing to these fluctuations the secondary and tertiary structure of the dissolved protein molecules are never quite homogeneous. Their changing structure can be considered as the result of two competing processes; one of these is the tendency of the amino acid side-chains to random motions, the second is the tendency of the water molecules to combine with each other or with polar residues and at the same time to drive the hydrophobic side chains of some amino acids into the globular parts of the protein molecule. Both processes involve an increase in entropy. As a result of these two processes the nonpolar side chains of phenylalanine residues move only ± 0.6 Å from the center of their side chain, whereas the "floppy," elongated side chains of lysine residues move up to ± 2.3 Å from the center of their side chains.[31] Since only the surface of the dissolved and hydrated protein molecules can act as an antigen, more information concerning the surface groups of the dissolved protein molecules can be obtained by the production of antibodies against the antigenic parts of these molecules. The interaction of these antibodies directed against antigenic portions of the protein surface is specifically inhibited by synthetic peptides whose amino acid sequence is identical with that of the surface area of the antigenic protein molecule.[32,33] In this way it can be found out which parts of the globular protein molecule are in its surface.

Obviously, the conformation of a dissolved protein molecule depends also on the presence of other protein molecules. Thus the conformation of the light chain of an immunoglobulin in its dimers (L_2) is different from its conformation in the complex LH in which it combines with a heavy chain.[34]

THE SIGNIFICANCE OF HETEROGENEITY AND ITS LIMITS

The title of my topic was "Heterogeneity, its History, its Bases, and its Limits." I hope to have shown you its history; twenty years ago I would have shown you quite convincingly its bases and its limits, namely, the Central Dogma: DNA → RNA → Protein as final solution of the problem of protein biosynthesis. However, the action of the genome is much more complicated than that of a template. The genome codes not only for the production of a definite amino acid sequence, but also for the simultaneous production of enzymes that modify the primary genome by translocations and thus give rise to protein molecules that do not reflect completely the trinucleotide sequence of the original genome. Moreover, the genome contains also the codes for the production of enzymes that modify the protein molecules produced by the translation of mRNA. Other enzymes are involved in the biosynthesis of carbohydrates, lipids, and also nucleic acids. Their products, particularly carbohydrates and lipids, seem to be essential for all living cells. The carbohydrates form the backbone of all nucleic acids, and the lipids the backbone of the cell membranes.

Proteins, carbohydrates, and lipids are thus members of a complicated chain of interdependent reactions, each of them forming a link in this chain. In the living organism these links of a chain undergo continual changes and thus must be heterogeneous. They may become homogeneous when we isolate them by removing the adjacent links in the chain of reactions. However, they have then lost their ability of changing continually their conformation. Life involves heterogeneity. If we prevent heterogeneity and isolate homogeneous proteins, we may obtain beautiful crystals. However, they lack the functional reactivity which they display in the living cells.

If we dissolve these isolated homogeneous proteins in a physiological salt solution, they acquire some of their mobility, their ability to change their secondary and tertiary structure. If the dissolved protein happens to be an enzyme and if we add to its solution some of its substrate, two links of the chain, enzyme and substrate, begin to come alive and a metabolic reaction may take place. It is clear from this example that the ability of the proteins to undergo changes in their secondary and tertiary structure is essential for life. Obviously an excess of heterogeneity such as that produced by raising the temperature of an aqueous protein solution beyond a certain limit causes irreversible changes in the structure and in the functions of a protein. The limits of heterogeneity, which I mentioned in the title to this article, separate the functional state of the natural proteins in the biological environment from excessive heterogeneity on the one hand, and from absolute purity and homogeneity on the other hand. In both of these states the protein molecules lose their natural fine structure and function.

REFERENCES

1. Cohnheim, O. 1900. Chemie der Proteine. Vieweg and Sons. Braunschweig.
2. Kühne, W. & R. H. Chittenden. 1884. Z. Biologie *20*: 11.
3. Hofmeister, F. 1889. Z. Physiol. Chemie *14*: 165.
4. Sörensen, S. P. L. 1930. Kolloid-Zeitschrift *53*: 102.
5. Hüfner, G. & E. Gansser. 1907. Arch. f. Anat. u. Physiol. *31*: 209.
6. Adair, G. S. 1928. Proc. R. Soc. Ser. A. *120*: 573.
7. Svedberg, T. & J. B. Nichols. 1927. J. Am. Chem. Soc. *49*: 2920.
8. Singer, S. J. & H. A. Itano. 1959. Proc. Nat. Acad. Sci. U.S.A. *45*: 174.
9. Darnell, D. W. & I. M. Klotz. 1975. Arch. Biochem. Biophys. *166*: 651.
10. Kegeles, G. 1977. Arch. Biochem. Biophys. *180*: 530.
11. Midelfort, C. F. & A. H. Mehler. 1972. Proc. Nat. Acad. Sci. U.S.A. *69*: 1816.
12. Nisonoff, A. & J. L. Palmer. 1964. Science *143*: 3604.
13. Körber, E. 1866. Über Differenzen des Blutfarbstoffes; Inaugural dissertation at the University of Dorpat. Quoted by Krüger, F. v. 1932. Z. f. Vergleich. Physiol. *17*: 337.
14. Krüger, F. von. 1925. Z. Vergleich. Physiol. *2*: 254.
15. Haurowitz, F. 1929. Hoppe Seyler Z. Physiol. Chem. *183*: 78.
16. Wakulenko, J. L. 1910. Communications from the med.-chem. laboratory of the University of Tomsk (Siberia). Quoted by Degel, P. 1932. Z. Vergleich. Physiol. *17*: 337.

17. HAUROWITZ, F. 1930. Hoppe-Seyler Z. Physiol. Chem. *186*: 141.
18. HAUROWITZ, F. 1935. Hoppe-Seyler Z. Physiol. Chem. *232*: 125.
19. BETKE, K. & E. KLEIHAUER. 1958. Blut *4*: 241.
20. NOSSAL, G. J. V., A. SZENBERG, C. L. ADA & C. M. AUSTIN. 1964. J. Exp. Med. *119*: 485.
21. CHOU, F. C. H., C. H. J. CHOU, R. SHAPIRA & R. F. KIBLER. 1977. J. Biol. Chem. *252*: 2671.
22. QUINN, P. S., M. GAMBLE & J. D. JUDAH. 1975. Biochem. J. *146*: 389.
23. STRAUSS, A. W., C. D. BENNETT, A. M. DONOHUE, J. A. RODKEY & A. A. ALBERTS. 1975. J. Biol. Chem. *252*: 6846.
24. SANGER, F., S. NICKLEN & A. R. COULSON. 1977. Proc. Nat. Acad. Sci. U.S.A. *74*: 5462.
25. TONEGAWA, S., A. MAXAM, R. TIZARD, O. BERNARD & W. GILBERT. 1978. Proc. Nat. Acad. Sci. U.S.A. *75*: 1485.
26. GILBERT, W. 1978. Nature *271*: 501.
27. TONEGAWA, S., N. HOZUMI, G. MATTHYSSENS & R. SHULLER. 1976. Cold Spring Harbor Symp. Quant. Biol. *41*: 877.
28. BRAND, L. & J. R. GOHLKE. 1972. Ann. Rev. Biochem. *41*: 843.
29. YGUERABIDE, J., H. F. EPSTEIN & L. STRYER. 1970. J. Mol. Biol. *51*: 573.
30. KANEHISA, M. I. & A. IGEKAMI. 1976. Biophys. Chem. *5*: 131.
31. CRAMPIN, J., B. H. NICHOLSON & B. ROBSON. 1978. Nature *272*: 558.
32. BALDWIN, R. L. 1975. Ann Rev. Biochem. *44*: 453.
33. ANFINSEN, C. B. & H. A. SCHERAGA. 1975. Adv. Protein Chem. *29*: 205.
34. FIRCA, J. H., K. R. ELY, P. KREMER, F. A. WESTHOLM, K. J. DORRINGTON & A. B. EDMUNDSON. 1978. Biochemistry *17*: 148.

DISCUSSION OF THE PAPER

CARTER: I am glad to see that somebody in my generation understands a little of the complexity of gene expression, to which I have been exposed for the last six months in seminars.

FRAENKEL-CONRAT: I have both a comment and a question. The comment is that I think that Dr. Haurowitz has overplayed his concern regarding recent developments in molecular biology. The splicing that he refers to is apparently real and very widespread, but the fundamental fact remains: the central dogma, that the sequence of the nucleotides in DNA is transcribed into the sequence of nucleotides in RNA and codes for a specific protein sequence, is still true. The fact that there are gaps in between and that the situation is becoming much more complicated does not rule out the fact that there is an absolutely reliable relationship between nucleotide sequence and amino acid sequence. No variations in protein sequences result from the newly found complexity.

My question is a historical one. I vaguely remember that heterogeneity of proteins was suggested during the era when proteins were not yet recognized to be decent well-behaving molecules. They were still questionable and the vitalists, so to speak, wanted to have proteins behave quite differently from ordinary molecules and not follow rules. Somebody suggested that every protein molecule in solution might be different from its neighbors. That is, there was random variation of amino acid sequences, and therefore proteins should never be called pure because they would have different sequences. That represented a defeatist attitude, if I remember it right, and I hope that somebody else either confirms or corrects me that this micro-heterogeneity conception raised the question whether proteins sequences would ever be known.

HAUROWITZ: Well, this is true and maybe I presented it too strongly but I could give a few examples. A protein that we had considered absolutely homogeneous was discovered to be heterogeneous. Serum albumin was crystallized by Hofmeister many decades ago and was considered an example of a homogeneous protein. And yet there was found a precursor of serum albumin, and likewise with insulin, pro-insulin was found. Well, then we had two proteins, pro-albumin and albumin. But now just quite recently a precursor for pro-albumin was found, it was called pre-pro-albumin and I do not know how many pre-pro-pre-pro and so forth will finally result. But this is the same thing you find for instance in lactate dehydrogenase, an enzyme made up of A and B subunits; you can have either A_4 or A_3B_1, A_2B_2, A_1B_3, and B_4 so this is again a mixed up substance which apparently was homogeneous but was later found to be heterogeneous. I think this field is still developing and we have no final answer yet.

DISCHE: I think Prof. Haurowitz was very right to draw our attention to this difficulty in stabilizing the physiologically essential units in such molecules as DNA and proteins in some organs. For instance, in the ocular lens, posttranslational changes in proteins are of fundamental importance for the structure of the organ and for its function, and we do not understand at all how such processes are really regulated. Then there is this phenomenon that only a small portion of the DNA nucleotides (3%–6%) is used as determinants for the differentiation of practically all organs of the vertebrate body except the brain for which up to 12% are used. What the function of the rest of the DNA is we have really no idea. And we must remember that in certain vertebrates like the amphibians there is eight times as much DNA as in humans. So we are really at the beginning of the story.

KARLSON: First a comment to Dr. Fraenkel-Conrat: It is indeed true to my knowledge, and I remember the discussions with physiologists in the late 40s and early 50s. There was a general belief for some time that proteins are not homogeneous and will never be obtained in homogeneous forms, but the proponents of these theories also denied the general concept of polypeptides, or at least of one polypeptide chain with one definite sequence. They promoted the concept of colloid chemistry, that there are units of different sizes combining and recombining to various aggregates, and that was the main reason why some people believed in heterogeneity. Now I want to come back to the lecture. I want to make two points.

Homogeneity is at first something which you achieve when you want to purify a protein from a solution that is from authentic material, from biological material. In that case you first want to have criteria of purity, that is, criteria for the process in which you remove the impurities and finally end up with a protein which you can tell is pure.

When I was brought up as a chemist I learned that the criterion for homogeneity with respect to molecular size is determined by the ultracentrifuge. That was around 1940. The second criterion was obedience to the Gibbs phase rule. In this respect I wanted to ask, would albumin, as in your drawing there, obey the phase rule since there is no clear cut evidence regarding solubility?

Well nowadays of course we have other methods, especially chromatographic methods, to check the homogeneity of protein preparations and I would like to ask you, and perhaps others in the audience, what would be the most efficient method nowadays to really define this purity of a protein preparation? That is, the homogeneity of a preparation?

My second point concerns heterogeneity as it occurs in the body, and I would like to stress one point already made by Dr. Haurowitz, that is the posttranslational modification of the molecule. We have many examples and I may mention a few more. A molecule, after it has been translated from a messenger RNA in the form of a polypeptide and chopped to the proper size by removal of the N-terminal and C-terminal residues, can be further modified by acetylation, by methylation in the histone series, phosphorylation, adenylation, and by carboxylation in the case of blood clotting factors and maybe other molecules. And these modifications do not always lead to the conclusion that one has a mixture of protein molecules as a result of posttranscriptional events.

HAUROWITZ: We have to fight here also with linguistics. Let's take a molecule of hemoglobin; when we say a molecule of hemoglobin it just means the tetramer consisting of two alpha and two beta chains. But according to classical chemistry each of these is a molecule; we have four separate molecules which combine to form a polymer. So you see even the idea of a molecule is chemically not quite clear.

CARTER: I think it is a marvelous contribution to the sophistication which we have achieved that our discussion on purity is approaching the philosophical.

EMIL SMITH (*University of California, Los Angeles, Calif.*): It seems to me that the discussion on homogeneity or inhomogeneity of proteins has undergone a whole series of revolutions and one must keep clearly in mind the various distinctions. When in the 1930s Andrée Roche in Paris published a small book on this, she meant that proteins never had a defined composition, and that you could even modify the amino acid composition of the protein by the diet fed the animal. Clearly this notion does not accord with modern views of the genetic determination of protein sequence, and is completely incorrect.

Now in regard to the heterogeneity in part that Dr. Haurowitz referred to in terms of the precursors of active proteins, pro-proteins and pre-pro-proteins, these are definite physiological states in the synthesis of a protein. That does not say that insulin does not have a defined sequence, as Sanger very well proved. And this is certainly now true of a large number of definitely secreted proteins all of which are formed initially as larger precursor molecules.

I think the other thing that we must realize is that proteins are active biological substances existing in an active biological environment, and they are subject to enzyme degradations and modifications during their lifetime in the cell. That does not mean that the protein did not have a well-defined sequence to begin with. We know, for example, from the work of Horecker and his co-workers on aldolase that if you take aldolase out of the muscle of an old rabbit you find that one residue of glutamine or asparagine, I have forgotten which, has become deamidated, whereas if you get aldolase out of the muscles of a young rabbit it is a perfectly homogeneous protein. That deamidation can occur within the protein within the cytoplasm of a living cell, by all means we accept. In the same way the posttranslational modifications which we discovered in the histones, for example, offer a good object lesson. When one of my colleagues found that we were working on the amino acid sequences of histones he thought we were insane because, he said, one could never obtain a histone chromatographically pure. Yet the reason we started is that the amino acid composition gave beautiful stoichiometry and it permitted us to discover the methylation, phosphorylation, and acetylation of histones. There are posttranslational modifications not only of histones but of a variety of enzymes as a part of their physiological role. And one must remember that proteins are dynamic substances and that many of them take part in highly dynamic processes and it may be a modification of a single residue or many residues; and a great deal of the heterogeneity that one is talking about now is a functional heterogeneity, which is extremely important in understanding the role of these enzymes.

JOHN T. EDSALL received his A.B. from Harvard College in 1923. He studied biochemistry and pathology for two years in Cambridge, England, and returned to Harvard Medical School, where he received the M.D. degree in 1928. He continued on the faculty there until 1954, when he moved to Biochemistry in Harvard College, where he is now Professor Emeritus. He has been involved in many different areas of research, ranging from amino acids to war-related work with E. J. Cohn in the Plasma Fractionation Program.

Dr. Edsall has been Editor of, among other publications, the *Journal of Biological Chemistry* and *Advances in Protein Chemistry*. He served as president of the Sixth International Congress of Biochemistry. In addition, he is a member of the National Academy of Sciences and the American Philosophical Society and is involved in various programs on the history of biochemistry, including the Survey of Sources for the History of Biochemistry and Molecular Biology.

The Development of the Physical Chemistry of Proteins, 1898-1940*

JOHN T. EDSALL
Biological Laboratories
Harvard University
Cambridge, Massachusetts 02138

THE USE OF physical chemistry in the study of proteins is only one approach to the understanding of these substances—substances that were recognized from the beginning to be of central importance for living organisms. The physical chemistry of proteins is inseparable from the work of the analytical and structural chemists, so that what I have to say today is closely interwoven with the themes developed by the other speakers, notably with Dr. Fruton's and Dr. Pirie's contributions. Moreover it would be impossible to develop my story without some reference to the immense influence of the ultracentrifuge, and of x-ray diffraction studies, on the study of proteins by physical chemists. Dr. Williams and Dr. Hodgkin will develop those subjects in further detail, and with the personal insight that comes from their outstanding mastery of those great fields of research.

The period I will discuss includes the first forty years of the twentieth century. I have chosen an opening date in 1898, when two Hungarian investigators applied reversible electromotive force measurements in a study of the binding of ions to proteins. That was elegant and solid work, to which I return later. My closing date of 1940 marks not only the onset of a cataclysmic period of world history, but it is in a way the end of an epoch in protein chemistry. By then the nature of proteins as definite molecules was generally accepted, the molecular weights of many proteins were fairly accurately known, and something was known about their general shapes. Great new scientific insights, and rapid and revolutionary advances in our understanding of proteins and their role in biology, were to follow in the years after the war; but that is not the theme of this paper.

The area of science termed physical chemistry was officially baptized in the last two decades of the nineteenth century, with the work of Arrhenius, van't Hoff, Wilhelm Ostwald, Nernst, and others. It centered

* This work was supported by the National Science Foundation (Grant No. SOC 72-05516).

0077-8923/79/0325-0053 $01.75/0 © 1979, NYAS

largely on the thermodynamics and kinetics of substances in solution, and on gaseous and solid phases in equilibrium with liquid solutions. A decade earlier, in 1876, Willard Gibbs had formulated the thermodynamics of multicomponent systems more rigorously, and with deeper insight, than any of his immediate successors; but his papers made hard reading for chemists, and his insight was not fully appreciated for another generation or more. Much that we would call physical chemistry had indeed preceeded Gibbs—think for instance of the work of Dulong and Petit on heat capacities in 1819, or of the formulation of the law of mass action by Guldberg and Waage in 1867. After about 1885, however, the subject acquired a new momentum, manifested by the founding of the *Zeitschrift für Physikalische Chemie* by Ostwald in 1889. Shortly thereafter the new developments began to influence the work of protein chemists. To appreciate this influence, however, we must consider the state of knowledge of proteins at the end of the nineteenth century.

PROTEIN CHEMISTRY AT THE END OF THE NINETEENTH CENTURY

Around the year 1900, and indeed for more than twenty years thereafter, the techniques available for separating, purifying, and characterizing proteins were very limited. Proteins were known to need careful handling, with mild reagents; in general they were altered by heating, even at temperatures well below the boiling point of water, and underwent seemingly irreversible coagulation on subsequent cooling. Exposure to acid or alkali of rather moderate concentration produced similar damage; titrating the solution back to somewhere near neutrality also produced a coagulum, or at least a solution that lacked some of the properties of the original protein. Moreover, protein solutions commonly proved to be good culture media for bacteria. Attempts to separate individual proteins from the complex mixtures usually found in nature commonly involved the use of neutral salts. Shortly after 1850, Panum,[1] Virchow,[2] and Claude Bernard, as reported by Robin and Verdeil[3] had developed the technique of salting out, whereby a protein is precipitated by adding a salt at high concentration—commonly sodium chloride or sodium sulfate; later ammonium sulfate became a favorite salting-out agent because of its high solubility in water. The higher the salt concentration the lower of the solubility of the protein; but the process was reversible. On dilution of the protein precipitate with water, to lower the salt concentration, the protein readily redissolved. Different proteins required different concentrations of salt to precipitate them, so that separations of different protein fractions were possible. The separations, however, were generally not clear cut, and there was seldom any criterion for judging that a given fraction did or did not constitute a pure protein.

P. S. Denis, in his memoir on blood in 1859,[4] made use of the salting out process for plasma proteins, but he also found that some proteins of blood plasma, while soluble in 10% sodium chloride, precipitated reversibly when the solution was diluted with water. These were later denoted as euglobulins, relatively insoluble in water or very dilute salt, but soluble in somewhat higher salt concentrations—concentrations still too small to produce a salting out effect.

The addition of organic liquids, such as ethanol or acetone, to aqueous protein solutions, was another powerful method of protein precipitation. Often, with these reagents, the precipitation was irreversible, but with certain proteins, and with careful addition of reagents in the cold, one might obtain an apparently unaltered, indeed even a crystalline protein preparation. In 1871, Preyer[5] described the crystallization of hemoglobin from some 40 species of mammals, birds, fish, reptiles, and amphibians. Although he employed a variety of methods, he strongly favored the addition of alcohol in the cold as a method of inducing crystallization. He was by no means the first to crystallize hemoglobin; crystals had been obtained from the red cells of some species by 1845 or earlier.[6]†

Hemoglobin was not unique in this respect. Indeed, by the year 1900 an impressive array of other protein crystals had been obtained. In 1889 Hofmeister had crystallized ovalbumin from egg white by salting out, and Hopkins in 1898 had greatly improved the process, by careful acidification of the solution, by adding acetic or dilute sulfuric acid, at high salt concentration. H. Ritthausen in Germany, and later T. B. Osborne in New Haven, had crystallized a whole series of globulins from plant seeds. In 1901, F. N. Schulz[7] gave a valuable survey of the many protein crystals that had been obtained up to that time. I must refer to him, and to a briefer discussion of my own[8] for further discussion of these studies on protein crystals.

LABORATORY EQUIPMENT AND TECHNIQUES IN THE EARLY TWENTIETH CENTURY

It is well to recall briefly the conditions of laboratory research on proteins around the turn of the century. We have already seen some-

† Mr. Horace Judson has since called to my attention the work by F. L. Hünefeld (1840. Der Chemismus der Thierischen Organisation. :160–161. R. A. Brockhaus. Leipzig). Hünefeld observed under a microscope crystalline tabular (tafelförmig) crystals to separate from human and pig blood when the blood was placed between glass plates and partially dried in a desiccator. This may be the first report on crystalline hemoglobin.

thing of the techniques for separating and purifying proteins. For the young workers of today who try to cast their minds back to that early era, it is important to remember not only what was present, but what was lacking. There was of course no ultracentrifuge, no chromatography, no photoelectric spectrophotometry. Biochemists had indeed used spectroscopes since about 1860; one inserted the absorbing solution in the light path, and observed the absorption bands by eye in the visible spectrum. Observations in the ultraviolet required photographic studies; they were technically difficult and tedious, and in general only a few physicists attempted such measurements. No biochemist in those days even dreamed of using infrared radiation to study his materials. Electromotive force measurements were already being done on proteins by 1898, but Sørensen's development of the concept of pH, and of its experimental measurement, was not to come until 1909.[9]

There were indeed centrifuges—very few indeed about 1900, considerably more by 1920—but they were of low power and usually of small capacity. By the time I started working in Edwin Cohn's laboratory at Harvard Medical School in 1926, there were good electrically driven centrifuges that could handle one or two liters of material at a time, and concentrate ordinary protein precipitates in a reasonable time. By that time, of course, Svedberg had developed his first ultracentrifuge in Uppsala, and was getting important results, but we had no ultracentrifuge until 13 years later, and what we then got was not the Svedberg machine, but the air driven model designed by Beams and Pickels, which later evolved into the Spinco instrument.

One important technical problem that could not be adequately handled 75 years ago was refrigeration. When I started working in Cohn's laboratory, there was already a fair-sized cold room in operation. Some of the chemical operations took place there, and that was where we kept our protein solutions, to protect them from denaturation and bacterial contamination, when they were not being actively worked on. The Department of Physical Chemistry at Harvard Medical School—the only department of its kind in any medical school, I believe—was six years old at that time. I think that the cold room had been installed at a very early stage; Cohn certainly regarded it as an imperative necessity. More and better cold rooms were added as the laboratory grew. But in other laboratories it was different. Certainly the Sir William Dunn Institute of Biochemistry in Cambridge, England, was one of the great biochemical institutes of the world, when it opened its doors in 1924. However, Malcolm Dixon, in a gathering held to commemorate that event fifty years later, has written: "You could not buy biochemicals then—you had to make them; so the isolation of aminoacids was very important. So the building had two hot rooms (for running tryptic digests) but no cold room; many incubators but (as far as I can remember) not a single refrigerator! When we wanted ice we sent out to the fishmonger for a great block about two feet high, which had to be attacked with hammer and chisel."[10] J. B. Sumner, in his Nobel prize address in 1946, recalled his early endeavors to achieve crystallization of urease from jack bean meal at Cornell about 1920. He made unsuccessful attempts by adding ethanol to the aqueous extract. "At that time" he wrote "we had no ice chest in our laboratory and we used to place cylinders of 30% alcoholic extracts on our window ledges and pray for cold weather."[11] Later he found that addition of acetone, to 30% by volume, was more promising and he noted: "The acetone extract was chilled in our newly acquired ice chest overnight," and numerous fine crystals of urease appeared by the next morning. This was in 1925 or 1926. I suspect that Sumner may have enjoyed dramatizing the primitiveness of the equipment with which he did the work that started a revolution in enzyme chemistry; but there is no question that laboratories were primitive in those days, by the standards to which the workers of today have become accustomed.

Having taken a look at the background, I turn now to some of the major problems with which the protein chemists of the early twentieth century were grappling. I will take them under four heads: (1) the ntaure of protein molecules, or particles; (2) the electrical properties of proteins, and their behavior as acids and bases; (3) the solubility of proteins, and (4) the denaturation of proteins and its reversal.

PROTEINS AS MOLECULES OR AS AGGREGATES?

In 1903, two years after his review on protein crystals, F. N. Schulz published an extensive discussion concerning what was then known of the size of protein molecules.[12] In contrast to the impressive achievements in protein crystallization, the evidence concerning the size of protein molecules was in general uncertain and confusing. There were some bright spots in a generally cloudy picture, notably hemoglobin. In 1885 Zinoffsky, in the laboratory of Gustav Bunge, had determined the iron and sulfur content of horse hemoglobin, and found the S:Fe atomic ratio to be quite accurately 2 to 1; a figure fully in accord, by the way, with the most recent measurements on this protein. The minimum molecular weight, as judged by the iron content of 0.336%, was close to 16,700.[13]

Closely similar findings were obtained on other hemoglobins. Hemoglobin certainly appeared to be a large molecule, phenomenally large indeed by the standards of the time. For protein molecules in general, however, there was no similar basis for accurate analytical determinations that could fix even a minimum molecular weight in terms of some well-defined constituent. Amino acid determinations, after hydrolysis of the protein, were necessarily of low accuracy.

The physical techniques for molecular weight determination, such as melting point depression or lowering of the vapor pressure of the solvent, would be unreliable for solutes of high molecular weight. There was indeed one remarkable report[14] of melting point depression in carefully dialyzed egg albumin solutions, from which the authors calculated a molecular weight of 14,000; but this was a tour de force, never to my knowledge repeated on any other protein. As we know from later work, even this value was low by a factor of three. As Schulz surveyed the available data he could only conclude: "Im Ganzen ist das Bild ein wenig erfreuliches."[12]

One really hopeful line of attack was through osmotic pressure studies. Given a suitable membrane, with pores small enough to be impermeable to the protein, but large enough to be permeable to water, salts, and small organic molecules, van't Hoff's calculations had shown that the osmotic pressure of a solute of molecular weight of 100,000 or more should be measurable with fair accuracy. Starling[15] had measured the osmotic pressure exerted by the blood plasma proteins, in connection with his studies of the exchange of fluids across the capillary wall between the blood and the tissues. E. W. Reid in Scotland took up the study of individual proteins, and surprisingly found no detectable osmotic pressure in solutions of egg and serum albumin[16] though his method was apparently sensitive enough to have measured the pressure due to solutes as large as 300,000 daltons. A year later. however, Reid reported studies on dog hemoglobin that pointed to a molecular weight of 48,000, about three times the minimum value inferred from Zinoffsky's and other analytical measurements.[17] In the light of later events this was an impressive achievement; Reid's value was still too low, but it was better than any other obtained around that time, or indeed until the work of Adair from about 1920 on. Such judgment by hindsight, however, can be misleading. The work made little impression on Reid's contemporaries, who generally continued to believe that the minimum molecular weight of about 16,700 was the true value. Reid's earlier puzzling findings on egg and serum albumins tended to weaken confidence in his findings on hemoglobin.

Moreover, the concept that proteins could be characterized by definite molecular weights was under attack by some of the colloid chemists. Although in the light of later events they turned out to be wrong on this important issue, one should not underestimate the importance of some of their discoveries. Thomas Graham, a generation earlier, had made pioneer observations on substances of colloidal dimensions, emphasizing in particular their slow diffusion rates and their inability to pass through membranes that allowed smaller molecules to pass freely. The invention of the ultramicroscope by Siedentopf and Zsigmondy, at the beginning of the twentieth century, furnished a powerful impetus to the study of colloids. Particles, formerly invisible, could now be seen, and their motion could be followed. Svedberg,[18] writing in 1946, noted the immense impression that Zsigmondy's book *Zur Erkenntnis de Kolloide* (1905) had made upon him as he started his own research. His own early work was mainly with the ultramicroscope, and the objects of study were generally inorganic colloids—gold sols, mercury sols, sulfur sols. The particles in such preparations were always heterogeneous—their sizes had to be characterized by a distribution function, although the preparations could be made more uniform by careful fractionation. Inevitably investigators, whose first experience of colloids came through the study of such systems, thought about proteins in similar terms. Indeed, in the first decade of the twentieth century, and for some years thereafter, the situation regarding proteins was so complex and confused that it was not unreasonable to regard them also as capable of forming ill-defined aggregates of various sizes and shapes.

The work of Einstein[19] and Smoluchovski[20] on the kinetics of diffusion and the theory of the Brownian movement opened up new vistas. The motion of particles that were visible in the ultramicroscope provided direct evidence to interpret the kinetic behavior of smaller molecules that could not be directly seen. Experimentalists, among whom Jean Perrin[21] was outstanding, soon verified Einstein's predictions, and studied both translational and rotatory diffusion with particles of known dimensions. Such particles were indeed far larger than protein molecules—proteins were much too small to be seen in the ultramicroscope. The available methods of that time could tell nothing of the uniformity or nonuniformity of protein molecules, but the calculations of Einstein, and their verification by Perrin, certainly strengthened the enthusiasm of the colloid chemists and their belief that a new world of hitherto unexplored particle dimensions was opening before them.

It would be a mistake, however, to suppose that the colloidal school ever really dominated the thinking of protein chemists. Workers like T. B. Osborne[22] who had prepared dozens of crystalline proteins and carefully studied their chemical composition and their physical properties, continued quietly to proceed on the assumption that proteins were definite large molecules.

In 1908 Henderson[23] formulated the principles of buffer action, and Sørensen[9] developed the concept of pH and showed how to measure it. In 1911 Donnan[24] treated the thermodynamics of systems containing electrically charged particles enclosed in a membrane impermeable to them, but permeable to all small ions. He showed that the diffusible ions must be distributed unequally on the two sides of the membrane, and that this will give rise to an additional osmotic pressure, beyond that due to the colloidal ion itself.‡

These two developments made possible, for the first time, reliable and reproducible osmotic pressure studies on proteins. By pH and titration measurements one could determine the charge on the protein, and with the aid of Donnan's equations one could correct the data for the effect of the diffusible ions on the osmotic pressure. It was indeed Sørensen, after his fundamental work on the pH scale, who was the first to do osmotic pressure studies, on egg and serum albumins,[27] that carefully took account of these factors. The molecular weights he derived from his data—33,000 for egg albumin, 45,000 for serum albumin—seemed incredibly high to some of the colloid chemists like Wilder Bancroft, who suggested that Sørensen must have been studying aggregates, not real protein molecules; yet later a critical re-evaluation of Sørensen's data, by his younger associate Linderstrøm-Lang, indicated that the true molecular weight values were higher still—45,000 and 70,000 respectively.

A few years later, Adair[28] in Cambridge, England, established the molecular weight of hemoglobin, by osmotic pressure studies, as 67,000, four time the minimum molecular weight deduced from the iron content. Some of the older chemists found Adair's claim at first hard to believe, but all doubts were soon swept away when Svedberg and Fåhreus[29] reported the same value quite independently, from measurements of sedimentation equilibrium in the ultracentrifuge. Moreover, the concentration gradient at equilibrium indicated that the molecular weight was the same at all points in the cell, and therefore that the molecules were all of the same size. This was a point that could not be established from osmotic pressure alone, for this yielded only what we would now call a number average molecular weight. A year later Svedberg and Nichols[30] studied hemoglobin by the sedimentation velocity method, which provided an even more rigorous test of homogeneity by the observation of the form of the single sedimenting boundary.

As mentioned above, Svedberg, before he started to work on proteins, had been concerned with inorganic colloids, which were invariably heterogeneous mixtures of particles of different sizes. He began his work on proteins in the belief that they too would be similar mixtures. This

‡ As pointed out later by Adair,[25] Willard Gibbs had formulated the fundamental principle of the Donnan equilibrium in his great treatise of 1876,[26] but the formulation of Gibbs was so abstract that no one had perceived its significance.

indeed proved to be true for certain protein preparations, such as casein; but the large majority of the proteins he studied soon turned out to be either completely homogeneous, like hemoglobin, or composed of a small number of distinct components.

Having started as a skeptic concerning the concept that definite protein molecules existed, Svedberg became the outstanding champion of the view that proteins were indeed well defined molecules. In a discussion on "The Protein Molecule" held by the Royal Society in November 1938, he concluded that "the proteins are built up of particles possessing the hallmark of individuality, and therefore are in reality giant molecules. We have reason to believe that the particles in protein solutions and protein crystals are built up according to a plan which makes every atom indispensable for the completion of the structure."[31] His faith in the characteristic individuality of protein molecules did indeed go well beyond the then known facts, though later events have abundantly justified it. It was primarily the ultracentrifuge work in his own laboratory, and the beautiful x-ray diffraction patterns of crystalline proteins, first revealed by Bernal and Crowfoot,[32] which had led him to a conclusion so strikingly different from his early views on the colloidal nature of proteins.

EARLY WORK ON BINDING OF IONS AND ELECTRICAL MOBILITY OF PROTEINS

In 1898 two investigators in the Imperial Veterinary College in Budapest, Bugarsky and Lieberman,[33] made what I believe was the first electrometric study of the binding of ions to a protein, egg albumin. They employed the cell:

$$\mathrm{Pt} \mid H_2 \quad 0.05\mathrm{N\ HCL} \parallel 0.05\mathrm{N\ HCL} \ + \ \mathrm{albumin} \mid H_2 \mid \mathrm{Pt}.$$

This cell contained a liquid junction between the dilute HCL and the acidic protein solution. Bugarsky and Lieberman were well aware of the work of Nernst and Planck, and the problem of such liquid junctions. Nevertheless they concluded reasonably that the potentiometric voltage measurements could be interpreted in terms of the Nernst equation:

$$E = (RT/F)\ell n\left(\frac{C_H(0.05N)}{C_H(\mathrm{protein})}\right) = 0.0590 \ \log_{10}\left(\frac{C_H(0.05N)}{C_H(\mathrm{protein})}\right) \text{ at } 25^{\circ}\mathrm{C}. \qquad (1)$$

Here E is the electromotive force in volts, R is the gas constant, T the absolute temperature, and F the Faraday equivalent.

They varied the concentration of albumin stepwise from zero to 12.8 g/100 ml, calculated C_H in the solutions containing albumin from the value of E, and calculated the number of hydrogen ions bound per gram albumin from the difference between C_H in the albumin solution and the total added acid (0.05N). Their results stood up well in comparison with those of later workers, as may be seen from the composite titration curve of egg albumin, constructed in 1925 by E. J. Cohn[34] from the work of several later investigators (FIGURE 1). I return to this titration curve below.

In the next year W. B. Hardy in Cambridge, England made a fundamental observation. In his own words: "Under the influence of a constant current the particles of proteid in a boiled solution of egg-white move with the negative stream if the reaction of the fluid is alkaline; with the positive stream if the reaction is acid."[35] At an intermediate point, denoted by Hardy as the isoelectric point, the particles did not move in either direction. With this paper Hardy, already distinguished as a zoologist and histologist, moved into the study of proteins. I have no room here to discuss his important later contributions, and can merely note that he was one of the great pioneers of protein physical chemistry. Wolfgang Pauli showed a few years later that soluble proteins behaved similarly to Hardy's coagulated egg white; and Leonor Michaelis, after Sørensen's work on pH, was the first to define the pH value of the iso-

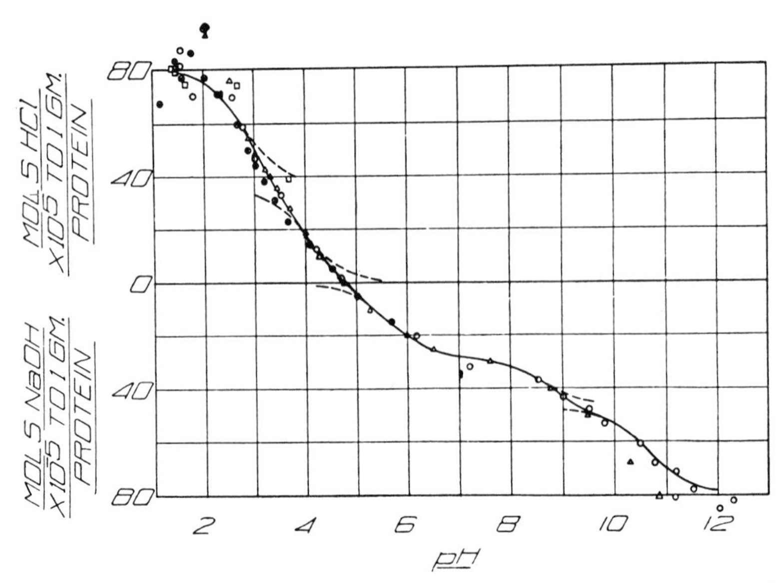

FIGURE 1. Titration curve of egg albumin, as given in 1925 by Cohn.[34] The data of Bugarsky and Lieberman are indicated by open circles. Other data are from Loeb, Hitchcock, and Cohn and Berggren.

electric points of certain proteins, for instance pH 4.7 for serum albumin.[36]

Several workers, notably Pauli and his collaborators in Vienna, studied the titration curves of proteins; it was E. J. Cohn in 1925 who made the first systematic attempt to correlate the titration curves with what was known of the amino acid composition of the proteins.[34] The maximum proton combining capacity should correspond with the sum of the histidine, arginine, and lysine content, plus any terminal α-amino groups present. What Cohn called the maximum base-combining capacity—that is, the maximum number of protons removable from the isoelectric protein—should correspond with the number of ionizable carboxyl and phenolic hydroxyl groups. There were some uncertainties in the analytical data. Nevertheless the correlation between the two sets of data was highly encouraging, and gave promise that Cohn's assumptions were fundamentally correct. He also attempted to correlate the steepness of the titration curve, in various pH regions, with the ionization constants of the groups that were expected to ionize at those pH values from studies on simpler compounds. This was a first step in a long series of researches, in many laboratories. It was gradually recognized that some potential ionizing groups in native proteins were buried inside the molecule, and were released only when the molecule unfolded during denaturation. Crammer and Neuberger[37] demonstrated this most impressively in the case of egg albumin, for which they showed by ultraviolet spectroscopy that the tyrosyl phenolic groups could not be titrated in the native, but only in the denatured protein, about pH 12–13.

One early triumph in the theory of protein titration was the work of Linderstrøm-Lang,[38] who applied the Debye-Hückel theory of interionic attraction, only a year after its publication, to describe the effect of ionic strength on the form of protein titration curves. Linderstrøm-Lang remained a central figure in protein chemistry until his death in 1959, but in this brief account I can mention only this one example of his brilliant contributions.

THE SOLUBILITY OF PROTEINS

At equilibrium between a solution and a solid phase, Gibbs showed that the chemical potential of each component of the system must be the same in both phases. A solution of a single protein in an aqueous salt solution, containing some excess acid or base, may be regarded as a system of four components: water, protein, salt, and acid or base. The pH of the solution may be taken as an index of the acid or base present. (If the protein is strictly isoelectric, only three components are present.)

If salt concentration, pH, pressure, and temperature are all fixed, the chemical potential of the protein should also be determined and it should give a definite solubility, independent of the amount of protein in the crystalline solid phase, according to the phase rule of Gibbs.

Sørensen and Høyrup[39] were the first to carry out such solubility studies, on egg albumin, and they concluded that the albumin could be considered as a pure chemical component of a true solution, in contrast with the views of some colloid chemists who held that so large a molecule as a protein must be regarded as a separate phase, suspended in the water.

A few years later Cohn[34] pointed out that the solubility data of Sørensen and Høyrup, and earlier data of Chick and Martin[40] on egg albumin, in concentrated salt solution, could be fitted by a simple equation:

$$\text{Log}\, S = \beta - K_s M_{\text{salt}} = \beta - K_s' \mu . \tag{2}$$

Here S is the protein solubility, K_s and K'_s are "salting-out constants," μ is the ionic strength, and β is the intercept obtained by extrapolating the data for log$S$ back to zero salt concentration. FIGURE 2 shows Cohn's plot of the data. The fit is remarkably good, even for serum pseudoglobulin which in fact was a mixture of components and not a pure protein at all. Naturally Cohn was eager to test this relation further, and at his suggestion Dr. Arda A. Green, who had recently arrived in his laboratory, undertook what became an outstanding set of solubility studies on

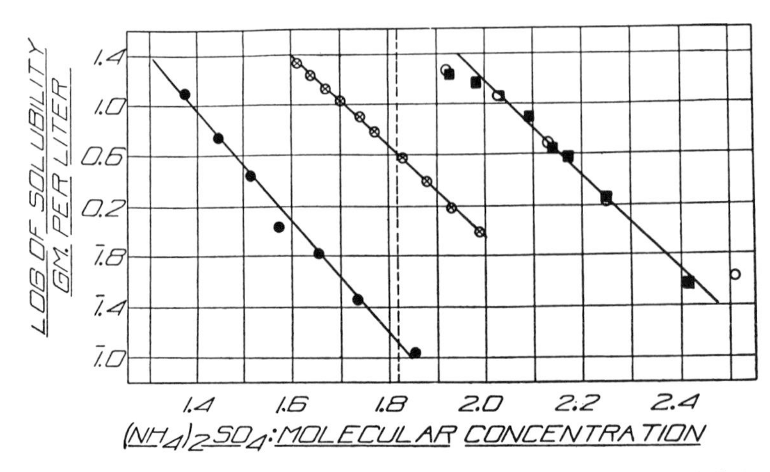

FIGURE 2. The salting out curve of pseudoglobulin from Sørensen (left-hand line), egg albumin from Sørensen and Høyrup (center line), and egg albumin from Chick and Martin (right-hand line). The data of Chick and Martin were obtained at a different pH from those of Sørensen and Høyrup. Data as plotted by Cohn.[34]

horse carboxyhemoglobin, over a range of pH, salt concentration, and temperature,[41] She studied salting out in concentrated phosphate buffers, and showed that the salting out constant K'_s was unchanged by variation of temperature between 0° and 25°C, or by variation of pH between 6.6 and 7.4 at 25°. The parameter β, on the other hand, varied markedly with pH and temperature, as seen in FIGURE 3. The solubility is a minimum near the isoelectric point, and rises as the pH becomes more acid or more alkaline. If the crystalline phase is the isoelectric protein, this is to be expected. The concentration of this form should be essentially independent of pH; but as the pH deviates from that of the isoelectric point, the concentration of charged forms of the protein increases, and so does the total solubility.

Various salts differ greatly in their effectiveness in the salting out of a given protein, as Hofmeister[42] had recognized long ago, when he formulated the well-known Hofmeister series. Hofmeister listed the effectiveness of certain salts in salting out, in the following order:

$$Na_2SO_4 > K_2HPO_4 > Na_3 \text{ citrate} > (NH_4)_2SO_4 > MgSO_4 > NaC1 > KCl$$

This series corresponds to Green's findings, and to the practically universal experience of protein chemists.

Indeed, the salting-out effect is far more general; the great majority of small molecules containing nonpolar groups become less soluble in aqueous solutions, on adding most salts, and the Hofmeister series is found here, as it is in proteins. For these systems an equation similar to (2), but even simpler, applies:

$$\log(S/S_0) = -K_s M_{salt} = -K'_s \mu \qquad (3)$$

Here S_0 is the solubility in the absence of salt, and the simple linear relation for $\log S$ often holds over a wide range of salt concentrations. A good example from the great mass of available data is shown in FIGURE 4, for the salting out of acetone in various salts.[43] The same series of salt effects found in proteins is immediately apparent, although the magnitude of the salting-out constants is far lower than in proteins. The bottom part of the figure shows by contrast the "salting-in" of HCN by the same salts. HCN has a very high dipole moment, and the pure liquid has a dielectric constant higher than that of water. Qualitatively salting-out represents an attraction of the highly polar water molecules for the ions of the salt; the water molecules crowd around the ions, and "squeeze out" less polar molecules like acetone. HCN, by contrast, being even more polar than water, tends to crowd the water molecules out. Qualitatively this general picture seems essentially correct[44] but a good quantitative theory has not, even yet, been developed.[45]

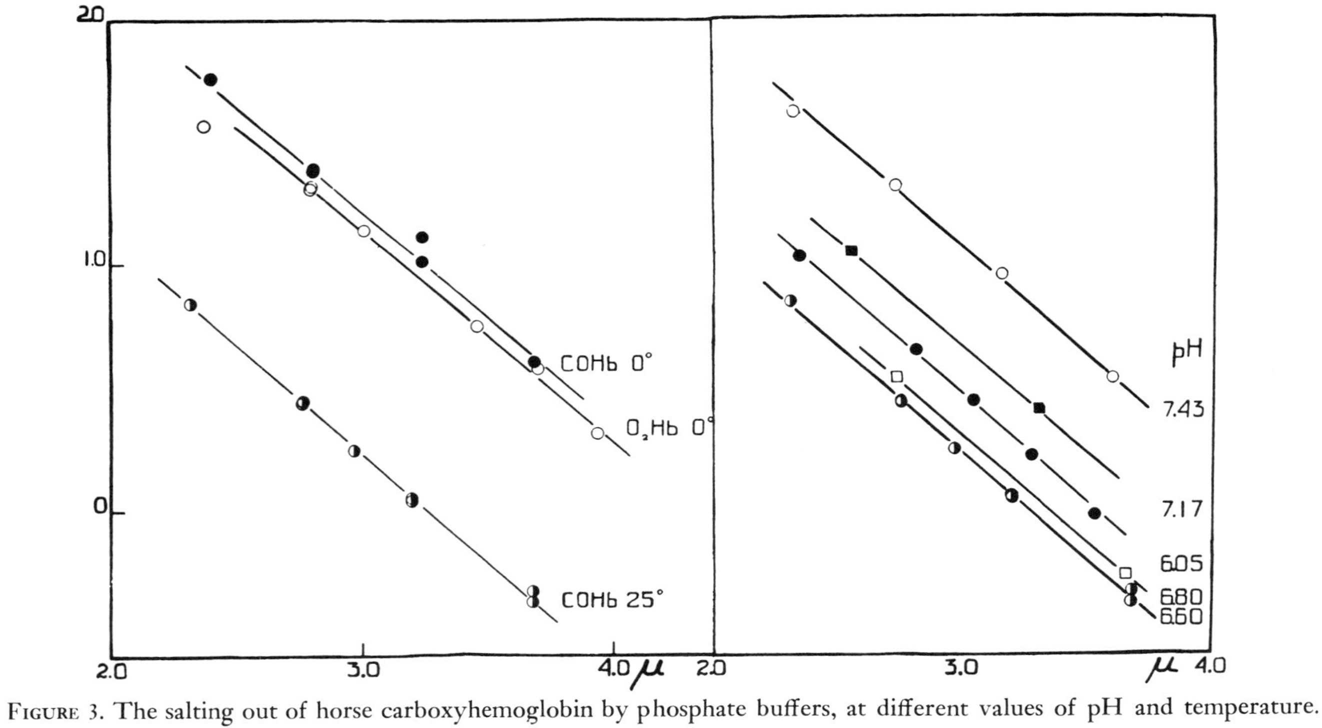

FIGURE 3. The salting out of horse carboxyhemoglobin by phosphate buffers, at different values of pH and temperature. The symbol μ denotes ionic strength. From Green.[41]

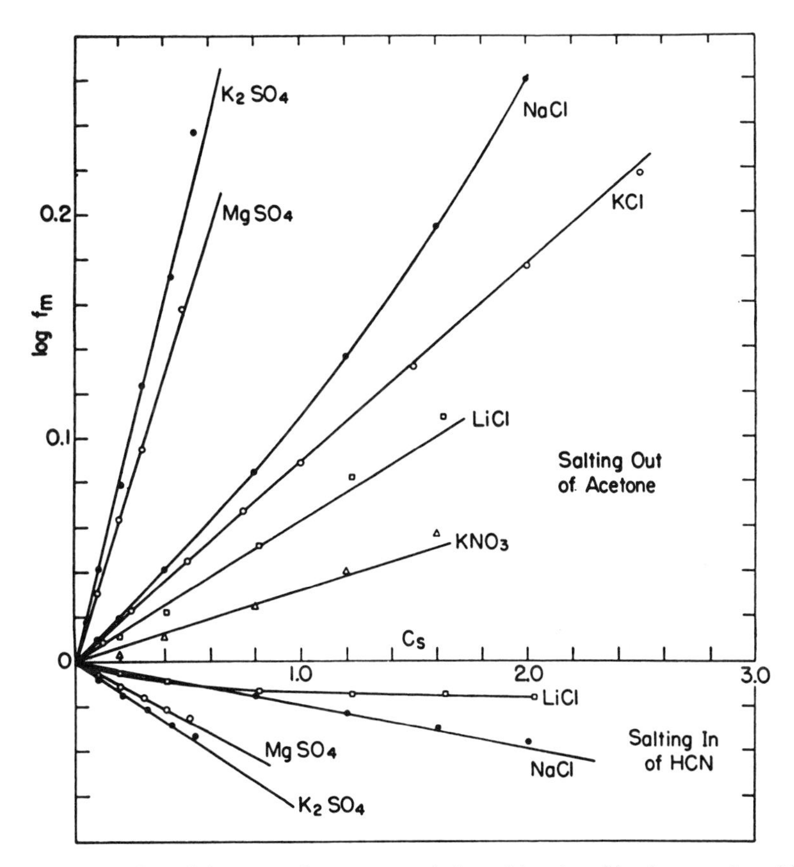

FIGURE 4. The salting out of acetone and the salting in of hydrocyanic acid, by various salts. The symbol C_S denotes molar salt concentration. Data from Gross *et al.*[43] as plotted by Edsall and Wyman.[60]

The action on proteins of salts at *low* concentrations is very different. It had long been known that some proteins, later termed globulins or euglobulins, will dissolve in dilute salt solutions but will then precipitate reversibly on dilution with distilled water. Such observations go back to the early work of Panum[1] and Denis.[4] Green studied these effects quantitatively in horse carboxyhemoglobin, and some of her results are shown in FIGURE 5. The protein has a well-defined solubility in water at zero ionic strength, but all the four salts studied increase its solubility at low ionic strength. In 3M sodium chloride it is about 15 times as soluble as in water; potassium chloride has a similar action, but is slightly less effective as a solvent. Ammonium and sodium sulfates have a moderate solvent action at low concentration, but as more salt is added the solubility goes

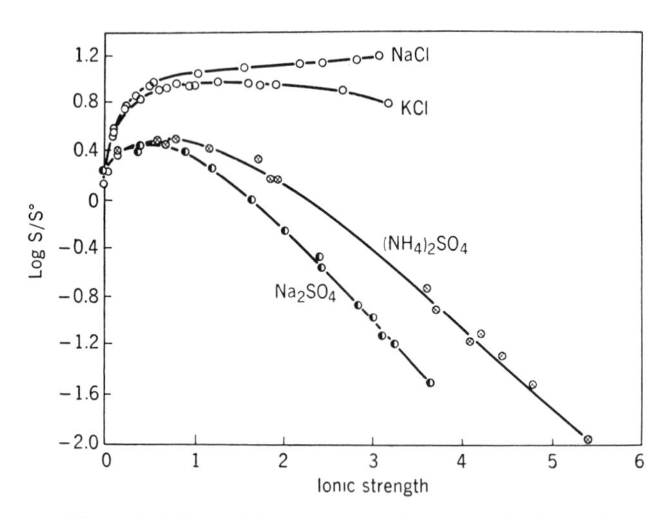

FIGURE 5. The solubility of horse carboxyhemoglobin in various salt solutions. From Green[41] as plotted by Cohn and Edsall. Temperature 25° C, pH near 6.6.

through a maximum and the salting-out effect becomes predominant. At an ionic strength of 5 M in ammonium sulfate the hemoglobin is a hundred-fold less soluble than in water.

The solvent actions of salts at low ionic strength are much greater for some other proteins. For instance isoelectric β-lactoglobulin is about 100 times as soluble in 0.1 M sodium chloride as in water.[46] Such phenomena were recognized by Edwin Cohn[47] and by his colleague at MIT, George Scathard[48] as related to the action of salts in dissolving other slightly soluble salts, such as thallous chloride or barium iodate—actions that were quantitatively interpreted in terms of interionic forces by the Debye-Hückel theory. It was therefore natural to look to simple amino acids and peptides, in interaction with salts, as possible models for the action of salts upon proteins. Isoelectric amino acids and peptides had been recognized by Adams[49] and Bjerrum[50] as *Zwitterionen* or dipolar ions, bearing positive charges on their amino groups and negative charges on their carboxyl groups. Qualitatively, therefore, they would interact rather like ions with the ions of neutral salts, and would be more soluble in dilute salt solutions than in pure water. The theory of these phenomena was worked out by Scatchard, and especially by his younger colleague John G. Kirkwood, as the result of a long series of discussions, interwoven with experimental studies, by Cohn, T. L. McMeekin, J. P.

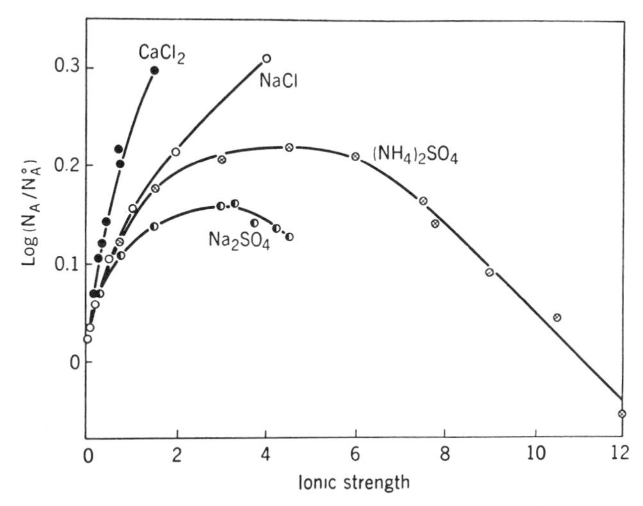

FIGURE 6. The solubility of cystine in aqueous salt solutions. The term N_A denotes mole fraction of cystine. Data of McMeekin, Cohn, and Blanchard[52] as plotted by Cohn and Edsall.[51]

Greenstein, Jeffries Wyman, myself, and others. It is impossible here to attempt a picture of those developments, which were central in my own scientific life for a period of nearly ten years; the results were set forth comprehensively in a book by Cohn and myself.[51] I will merely illustrate them with one example: the solubility of cystine in salt solutions, as determined by McMeekin, Cohn, and Blanchard[52] and shown in FIGURE 6. Qualitatively it bears a striking family resemblance to FIGURE 5 for hemoglobin. The pattern is like that of hemoglobin in minature, with the increase of solubility at low ionic strength, and the salting out by sodium and ammonium sulfates at high ionic strength. The order of the various salts is the same, but the magnitude of the effects is very small indeed, compared to the effects seen in FIGURE 5. Note the difference in scale for both abscissa and ordinate in comparing FIGURES 5 and 6. This illustrates, nevertheless, the fundamental point that there are common principles that underlie the solubility behavior of the simple amino acids and the very complex proteins.

We also observed, in the course of those studies, what would now be termed hydrophobic interactions: every added nonpolar group on the side chain of an amino acid increased its solubility in organic solvents, relative to its solubility in water, and we formulated these relations in quantitative terms. However, we certainly did not then perceive the fundamental importance of such interactions in determining the structure of native proteins. At that time, indeed, I do not think that anyone had grasped this essential conception.

DENATURATION OF PROTEINS AND ITS REVERSAL

The coagulation of proteins on heating had long been known; anyone who has boiled an egg will be familiar with the phenomenon. It was an important advance in 1910 when Harriette Chick and C. J. Martin[53] at the Lister Institute in London, first drew a clear distinction between the initial process of denaturation and the subsequent coagulation of the denatured protein. A denatured protein may still remain in solution, if it is far from its isoelectric point, yet its properties are quite different from those of the native protein. If, for instance, it is an enzyme its activity is lost. At the isoelectric point it coagulates. Chick and Martin measured the kinetics of heat denaturation of egg albumin and hemoglobin and showed that in both cases the reaction proceeded as a first-order process if the pH was controlled. Moreover, the process was characterized by a very high temperature coefficient; a rise of 10°C, in the case of hemoglobin, accelerated denaturation by a factor of 15 or more. For egg albumin the temperature coefficient was substantially higher.

About 1925 it began to be realized that denaturation might be a reversible process. The first study of reversal was apparently that of Mona Spiegel-Adolf[54] in Pauli's laboratory in Vienna, who found that serum albumin, after heat coagulation in boiling water and subsequent cooling, could be brought into solution by careful addition of dilute alkali. The optical rotation, and the specific precipitation by antibody, corresponded to the values for the native protein; and the "reversed" protein underwent heat coagulation like the native protein.

Shortly thereafter Anson and Mirsky[55] succeeded in reversing the denaturation of hemoglobin that had been completely coagulated, by letting it stand in slightly alkaline solution under conditions they defined. Only a fraction of the original hemoglobin was regenerated, but this was indistinguishable from native hemoglobin in solubility, absorption spectrum, and oxygen-combining capacity. In later work Anson and Mirsky improved the conditions for regenerating the protein, and obtained better yields.

One of their most striking results was obtained with Northrop's crystalline trypsin[56, 57] for which they demonstrated the existence of a reversible equilibrium between the native and the denatured protein, varying with temperature. At 42°C, 32.8% of the enzyme was denatured; at 50°C, 87.8%. The data indicated an enthalpy (ΔH) of reaction equal to

67 kcal/mole, the formation of the denatured protein being an endothermic process.

Thus, there was clear evidence by the early 1930s that the denaturation of some proteins was reversible. Some thirty years later the interest in such phenomena revived and was greatly intensified when it was realized that the amino acid sequence of the protein, in a suitable solvent, essentially determined the three-dimensional structure—but that is another story.

Finally I would mention two important early papers on the interpretation of denaturation. The first was by the Chinese biochemist Hsien Wu[58] who concluded that the compact form of the native protein must be maintained by interactions weaker than covalent bonds, which he called "secondary linkages," and that the process of denaturation involved the disruption of these linkages into a random disorganized unfolded structure. Familiar as this concept is today, it represented a major conceptual advance at the time, and Wu's achievement should not be forgotten.

Five years later Mirsky and Pauling[59] considered protein denaturation, and came to conclusions very similar to those of Wu. They did not refer to his paper, and presumably were unaware of it. They stressed the very large increase of entropy that accompanies denaturation, as evidence of an order–disorder transition, and they emphasized the importance of hydrogen bonds in maintaining the structure of the native protein. Like others at that time, they did not consider hydrophobic interactions. Later work has shown that hydrogen bonds, though important, contribute much less to the stabilization energy of the protein than they thought.

These two papers may be said to have launched modern concepts of protein denaturation. The contribution of Mirsky and Pauling is more widely known, but Wu's paper clearly ranks as the first statement of the fundamental picture of the denaturation process.

REFERENCES

1. Panum, P. L. 1852. Virchow's Arch. Path. Anat. *4*: 419–467.
2. Virchow, R. 1854. Virchow's Arch. Path. Anat. *6*: 572–579.
3. Robin, C. & F. Verdeil. 1853. Traité de Chimie Anatomique *3*: 299.
4. Denis, P. S. 1859. Memoire sur le Sang. Paris.
5. Preyer, W. 1871. Die Blutkrystalle. Mauker's Verlag. Jena.
6. Edsall, J. T. 1972. J. Hist. Biol. *5*: 205–257.
7. Schultz, F. N. 1901. Die Krystallisation von Eiweisstoffen und ihre Bedeutung für die Eiweisschemie. Gustav Fischer. Jena.
8. Edsall, J. T. 1962. Arch. Biochem. Biophys. (Suppl.) *1*: 12–20.
9. Sørensen. S. P. L. 1909. Biochem. Z. *21*: 131–304; *22*: 352–356.

10. DIXON, M. 1974. *In* Sir William Dunn Institute of Biochemistry, 1924–1974." Fiftieth Anniversary of the Opening of the Institute, 9 May 1974 (Sir Frank Young, presiding). :10.
11. SUMNER, J. B. 1946. *In* Nobel Lectures in Chemistry, 1942–1962. :116. Elsevier Publishing Co. 1964. New York.
12. SCHULZ, F. N. 1903. Die Grösse des Eiweissmolecüls. :101. Gustav Fischer Verlag. Jena.
13. ZINOFFSKY, O. 1885. Z. Physiol. Chem. *10*. 16–34.
14. SABANEJEFF, AL & N. ALEXANDROW. 1891. J. Russ. Phys-Chem. Soc. *1*: 7.
15. STARLING, E. H. 1899. J. Physiol. *24*: 317–330.
16. REID, E. W. 1904. J. Physiol. *31*: 438.
17. REID, E. W. 1905. J. Physiol. *33*: 12–19.
18. SVEDBERG, T. 1946. Forty years of colloid chemistry. *In* Harold Nordensen 60 år. :340–349. Stockholm.
19. EINSTEIN, A. 1905. Ann. Phys. (Leipzig) *17*, 549; 1906. *19*: 371.
20. SMOLUCHOWSKI, M. 1906. Ann. Phys. (Leipzig) *21*: 756.
21. NYE, M. J. 1972. Molecular Reality: A Perspective on the Scientific Work of Jean Perrin. Science History Publications. New York.
22. OSBORNE, T. B. 1924. The Vegetable Proteins. 2nd edit. Longmans Green. London.
23. HENDERSON, L. J. 1908 Amer. J. Physiol. *21*: 173–179, 427–448; 1909. Ergebnisse der Physiol. *8*: 254.
24. DONNAN, F. G. 1911. Z. Elektrochem. *17*: 572–581.
25. ADAIR, G. S. 1923. Science *58*: 13.
26. GIBBS, J. W. 1931. Collected Works. Vol. *1*: 82–85. Longmans Green. New York. Reprinted from 1876. Trans. Conn. Acad. Sci. *3*: 108–248.
27. SØRENSEN, S. P. L. 1917. C. R. Trav. Lab. Carlsberg *12*: 262–372.
28. ADAIR, G. S. 1925. Proc. R. Soc. London *A108*: 627–637.
29. SVEDBERG, T. & R. FÅHREUS. 1926. J. Amer. Chem. Soc. *48*: 430–438.
30. SVEDBERG, T. & J. B. NICHOLS. 1927. J. Amer. Soc. *49*: 2920–2934.
31. SVEDBERG, T. 1939. Proc. R. Soc. London *B127*: 1–17; also in Proc. R. Soc. London *A170*: 40–56.
32. BERNAL, J. D. & D. CROWFOOT. 1932. Nature *133*: 794–795.
33. BUGARSKY, S. & L. LIEBERMAN. 1898. Pflüger's Arch. Ges. Physiol. 72: 51–74.
34. COHN, E. J. 1925 Physiol. Rev. *5*: 349–437.
35. HARDY, W. B. 1899. J. Physiol. *24*: 288–304; also 1936. Collected Scientific Papers of Sir William Bate Hardy. :294–307. Cambridge University Press. London.
36. MICHAELIS, L. & P. RONA. 1910. Biochem. Z. 72: 38–50.
37. CRAMMER, J. L. & A. Neuberger. 1943. Biochem. J. *37*: 302–310.
38. LINDERSTRØM-LANG, K. 1924. Compt. Rend. Trav. Lab. Carlsberg. *51*: No. 7, 1–29.
39. SØRENSEN, S. P. L. & HØYRUP. 1917. Compt. Rend. Trav. Lab. Carlsberg. *12*: 213–261.
40. CHICK, H. & C. J. MARTIN. 1913. Biochem. J. 7: 380–398.
41. GREEN, A. A. 1931. J. Biol. Chem. *93*, 495–516, 517–542; 1932. *95*: 47–66.
42. HOFMEISTER, F. 1887–1888. Arch. Exp. Pathol. Pharmakol. *24*: 247; 1888–189. *25*: 1; 1890. *27*: 395.
43. GROSS, PH. & M. ISER. 1930. Monatsh. Chem. *55*, 329–337; GROSS, PH. & K. SCHWARZ. *ibid*. *55*: 287–306.
44. DEBYE, P. 1927. Z. Phys. Chem., Cohen Festband :56–64.

45. Scatchard, G. 1976. Equilibrium in Solutions. :136–142. Harvard University Press. Cambridge, Mass.
46. Palmer, A. H. 1934. J. Biol. Chem. *104*, 359–372.
47. Edsall, J. T. 1961. Edwin Joseph Cohn 1892–1953. Biog. Mem. Nat. Acad. Sci. U.S.A. *35*: 47–84.
48. Edsall, J. T. & W. H. Stockmayer. In press. George Scatchard 1892–1973. Biog. Mem. Nat. Acad. Sci. U.S.A.
49. Adams, E. Q. 1915. J. Amer. Chem. Soc. *38*: 1503–1510.
50. Bjerrum, N. 1923. Z. Phys. Chem. *104*: 147–173.
51. Cohn, E. J. & J. T. Edsall. 1943. Proteins, Amino Acids and Peptides as Ions and Dipolar Ions. Reinhold. New York.
52. McMeekin, T. L., E. J. Cohn, & M. H. Blanchard. 1937. J. Amer. Chem. Soc. *59*: 2717–2723.
53. Chick, H. & C. J. Martin. 1910. J. Physiol. *40*: 404–430; 1911–12. *43*: 1–27; 1912–13. *45*: 61–69, 261–295.
54. Spiegel-Adolf, M. 1926. Biochem. Z. *170*: 126–172.
55. Anson, M. L. & A. E. Mirsky. 1929. J. Gen. Physiol. *13*: 133–143.
56. Anson, M. L. & A. E. Mirsky. 1934. J. Gen. Physiol. *17*: 393–398.
57. Northrop, J. H., M. Kunitz & R. M. Herriott. 1948. Crystalline Enzymes. 2nd edit. :138–139. Columbia University Press. New York.
58. Wu, H. 1931. Chinese J. Physiol. *5*: 321–344.
59. Mirsky, A. E. & L. Pauling. 1936. Proc. Nat. Acad. Sci. U.S.A. *22*: 439–477.
60. Edsall, J. T. & J. Wyman. 1958. Biophysical Chemistry. Vol. 1: 271. Academic Press. New York.

DISCUSSION OF THE PAPER

D. Kirschenbaum (*Downstate Medical Center, Brooklyn, N.Y.*): Dr. Pirie alluded to the lack of people who would make a centrifuge for him, and Dr. Edsall alluded to the lack of certain equipment and lack of chemicals and I would like to suggest to the Academy of Science that perhaps at another meeting on aspects of protein biochemistry they might send an invitation for a one-day symposium to the producers of chemicals that have made modern biochemistry what it is. I am certain that there are many people who have had their theses because Mann happened to have the chemicals they needed or else they would still be making the starting material and that perhaps Sigma might have contributed to the thesis. A one-day or two-day symposium of the people who made our degrees possible might be very useful. I know that Fred Mann is still alive because I have lunch with him occasionally and it might be of interest to see some of these other people down here and certainly learn more about how or why we got our theses.

Carter: If there were some source of funds to pay for it. I will have a comment or two to make about the amino acid situation in my paper.

COMMENT: I think it is a very good idea to have a symposium of that kind. I have often wanted to suggest to the companies who make the compounds that they should all along make gifts of the compounds that they market to research laboratories, because the knowledge that they have used to make those compounds comes from research papers. I have often felt that it is very inequitous.

CARTER: Equity has different meanings depending upon whether you are turning to NSF or NIH for your funds I think.

FRUTON: I wonder if Dr. Edsall would care to expand just a bit on one particular thing that happened during the 1920s that had considerable influence on those of us who entered protein chemistry ten years after he did? And that was the appearance of the book of Lewis and Randall, which, at least to some of us, had a very important educational influence in that he went around saying, from thermodynamic considerations, that follows that, and I wonder if you would care to add some more?

EDSALL: I would be glad to comment on that. I graduated from Harvard College in 1923 and in my senior year I had a course in thermodynamics given by a distinguished physicist which left me somewhat baffled and unsatisfied. Just after that the Lewis and Randall book appeared and I bought a copy of it promptly, began to study it, and continued to assimilate Lewis and Randall over the course of the next few years. I remember that when I was first working in Cohn's laboratory, about 1927, there was a small group of us which included Jeffries Wyman, myself, and Magnus Gregersen and a couple of other people. We used to meet one evening a week to discuss various parts of the Lewis and Randall book and work through some of the problems together. This had a tremendous impact on the thinking of all of us. Lewis and Randall put together the fundamental and rigorous thermodynamics in a form that could be assimilated by physical chemists and biochemists generally. No such book had been available before and I think the influence in that book was indeed profound.

I might add by the way, regarding the Donnan equilibrium, that, as some of you know, Adair was one of the few people among the biochemists in the 1920s who could and did read Willard Gibbs. Adair pointed out in a note in *Science* that Willard Gibbs had actually formulated the Donnan equilibrium in his great treatise of 1876, 35 years or so before Donnan. However, the form in which Gibbs stated it was so abstract that nobody before Adair had ever realized the relation between Gibbs and Donnan.

J. W. Williams received his B.S. degree in 1921 from Worcester Polytechnic Institute and his Ph.D. in 1925 from the University of Wisconsin-Madison. In 1973 he received the degree of Sc.D. (hon.) from W.P.I.

Dr. Williams is considered one of the world's greatest authorities on the ultracentrifuge, being involved in it practically from the beginning. He was at the University of Wisconsin in 1923 when Svedberg spent a year there while he was still developing the ultracentrifuge. Dr. Williams was also closely associated with Peter Debye and his work on dipole moments. The ultracentrifuge, however, has remained his primary interest, and his laboratory at the University of Wisconsin is regarded as one of the great centers for ultracentrifuge work.

The Development of the Ultracentrifuge and Its Contributions

J. W. WILLIAMS

Department of Chemistry
University of Wisconsin-Madison
Madison, Wisconsin 53706

DEVELOPMENT

FOR SOME TIME prior to his development of the ultracentrifuge the Swedish physical chemist The Svedberg had been interested in what was then called Colloid Chemistry. His dissertation, presented in 1907 to the Faculty of the University of Uppsala, bears the title "Studien zur Lehre von den Kolloiden Lösungen." In it one finds discussions of two topics: the formation of colloidal systems and the Brownian motion. His experiments with the second of these subjects were claimed to have provided the first verification of the Einstein-von Smoluchowski theory of this motion, but the experiments and their interpretation have not met with universal acceptance; at stake was the then perplexing question of the existence of molecules.

His later proposal to utilize centrifugal force for the investigation of fine-grained systems and of dissolved macromolecules turned out to be a different story, but even here it was some time before acceptance. Thinking about it today it seems strange that the new instrument, the ultracentrifuge, so simple in physical principle, could have become a matter of any contention.

By 1924, the work with the Madison prototype of 1923, and then the first Uppsala ultracentrifuge (equipped with the sector-shaped cell), had reached the point where it could be described in the literature. In Uppsala soon thereafter there had come into being two types of ultracentrifuge, the large oil-turbine, high-velocity machine, and the small, electrically driven equilibrium apparatus. (In the modern commercial instruments the two distinct purposes are served by a single assembly; they are powered by electric-motor, gear-drive combinations, not too different in principle from the drive mechanisms adopted for the first sedimentation equilibrium ultracentrifuges that were designed and constructed in Uppsala.)

In my opinion it was the complexities of installation that discouraged any wide-spread use of the oil-turbine machine in the United States; It

0077-8923/79/0325-0077 $01.75/0

was not, as generally believed, the cost of the apparatus. Two of these instruments were installed in this country during the mid-1930s, one of which went to an industrial laboratory, and the other to the Department of Chemistry of the University of Wisconsin. The Madison machine was in almost continuous daily use for a quarter of a century, 1937–1962.

Space and time do not permit any detailed account of the far-reaching Uppsala researches on proteins; nor is it necessary because a fascinating account of them is to be found in the Svedberg-Pedersen monograph *The Ultracentrifuge.*[1]

At the time the program with the proteins was begun, macromolecules were not supposed to exist; they were almost universally believed to be composed of clusters of ordinary small molecules which formed particles of undefined mass. As the Second World War approached, it became necessary for Svedberg to turn his attention to other channels, but following a "golden period," which extended over the years 1924–1938, he wrote: "We have reason to believe that the particles in protein solutions and protein crystals are built up according to a plan which makes every atom indispensable for the completion of the structure." In passing, it is of interest to note that Svedberg even went so far as to predict: "The x-ray (and possibly the electron-ray) analysis will ultimately give us a complete picture of the architecture of the protein molecule as it exists in crystals."

The Svedberg-Pedersen monograph also contains detailed information about experimental details, the construction and operation of the two types of ultracentifuge, the optical systems, the computational procedures, and other items that continue to be of much value. In the early years of ultracentrifugation, there were those who could not believe that the soluble proteins were so well defined as to structure. And here was this machine, which seemed so often to produce a number from about 20,000 to 70,0000 for molecular weight of a protein! They believed that the real molecular weight was something on the order of 500 and that what was actually being studied in the aqueous systems was the behavior of aggregates formed by chance.

The famous experiments of Svedberg and his associates during the period 1924–1938 constitute a lustrous age for the beginnings of the subject of protein physical chemistry. Unfortunately, they had to be brought to a close as war came again to Europe, and they were not resumed following the restoration of peace.

During the mid-1930s a new field of research involving the "virus proteins" came into prominence. In 1935 Stanley had found it to be possible by chemical means to isolate a crystalline protein from mosaic-diseased tobacco plants that possessed the properties of this virus. At this time such substances were of unknown general nature—they were perhaps bacteria, perhaps protozoa, perhaps enzymes, and so forth. Tobacco mosaic virus (TMV) is a relatively stable molecule, and readily obtainable because of its high concentration in the host.

It was the original suggestion of Stanley that his crystals were "proteins," but by the close of the year 1936, Bawden, Pirie, and Bernal and Fankuchen, after identification, had recognized it to be a "liquid crystalline nucleoprotein."

The development of ultracentrifuges of a new type began with the isolation of many new "virus proteins" (as they were often called) and it was not long before attempts were begun to determine their molecular weights and other properties. Methods based on diffusion and on osmotic pressure, which had been used for other proteins, had failed, largely because of the great size of the molecule. However, the ultracentrifuge, in the hands of Svedberg and Eriksson-Quensel[2] soon provided the desired result for TMV: MW $= 15 - 20 \times 10^6$. The early preparations turned out to be quite inhomogeneous but as the methods were improved so was the quality of the material,[2] and it was not long before the purified substance was made available.

It became increasingly awkward, at best, to be required to send these new substances to the Uppsala Laboratory for molecular weight determinations. Thus, during this same period in the United States, several other types of ultracentrifuges were designed and constructed, largely for virus researches at Rockefeller institutions both in New York and Princeton, N.J. The drive mechanisms were adaptations of the air turbines earlier used for the "spinning tops" of Henriot and Huguenard[3] and of Beams[4] at the University of Virginia. Usually two air jets were used, one to lift the whole rotating assembly, rotor, drive suspension and turbine, (in this way to avoid mechanical bearings) and the other to supply the driving force. (In the post World War II years the air-turbine drive mechanisms were replaced with an electric motor and gear mechanisms; the apparatus became the highly successful Spinco Model E Analytical Ultracentrifuge.)

Actually, it was not a straightforward task to determine the molecular weight of the "virus proteins" because they were known to have a pronounced asymmetrical shape. The limiting value of the sedimentation coefficient alone was insufficient for the purpose; it had to be combined with additional data, normally the corresponding diffusion coefficient, a difficult quantity to obtain for such large molecules. Too, the ordinary asymmetry number, then used, is characteristic of an ellipsoid of revolution, which in this instance was not too bad as an approximation.

It was in 1938 when Stanley began to describe the viruses as being nucleoproteins (or "protein material"). Following the war the study of the architecture of these and many other viruses became a very active field and greatly expanded the subject of virology. In the final phases of the early TMV researches, scientists were able to achieve the separation of the virus into the component parts, protein and RNA, in such a way that the original virus could be reconstituted and to prove that it was after all, the pure virial RNA that carried the genetic information. The protein portion forms a coat over the nucleic acid. Among the other viruses studied by Stanley were those that cause influenza and polio.

The characterization of viruses as molecules provided a new experimental approach to the study of their nature; one of the great pioneers in the subject, Dr. Stanley,[5] at that time of the Rockefeller Institute for Medical Research at Princeton, wrote that "the rise of protein chemistry, as evidenced by the work of Sørensen, Osborne, Loeb, Northrop, Cohn, and especially the ultracentrifugal studies of Svedberg" has not only supplied the colloid chemist with a great group of macromolecules, but has forced down practically all the barriers between colloidal solutions and true solutions. Thus, the purpose of the Stanley report "was to describe the isolation and properties of a new group of macromolecules, the virus proteins."

In some instances self-association or dissociation reactions occur. However, most globular proteins in dilute aqueous electrolyte systems are, for practical purposes, thermodynamically ideal. (This is not true in the case of highly asymmetric molecules because their high excluded volume becomes involved.)

Whereas the early phenomenal progress of the Svedberg Laboratory may have been related to this fact, it was by no means the decisive factor. Nevertheless, to have been able to achieve so much in a period of fifteen years has always appealed to me as having been one of the great achievements in science. There were really two demands made of the ultracentrifuge: to provide satisfactory evidence of the homogeneity of a given protein, and following that to provide its molecular weight.

Actually, there were several problems. The most important ones were charge effects, the "Donnan effects," which were recognized almost at once. They were minimized by the presence of a buffer and the addition of a supporting electrolyte. Then, especially because of the sector-shaped cell that must be used in the transport experiment, there must be an involved theory for the interpretation and evaluation of the data.

The first completely general mathematical analysis descriptive of the equilibrium behavior in a multicomponent, nonideal system was presented by Goldberg[6] in 1953. To the sedimentation velocity case we owe an auspicious start to Hooyman *et al.*,[7] who made use of the thermodynamics of irreversible processes for the treatment of the subject. The utility of these extended theories is great; one of their attributes was that one sees how applicable for the properly conducted experiment were the earlier, simple, formulations of Svedberg, Tiselius, and Pedersen in the interpretation of the optical ultracentrifugal patterns for their protein systems; Svedberg himself had been remarkably prescient in the way in which the solutions for analysis were to be prepared. Further remarks concerning the development of the basic theory are out of place here, and we note only that there now exist adequate treatments for the interpretation and evaluation of the data.

By the mid-1930s interest in the general subject of ultracentrifugation had begun a rapid escalation. The protein physical chemists had not at that time realized that the smaller equilibrium apparatus was the better instrument for the purpose they had in mind; its cost was small. And, for example, it would have been ideal for sedimentation velocity analysis in the virus researches that were popular at the time.

SELECTED CONTRIBUTIONS

From the invitation to participate in this Conference, I seem to have come to the belief that the main consideration would be given to developments in protein chemistry during the period 1930–1950. In a 1953 review Professor Edsall has written: "The molecules most accurately studied hitherto include several plant viruses, several hemocyanins, and a few smaller molecules, such as fibrinogen, edestin, and catalase. There have also been extended studies on such fibrous proteins as collagen and on the structure of muscle and its component structural proteins, myosin and actin."

If we restrict ourselves to studies with the ultracentrifuges, one finds another impressive list; one which covers the period 1925–1939. In mind is Table 48, "Molecular Constants of Proteins," which is to be found in the Svedberg-Pedersen monograph of 1940.[1] In its index there are tabulated for consideration: "respiratory proteins, serum proteins, muscle proteins, milk proteins, chromoproteins from algae, oil-seed globulins, enzymes and hormones, plant virus proteins and other animal proteins and seed proteins."

Each of the above lists is a long one and what remains for me to attempt is to select a few topics for comment. The proteins selected are ones in which I have had more than ordinary curiosity; only one of them represents a subject studied at any length in Madison. In this instance the comments involve both the science and the history of the subject.

A few remarks have been already made about the use of the ultracentrifuge in the elucidation of virus structure, but with another purpose also in mind. I wish to attempt an outline account of researches with (1) the gamma globulins and antibodies, (2) insulin, and (3) the lipoproteins.

Obviously, this is a strictly limited coverage, even when one considers only contributions of the ultracentrifuge during a 20-year period.

GAMMA GLOBULINS AND ANTIBODIES

Begun in Madison in 1939 were studies on the molecular weights of diphtheria toxin, its antitoxin, and their reaction products. The antitoxin available, a gamma globulin of molecular weight 150,000, was but 43% specifically precipitable by the toxin and it was mildly heterogeneous as indicated by our experiments, but it was still quite suitable for the study of the reaction between the toxin and antitoxin. These materials were brought to Madison for the study of their reactions by Professor A. M. Pappenheimer, Jr., now of Harvard University. (All of us here profited much from his presence during the all too short visit.)

Mathematical theories to explain the antigen-antibody reaction, such as between toxin and antitoxin, were still in their very early stages of development; the more detailed one of Goldberg did not make its appearance until 1952. Even at this early date it was demonstrated that the molecular ratio of antigen to antibody in the reaction complex varies continuously, but starts and ends at specific ratios. Furthermore, for the flocculation it was assumed that both antigen (G) and antibody (A) are multivalent. Ultracentrifugal analyses of such reaction mixtures had been begun as early as 1936 by Heidelberger and Pedersen in Uppsala but the University of Wisconsin study was of a different type and more comprehensive in nature. It was possible for us to show that the empirical composition of the complex in the region of antigen excess varied continuously from AG to AG_2; as more antigen was added, the ratio approached AG_2. Similar observations in the region of antibody excess suggested the empirical formula A_7G. These data indicate that the number of combining sites on an antibody molecule (the "valence") is *two*, and not *one*, as many immunologists of the era had believed. The results were published in 1940.

During the progress of these studies, in 1936–1939 there were being conducted a number of investigations that concerned the action of proteolytic enzymes on antibody molecules. The objective was to "despeciate" for human use horse diphtheria antitoxin, horse tetanus antibody, and so forth, by treatments of the protein with several enzymes, such as pepsin, papain, and bromelin, with or without reducing agent, following which, by the application of heat, nonimmunologically active material could be removed from the system. These developments had taken place in pharmaceutical house laboratories.

With their interests being practical in nature, no attempt was being made to find out just exactly what had happened to the intact antibody molecule. The remaining biologically active portion had of necessity to be a smaller unit. During the properly controlled chemical reaction the mixture of molecular subunits had not lost any of its ability either to flocculate with antigen or to inhibit the precipitation reaction. It became of interest as well as of concern to us and others to ascertain whether these reactions had led to a discrete and reproducible fragmentation of the immunoglobulin (IgG) molecules to form true paucidisperse systems. The ultracentrifuge seemed to be the ideal tool for the purpose and by 1940 Tiselius in Uppsala, Northrop in New York, and we in Madison all had set out on this obvious quest. Our Madison group seems to have continued and extended such researches for a considerably longer period of time than did either Tiselius or Northrop, and we felt well rewarded for having done so.

It was possible for us to demonstrate that, under proper conditions, the enzymes pepsin, papain, and bromelin did indeed cause a limited and reproducible cleavage of the gamma globulins and antibodies from equine, bovine and human sera. It was possible for us actually to separate them in a physical way (ultracentrifuge patterns), and to characterize them, but not to do so in a chemical way. Parfentiev had used preferential heating to remove Fc.

In our 1940–46 reports we used the descriptions S_6 and S_4 components, also half and quarter molecules, for the fragments now designated as $F(ab)_2$, F(ab), and Fc, respectively. (At this early date the literature value for the molecular weight of the intact gamma globulins and antibodies was given as 170,000–190,000; it was because of this fact that the descriptive terms half and quarter molecules came into use. It was an unfortunate choice on our part.)

With reference to the description of the researches written by Professor Cohn,[8] and to our publications, it is evident that we had thought to ascertain and *did obtain quantitative immunological data* on whether the active antibody sites on the antibody had withstood the chemical reactions that were necessary for the preparation of the fragments. The assays were performed for the most part in the laboratory of Professor John Enders at the Harvard Medical School in Boston. We continue to be deeply indebted to him for them.

For an early, short, but precise description of the Madison researches that had to do with the enzymatic digestion of gamma globulin, we quote verbatim from an account that Professor Cohn[8] wrote in 1948.

Studies of the splitting of human γ-globulin by various enzymes were undertaken by Drs. Bridgman, Deutsch, and Petermann under Dr. Williams' direction at the University of Wisconsin in the interest of producing antibodies of smaller molecular size. Later these methods were turned toward increasing the yield of γ-globulin and diminishing its depressor effect. The course of the enzyme digestion was followed by ultracentrifugal studies, and the products obtained were studied by chemical and immunological methods.

The ultracentrifugal studies were of much value in determining the proper conditions for the digestion procedure. Immunological studies of the pepsin-digested products indicated little or no decrease in the titer of the antibodies usually studied and clinical tests proved their potency against measles. Traces of pepsin were detected in these preparations by the guinea pig anaphylaxis test, but no evidence of sensitization was observed in humans. Digestion by bromelin or papain led to a higher proportion of quarter molecules and to perhaps some destruction of certain of the antibodies. No appreciable destruction of the diphtheria antitoxin activity had been noted, however, and the deviations in the other tests were not much greater than might have been caused by experimental error in the limited series carried out. An inability of digested γ-globulin to fix complement was observed and is of interest."

At this point we insert TABLE 1, which we believe to be a complete list of the publications for the period 1940–46 that had to do with the use of the ultracentrifuge for the study of the action of proteolytic enzymes on gamma globulins and antibodies.

On a least two occasions Professor R. R. Porter has made reference to a report by Petermann (1946, TABLE 1, No. 6) that requires further scrutiny. Its subject is in part the digestion of human serum gammaglobulin antibodies by papain-cysteine. Our main problem was that our IgG had to be contained in a Cohn Fraction II from a "normal," not hyperimmune, pool. thus all antibody titers had to be very low. One finds in the Porter review[9] the statement: "and human γ-globulin digested with papain or bromelin gave either half or quarter molecules, the half molecules in some cases retaining antibody activity." On analysis, the digest system gave Fab (and Fc) 90%, F(ab)$_2$ 3%, intact γ-globulin 3%. In the Petermann article one finds in its section titled "Immunological Assays" the argument why "the activity *must* be contained in the *quarter*, (i.e., the Fab) fragments" (italics added). In his Nobel Lecture and with reference to exactly the same article, Porter has written that *no* immunological assays were performed!

There is, actually, another quotation which we may mention.[10] "Several new subfragments of IgG have been delineated since the first successful splitting of rabbit (Porter, 1959)[11] and human fragments by papain

TABLE 1
PUBLICATIONS ON THE ACTION OF PROTEOLYTIC ENZYMES ON GAMMA GLOBULINS AND ANTIBODIES 1940–1946

	University of Wisconsin, Madison
(1)	PETERMANN, M.L. & A.M. PAPPENHEIMER, Jr. 1941. The action of crystalline pepsin on horse anti-pneumococcus antibody. Science **93:** 458.
(2)	PETERMANN, M.L. & A.M. PAPPENHEIMER, Jr. 1941. The ultracentrifugal analysis of diphtheria proteins, J. Phys. Chem. **45**: 1.
(3)	PETERMANN, M.L. 1942. Ultracentrifugal analysis of pepsin-treated serum globulins. J. Phys. Chem. **46**: 183.
(4)	PETERMANN, M.L. 1942. The action of papain on beef serum pseudoglobulin and on diphtheria antitoxin. J. Biol. Chem. **144**: 607.
(5)	DEUTSCH, H.F., MARY L. PETERMANN & J.W. WILLIAMS. 1946. Biophysical studies of blood plasma proteins. II. The pepsin digestion and recovery of human γ-globulin. J. Biol. Chem. **164**: 93.
(6)	PETERMANN, M.L. 1946. The splitting of human gamma globulin antibodies by papain and bromelin. J. Am. Chem. Soc. **68**: 106.
(7)	BRIDGMAN, WILBER B. 1946. The peptic digestion of human gamma globulin. J. Am. Chem. Soc. **68:** 857.
	Institute of Physical Chemistry, Uppsala
(8)	TISELIUS, A. & O. DAHL. 1941. Some experiments on the enzymatic digestion of diphtheria antitoxic globulin. Arkiv Kemi, Minerol. Geol. Bd. 14B (No. 31).
	Rockefeller Institute (now University), New York
(9)	NORTHROP, JOHN H. 1942. Purification and crystallization of diphtheria antitoxin. J. Gen. Physiol. **25:** 465; also ROTHEN, A. 1942. *Ibid.* **25**: 487.

digestion (Edelman, *et al.*, 1960)."[12] The Porter article provides proper literature citation but such is not the case in the Edelman report.

There could be in this last quotation some question as to the meaning of the word "successful." In my own case we fail to understand from the sentence quoted above whether the word "first or the word "successful" is the more important to the author of this quote, but it is the second of the two words that troubles us. Wherein did we fail to cause fragmentation of the IgG molecule?

The 1946 article of Dr. Petermann (TABLE 1, No. 6) would seem to deserve more consideration. Its teaching to Professor Cohn has been already quoted. Edelman, *et al.*, describe their preparation as "twice recrystallized" (Mann Laboratories); papain was used to treat the γ-globulin samples under exactly the conditions described by Porter." The record of Dr. Porter could have been written in such a way as to indicate that his IgG-enzyme system corresponds closely to our enzyme system and that the conditions of reaction were otherwise substantially the same as used by us, except that he used crystalline papain. One of our earlier observations had been that "crystalline papain and crude papain split the (bovine) globulin in the same way" (cf. TABLE 1, No. 4).* We cannot escape the suspicion that in all three cases, (Petermann, Porter, and Edelman *et al.*) the *splitting* (or fragmentation) took place in exactly the same fashion. It was necessary in our experiments to use the ultracentrifuge to demonstrate the fact; the techniques of immunoelectrophoresis, suitable separation columns,[13] and so forth came much later. Our molecular weights require no modification.

It is true, of course, that the modern chemical era of antibody research did begin in 1959 with researches by Professor Gerald Edelman and associates at Rockefeller University. Their establishment of the fact that the antibody molecule is made up of multiple polypeptide chains rather than a single chain and the separation by them of the polypeptide chains without breaking them are of course masterpieces. (Note that this work of Professor Edelman is not the same as the enzymatic fragmentation.) Here the several polypeptide chains were separated without breaking them. From these researches there followed the four-chain model of Porter for the gamma globulins and antibodies; it is now commonly accepted.

INSULIN

Van Holde has written in the introduction to his chapter[14] on the "Sedimentation Analysis of Proteins" in the most recent edition of the Neurath and Hill "set":

> It is, perhaps, only a slight exaggeration to claim that the physical study of proteins really began with Svedberg's development of the ultracentrifuge. In the period of a few years, this remarkable scientist and his collaborators developed the principal techniques still used today and established two critical facts about proteins—that they were indeed macromolecules and that they were homogeneous.
>
> The first had already been suspected by many, but the demonstration that these molecules could be sedimented at appreciable velocities in moderate centrifugal fields made the fact vivid, and the application of absolute methods for the determination of particle weights provided a scale for comparison with smaller molecules. The demonstration of homogeneity (although crude at first) was more surprising and contained implications

* The activity of both the crude and the crystalline papain was measured by the hippurylamide method.[27] Reference to the Petermann article (Table 1, No. 4) is pertinent. The amount of enzyme in such units, per mg gamma globulin including IgG, was always carefully regulated.

concerning the biosynthesis of proteins that were not to be fully appreciated for many years.

One present day successor of the subject may be referred to as "associating systems"; they take two forms, self-associations and mixed associations. Both of them are often found in biochemical systems. The recognition that a reaction has taken place or that many proteins are made up of subunits has depended largely upon accurate molecular weight determinations. A very early example of a self-associating reaction is found in pH-stability diagrams for *Helix pomatia* hemocyanin of Svedberg and Heyroth.[15] The ultracentrifuge had amptly demonstrated its usefulness for this kind of study at a very early time.

Theoretical work on the self-association reaction at sedimentation equilibrium began with Tiselius.[16] However, the analysis was restricted to monomer-dimer self-associations that take place in thermodynamically ideal systems. It was a long time (early 1960s) before the theory was generalized and expanded to the point where, with certain simplifying approximations, it became possible to move rapidly ahead.

For a number of years the problems that are associated with the interpretation of the sedimentation equilibrium patterns for heterogeneous nonideal systems had been difficult to solve. For reactive systems the data obtained are apparent average molecular weights (as they are with any heterogeneous systems), and they must be adjusted to true values before any equilibrium constants, reaction mechanisms, and so forth can be deduced. To achieve this end, an approximation was made; it made use of a common second virial equation for all the reactants (which has turned out to have been a close and satisfactory approximation), but it does introduce an additional and unknown parameter into the analysis. This situation is now in a period of rapid change.

The protein systems selected for consideration in Madison have been, first of all, ones that can be readily "purified," with secondary importance being attached to their general overall interest to the biochemist. The effect of a relatively small amount of solute impurity in the study of chemically reacting macromolecular systems in the ultracentrifuge is the failure of apparent weight–average molecular weight curves to be continuous, single-valued functions of concentration for successive low-speed experiments at different initial concentrations. Such behavior has been observed in studies of ovine pituitary adrenocorticotropin, bovine insulin, muramidase, chymotrypsinogen A, and glucagon, for example.

An illustration of what can now be achieved is furnished in the study of self-associating reactions at sedimentation equilibrium; recent researches by Jeffrey, Milthorpe, and Nichol[17] and of Pekar and Frank,[18] who investigated the polymerization pattern of bovine zinc-free insulin at neutral pH, $I = 0.2$, and 25°C, may be cited. The sedimentation equilibrium data are shown to involve one of three different isodesmic indefinite self-association reactions, all of which involve specified oligomeric species, and to permit the computation of a set of equilibrium constants for each pattern. This conclusion is of interest in connection with the discussion of the structure of insulin which appears in a review article of 1972 by Blundell *et al.*[19]

In this same report interesting discussions are found relative to the aggregation of the insulin in solution and the crystal structure of both the natural dimer and of the hexamer.

Apparently the "natural insulin dimer" of weight of approximately 12,000 was first recognized by L. S. Moody,[20] but because of wartime conditions it was not described in the literature in the usual way. The protein was dissolved in 1.6% glycerine solution, which had been brought to pH 3.0 HCl. The observed sedimentation coefficient at pH 3.0 is 1.7S. With the measured diffusion coefficient of 15×10^{-7} cm^2/sec and assumed partial specific volume of 0.75, the molecular weight is in the neighborhood of 11,000; this value was repeatedly found later on by ultracentrifugation, by light scattering, and by osmotic pressure.

The discovery was made by Fredericq and Neurath[21] that this unit is actually a dimer. The monomer molecular weight of beef insulin was soon determined by Ryle *et al.*[22] to be 5734, this is an exact value because it was determined by its amino-acid sequence. Studies with the ultracentrifuge in *neutral solutions* give molecular weight values that vary between 35,000 and 45,000. Oncley and Ellenbogen[23] began the study of insulin as a self-association system, considering the insulin monomer to have a molecular weight of 12,000 and not about 6,000 as Fredericq and Neurath had claimed. They used observed weight average sedimentation coefficient data for their estimations of the equilibrium constants of the reaction $3I \rightleftarrows I_3$ as a function of several variables.

It was much later before the sedimentation equilibrium techniques came into use for the self-association systems. They, with the Rayleigh optical systems and improved methods for the evaluation of the data, have made it possible to extend the subject to a point where more complicated systems may be investigated with greater assurance that a proper recognition of the reaction mechanism may be deduced. For the case of bovine zinc-free insulin at pH = 7 and I = 0.2 at 25°C, Jeffrey *et al.*[17] have concluded from their sedimentation equilibrium experiments that the reaction involves the combination of the dimerization of monomeric insulin, and the isodesmic or indefinite self-association of the dimer, modifications of earlier interpretations by Jeffrey and Coates and of Pekar and Frank.

LIPOPROTEINS

Biochemical studies have provided explanations of how cholesterol is transported in the body. It seems to be generally accepted that virtually all of it, and the triglycerides as well, are bound into the complex macromolecules that are called lipoproteins.

A popular classification for these macromolecules is based on their buoyant densities. Gofman is probably the pioneer in using ultracentrifugation to separate them on the basis of these densities, using four classes for the purpose. They are the chylomicrons, the very low density lipoproteins, the low density lipoproteins, and the high density lipoproteins. Each one of these fractions has been shown to be heterogeneous with respect to size and composition, the latter is the factor that determines the density. Accordingly each system or group, in turn, becomes very difficult to characterize. Also, in many instances these macromolecules are subject to reversible self-association reactions, further to complicate matters.

The objective of the Gofman ultracentrifugal analysis was to determine whether there is a relationship between the relative amounts of one of the several lipoproteins and the tendency to arteriosclerosis with time. In other words it was sought to be able to predict the likelihood of a future onslaught of a heart attack in an individual by a size distribution analysis of the particular lipoprotein in the plasma.

The early results and claims created a sensation, but it soon became evident that they could not serve the purpose for which they were intended. One wonders to what extent these researches were predestined to failure. There are many habits in life that militate against success, but we shall mention only a technical one. It was (and to a lesser extent) still is, the question of the availability of suitable theory for guidance in the interpretation of the experimental records. The data had to be treated in an arbitrary way.

The original Baldwin-Williams[24] theoretical treatment of boundary spreading in sedimentation transport represented but a start with an involved subject; there are many complications, some of which are now understood. The distribution of the idealized sedimentation coefficients of a sample is usually represented by its normalized weight frequency function. Its determination for the lipoprotein system of blood plasma is very difficult because, in addition to the simpler transforms required for the observed boundary or boundary gradient curves of diffusion and of effects of concentration dependence on the sedimentation coefficient, one must make adequate allowances for the density variations of the lipoproteins in any one of the fractions or classes on both corrections.[25]

Actually it was the original purpose of the ultracentrifuge to provide this kind of information about the polydispersity of suspensions. However, the problems that accompany the interpretation of the data to this end were soon recognized to be extremely difficult to solve and it was only after experiences gained over a number of years with the simpler protein and organic high polymer systems, that some of them became at all manageable.

As an alternative, the technique of equilibrium banding in a buoyant density gradient has been suggested and used by Adams and Schumaker.[26] In principle it would seem to be advantageous in application. However, there is always the theoretical problem of the number of components which must of necessity be encountered. Then, there is the practical requirement of the use of long solution columns, and thus long times for the establishment of equilibrium in such a system.

In closing, may I remark that, over all, with its great successes and some failures, analytical untracentrifugation in one form or another is in considerable measure responsible for the fact that proteins are today recognized as being macromolecules. This discovery has had a number of far-reaching consequences; a discussion of some of them is one of the subjects of this Conference.

REFERENCES

1. Svedberg, T. & K. O. Pedersen. 1940. The Ultracentrifuge. Oxford University Press. Oxford.
2. Eriksson-Quensel, I. B. & T. Svedberg. 1936. J. Am. Chem. Soc. *58*: 1863.
3. Henriot, E. & E. C. Huguenard. 1925. Compt. Rend. *180*: 1389.
4. Beams, J. W. 1937. J. Appl. Phys. *8*: 795.
5. Stanley, W. M. 1938. J. Phys. Chem. *42*: 55.
6. Goldberg, R. J. 1953. J. Phys. Chem. *57*: 194.
7. Hooyman, G. J., H. Holtan, Jr., P. Mazur & S. R. de Groot. 1953. Physica *19*: 1095.
8. Cohn, E. J. 1948. *In* Advances in Military Medicine. Andrus *et al.*, Eds. Little, Brown & Co. Boston.
9. Porter, R. R. 1960. *In* The Plasma Proteins. F. Putnam, Ed. Vol. 1. Academic Press. New York.
10. Natvig, J. B. & H. G. Kunkel. 1973. Adv. Immunol. *16*.
11. Porter, R. R. 1959. Biochem. J. *73*: 119.
12. Edelman, G. M., J. F. Heremans, M-Th. Heremans & H. G. Kunkel. 1960. J. Exp. Med. *203*.
13. Peterson, E. A. & H. A. Sober. 1956. J. Am. Chem. Soc. *78*: 751.

14. VAN HOLDE, K. E. 1975. *In* The Proteins. Neurath & Hill, Eds. 3rd edit. Vol. 1, Ch. 4. Academic Press. New York.
15. SVEDBERG, T. & F. F. HEYROTH. 1929. J. Am. Chem. Soc. *51*: 550.
16. TISELIUS, A. 1929. Z. Phys. Chem. *124*: 449.
17. JEFFERY, P. D., B. K. MILTHORPE & L. W. NICHOL. 1976. Biochemistry *15*: 4660.
18. PEKAR, A. H. & B. H. FRANK. 1972. Biochemistry *11*: 4013.
19. BLUNDELL, T., G. DODSON, D. HODGKIN & D. MERCOLA. 1972. Adv. Prot. Chem. *26*: 280.
20. MOODY, L. S. 1944. Dissertation. University of Wisconsin, Madison.
21. FREDERICQ, E. & H. NEURATH. 1950. J. Am. Chem. Soc. 72: 2684.
22. RYLE, A. P., F. SANGER, L. F. SMITH & R. KITAL. 1955. Biochem. J. *60*: 541.
23. ONCLEY, J. L. & E. ELLENBOGEN. 1952. J. Phys. Chem. *56*: 87.
24. BALDWIN, R. L. & J. W. WILLIAMS. 1950. J. Am. Chem. Soc. 72: 4325.
25. ONCLEY, J. L. 1969. Biopolymers 7: 119.
26. ADAMS, G. H. & V. N. SCHUMAKER. 1969. Ann. N.Y. Acad. Sci. *164*: 130.
27. BALLS, M. & M. LINEWEAVER. 1939. J. Biol. Chem. *130*: 669.

DISCUSSION OF THE PAPER

KARLSON: I would like to make two comments and ask one question. The first comment is on Svedberg's work. I think his most important achievement is that he could show, quite unexpectedly, that proteins were monodispersed; that is, most proteins were monodispersed and not polydispersed as the colloid chemists had expected. And the system you mentioned briefly on hemocyanins showing oligodispersed preparations was quite consistent with this. He also proved that the ultracentrifuge could really measure associations and dissociations and thereby disprove the claims of some of the colloid chemists that this method was not of any use.

My second comment: in relation to tobacco mosaic virus I think that along with Stanley who was certainly the first we should not forget the contribution of the late Gerhard Schramm from Germany, a good friend of mine I should say, who in 1944 during the war detected the breakdown of tobacco mosaic virus in alkali into protein moieties which then could be reconstituted to form the tobacco mosaic virus protein molecule, either without RNA or with RNA; you can choose the conditions accordingly.

My question is: In the ultracentrifuge measurements the ratio of F/F_0 played a very important role especially during the early years. This has been explained as an axial ratio of a rotational ellipsoid. We know nowadays the dimensions of a protein molecule due to the x-ray work on many proteins. Does this axial ratio have any relation to the actual size and proportions of the molecule as evidenced by x-ray work?

WILLIAMS: This F/F_0 ratio depends on several factors.

HOWARD K. SCHACHMAN (*University of California, Berkeley, Calif.*): I am afraid we are getting into details. I think the question dealt with the reliability of interpreting shapes of molecules from frictional factors as evaluated from the ultracentrifuge, and it is an oversimplification to argue that the frictional factor gives you directly the axial ratio because Dr. Edsall, who is one of the pioneers in this field, and Dr. Williams himself will tell you that this involves the hydration and swelling of the molecule as well as the actual anisometry. So as a consequence it is not easy to interpret and unravel the distinctions between shape and volume. But insofar as we can interpret the data, the old numbers are very consistent. One of the triumphs of the hydrodynamic approach was that the size and shape of tobacco mosaic virus was predicted, largely by Max Lauffer, on the basis of hydrodynamics long before pictures were available through electron microscopy. So if you ask if the method is reliable, the answer is yes. Today we would use a newer method and will not consider any longer hydrodynamic treatment of ellipsoids but rather deal with much more sophisticated hydrodynamics following the methods of Bloomfield. So I hope that that summarizes very quickly the work of 15 years in that field.

KARLSON: One brief comment. When Prof. Williams attributed reconstitution to Fraenkel-Conrat and Stanley, I decided to keep quiet. But when later reconstitution was attributed to Schramm I decided I had to get up. I will later stress the importance of both of these workers in the field that we are discussing, and I give them much credit, but the reconstitution is mine.

EDSALL: Another historical note I think should be mentioned also, namely, that when Stanley first obtained the tobacco mosaic virus he called it a protein and did not detect the nucleic acid at that time. It was Bawden and Pirie who brought to light the fact that it was indeed a nucleoprotein.

A. H. GORDON left Cambridge in 1940 to join A. J. P. Martin and R. L. M. Synge at the Wool Industries Research Association at Leeds. While there he participated in the work that culminated in the introduction of paper chromatography of amino acids and peptides and also of electrophoresis in silica gel. He remained interested in electrophoresis and demonstrated the usefulness of agar for separation of proteins. Since 1950, at the National Institute for Medical Research, he has been especially concerned with purification and metabolism of plasma proteins.

Electrophoresis and Chromatography of Amino Acids and Proteins*

A. H. GORDON

National Institute for Medical Research
Mill Hill
London NW7 1AA England

THE DEVELOPMENTS that have led up the very many methods now available for separation by electrophoresis and chromatography of amino acids, peptides, and proteins have been described in considerable detail by Morris and Morris.[1] Although entitled *Separation Methods in Biochemistry*, a large part of their valuable book is concerned with methods of separation of amino acids, peptides, and proteins.

The purpose of the present contribution is no more than an attempt to identify those discoveries that seem to have contributed most to the present facility with which samples of very pure proteins are obtained. In addition, the most important advances that have made possible amino acid analysis and sequencing will be mentioned.

As shown in TABLE 1,[2-25] investigation of a particular protein may be divided into two stages; i.e., purification, and then structural analysis. Electrophoresis and chromatography have been the main methods employed in both stages. As is also shown in TABLE 1 the methods that have been devised for analysis and separation of proteins can be divided into those that are capable only of demonstrating the number and concentrations of the individual components of a mixture and those that lead to actual separation of the components. These in their turn can be divided according as to whether the separated components of the original mixture emerge as contiguous bands or whether, on the other hand, the components emerge some distance apart. Examples of the former type of separation are displacement development in chromatography and isotachophoresis, and, of the latter, elution development and zone electrophoresis.

* The section of this paper dealing with paper chromatography originally appeared in another article.[38] It is included here by permission of Elsevier North Holland Press and the International Union of Biochemistry.

0077-8923/79/0325-0095 $01.75/0

TABLE 1

LIST OF KEY DISCOVERIES

	Chromatography			Electrophoresis		
Demonstrations of complexity of mixtures:						
Proteins				U-tube method	Tiselius[3]	1937
				Immunoelectrophoresis	Grabar and Williams[4]	1955
Amino acids and peptides	Frontal analysis and displacement on active carbon	Tiselius[2]	1940			
Separation of proteins	Absorption	Tiselius[5]	1946	On filter paper	Konig[12]	1937
	Cytochrome C on IRC 50	Paleus and Nielands[6]	1950	Continuous flow	Svensson and Brattsten[13]	1949
	Ribonuclease (Partition)	Martin and Porter[7]	1951	Agar slab	Gordon	1949
	Plasma proteins (Ion exchange)	Sober and Peterson[8]	1954	Continuous flow	Grassmann and Hanig[15]	1950
	γ-globulin (Partition)	Porter[9]	1955	Cellulose	Porath[16]	1956

	Carbohydrates and small proteins on granular starch	Lathe and Ruthven[10]	1956	Starch gel	Smithies[17]	1955
	Plasma proteins on Sephadex	Porath and Flodin[11]	1959			
Separation of amino acids and peptides	Amino acids on silica gel (Partition)	Martin and Synge[19]	1941	Amino acids in a silica gel slab	Consden, Gordon, and Martin[23]	1946
	Amino acids and peptides on filter paper (Partition)	Consden, Gordon and Martin[20]	1944	High voltage paper electrophoresis	Michl[24]	1952
	Amino acids and peptides on granular starch	Synge[21]	1944	Finger printing	Ingram[25]	1958
	Amino acids (Ion exchange)	Stein and Moore[22]	1949			

ELECTROPHORESIS OF AMINO ACIDS, PEPTIDES, AND PROTEINS

After the development by Tiselius[3] of the U-tube method, which for the first time allowed quantitative electrophoretic analysis of mixtures of proteins, a further step forward was made by Coolidge,[26] who described a vertical tube apparatus. Because in this apparatus convection was prevented by the presence of glass powder, effective separation of albumin from the plasma globulins was achieved. At about the same time Von Klobuzitsky and Konig[27] reported a simple method which gave some separation of the constituents of snake venom by electrophoresis on a strip of filter paper. The fact that by 1942 Martin was well aware of the possibility of electrophoretic separation of amino acids on filter paper doubtless contributed to his ability to envisage partition chromatographic separations of amino acids on the same matrix. In fact, however, at the Wool Industries Research Association, silica gel in slab form was investigated as a matrix for electrophoresis of amino acids and peptides and was found to be a valuable technique.[23] These experiments were carried out just after the work on partition chromatography,[19] which led to the establishment of paper chromatography.[20] Despite the excellent separations into acidic, basic, and neutral amino acids achieved by electrophoresis in a silica gel slab, the method was never popular because only amino acids and rather small peptides can enter the pores of silica gel and because of the greater simplicity and convenience of paper chromatograms. However, these experiments did demonstrate a very simple electrophoretic method capable of complete separation of components, i.e., the so-called zonal method originated by Coolidge[26] was simplified and improved. Soon after, using almost identical apparatus but with agar gel, zonal separations of proteins was achieved.[14] Once again the method as described was less important than it was to become after a further important modification. This was achieved by Grabar and Williams[4] when they used electrophoresis in agar as the first stage in the method now known as immunoelectrophoresis.

At the same time that agar was under investigation as a matrix for zonal electrophoresis of proteins, Svensson and Brattsten[13] at Upsalla and Grassmann and Hannig[15] in Germany were developing continuous zone electrophoresis. Also at this time extensive studies of zone electrophoresis in vertical columns, packed initially with glass powder, were carried out at Upsalla by Hagland and Tiselius.[28] Comparison of many packing materials and numerous types of apparatus led to the widely used zonal column described by Porath.[16] With the introduction in 1955 of starch gel by Smithies,[17] zonal electrophoresis could again be conducted using only extremely simple apparatus. Because of the sieving effect of the starch gel, molecules of equal charge but different molecular size could now be separated by electrophoresis. The gradual replacement of starch gel by acrylamide initiated by Raymond and Weintraub[18] has led to a nearly optimal matrix with easily controllable pore size and many other advantages.

THEORIES OF CHROMATOGRAPHY

After 1944 chromatographic methods for separation of amino acids, peptides and proteins advanced rapidly. The most important theoretical contributions that made this possible were those of Tiselius,[2, 29] who distinguished between, and was able to demonstrate the existence of, three types of chromatographic system, viz., frontal analysis, elution development, and displacement development,[30] and that of Martin and Synge[19] who introduced into chromatography the concept of partition between a stationary and a moving phase. Earlier only Tswett (cf. Tsvet) (unlike most of his successors) had been clear about the countercurrent principles underlying his discovery.[30, 37] From the sound theoretical base thus established, the search for the most convenient and economical apparatus for chromatography of the vastly complex amino acid and peptide mixtures so readily obtainable from proteins could continue. In fact it was only during the sustained attempt made by Martin and Synge to find new ways of separating amino acids that their concept of the partition chromatogram was developed.

CHROMATOGRAPHY OF AMINO ACIDS AND PEPTIDES

In 1938, when Martin and Synge entered the field, the isolation of a single amino acid from the acid hydrolysate of a protein could take as long as two months. Their attack on the problem led them first of all to construct a countercurrent machine in which acetylamino acids were made to partition between an aqueous and an organic solvent phase.[19, 31] Their most important advance was their discovery that chromatography of acetylamino acids could be achieved in columns packed with granular silica gel. When in this type of chromatogram a moving phase consisting of water-saturated chloroform was used for elution, separations took place with discreet bands of individual acetylamino acid being made visible by the presence of a pH indicator in the stationary aqueous phase. Soon after this work, partition chromatography of unsubstituted amino acids on filter paper was found to be possible.[20] The technique was initiated in the following way: The experiments with silica gel chromatograms had indicated the need for a more sensitive detection system than that provided by the pH indicator. A different approach seemed to be required, both because the indicator failed to reveal very small quantities of acetylamino acids and because it would obviously be an advantage to be able to detect neutral molecules, such as unsubstituted amino acids and peptides. As a trial, several substances known to give colored reaction products were added to portions of the eluate from a chromatogram. The reactions all seemed insufficiently sensitive until ninhydrin was used. To obtain the color a drop of the organic solvent eluate was allowed to fall on to, and then evaporate from, a filter paper. When dry a dilute solution of ninhydrin was put on to the paper, which was then heated to 100°C. Preliminary experiments quickly showed that ninhydrin, used in this way, will give a purple color with microgram quantities of free amino acids and peptides. In fact, this detection system was not used for silica gel chromatograms because very soon after the great sensitivity of ninhydrin on filter paper as a color reagent for amino acids had been demonstrated, Martin made the essential intellectual jump and suggested that the filter paper itself might be a suitable chromatographic matrix. The first trials of this idea were promising, and the method never really looked back.[20]

There were, however, several important questions as to the mechanism on which the new method was based that had to be answered before publication could be contemplated. The most important was whether the paper chromatograms with which successful separations of amino acids were possible were acting as partition chromatograms. In order to resolve this point a number of weighings of filter paper, wet with one or both phases, were done. Happily, the amounts of water that were immobilized in the filter paper fitted well with the concept of a partition system. An unexpected problem, which emerged quite early on, was the curious appearance of the ninhydrin-colored spots, which was specially obvious after use of certain solvents such as phenol. The spots had "pink fronts." The leading edges of all the spots had a quite different color from the main areas, but the separation remained satisfactory. In the absence of any rational explanation of the pink front phenomenon there was a strong temptation to cease trying to understand the underlying cause and to pass on to other aspects of the new method. However, Martin took strong exception to such a course and stated his view that, if we failed to find the cause of the pink fronts, paper chromatography of the amino acids would never be successful. After quite a long period of frustration the answer was found. The pink fronts were the copper salts of the amino acids. Because these had slightly higher solubilities in the organic solvent phase they moved ahead of the amino acid spots themselves. We were very surprised to learn that there were sufficient amounts of copper in the reagents to show up in this manner. Actually much of the copper came from the commutator of the blower motor used to dry the sheets of filter paper. These particles of copper thrown on to the paper from the blower made it possible to identify the nature of the trouble. Proof of the hypothesis that the pink fronts were copper salts was soon obtained, because they disappeared when a complexing agent was added to the organic solvent phase.

The culmination of this work was the establishment of the sequence of the amino acids in the cyclic peptide gramicidin S.[32] Two-dimensional paper chromatography of partial hydrolysates of this peptide permitted separation of a sufficient number of dipeptides and tripeptides to make this possible. Confirmatory evidence was also obtained by analysis of fractions that had been separated by electrophoresis. Somewhat later, when Sanger[33] tackled the far more complex peptide mixtures derived by partial hydrolysis of insulin, two-dimensional paper chromatography by itself could not possibly have yielded sufficient information for establishment of the sequence. Thus, Sanger's success depended on preliminary fractionation by electrophoresis on paper, and it is interesting to note in retrospect that he paid attention to the different shades of color given by the various peptides when heated with ninhydrin.

As paper chromatography of amino acids and peptides became more and more widely used, many attempts to render the method quantitative were reported. None found wide acceptance. Fortunately the chromatographic separations obtained by Stein and Moore[22] of amino acids on ion-exchange resins provided an excellent alternative approach. Because the amino acids after separation on the chromatogram were made to react with ninhydrin under strictly controlled conditions of temperature and time, the color intensities thus produced were found to be extremely reproducible.

CHROMATOGRAPHY OF PROTEINS

The possibility that successful separations of proteins might be achieved by absorption chromatography became much more likely when in 1946 Tiselius[5] demonstrated that an increase in salt concentration led to greatly increased absorption of proteins by substances such as silica gel and that the effect is fully reversible. However, the earliest fully successful chromatographic separations of proteins were carried out on the ion exchange resin IRC 50 in 1950 by Paleus and Neilands[6] and by Hirs, Moore, and Stein,[34] but because only small basic proteins such as cytochrome-c and ribonuclease had been used, there was still doubt whether albumin and globulins could be separated chromatographically. Indeed, in 1952 Moore and Stein[35] sounded a cautionary note on this point. However, within three years Porter[9] was able to report successful chromatography of γ-globulin using a partition system with a kieselguhr matrix. The conditions under which such a system could be made to work were found to be extremely restricted and might not have been arrived at without much preliminary work with more stable proteins, including ribonuclease and insulin. The separation of antibody from nonantibody proteins thus brought about certainly marks a most important step forward in the methodology of protein purification. Almost equally important was the introduction in 1954 by Sober and Peterson[8] of an entirely new category of ion-exchangers, the substituted celluloses. Probably because ion-exchange chromatography of plasma proteins on these materials was technically rather simple, and because their load factor was much higher than the synthetic resins then available, with their introduction plasma protein fractionation advanced very rapidly.

Columns packed with granular potato starch eluted with butanol were first used by Synge[21] for separation of amino acids and peptides derived from gramicidin. For several years after 1944 the question of whether starch used in this way constituted a true partition chromatogram remained unsolved. Only with the work of Lathe and Ruthven[10] did the importance of the molecular size of the substances undergoing separation on such chromatograms become truly apparent. They showed that with untreated starch, molecules larger than 5000 MW were completely unable to enter the swollen grains. Furthermore, smaller molecules, which were able to enter the starch grains, separated chromatographically in correspondence with their molecular size rather than their partition coefficients between the mobile phase and the complex water/polysaccharide stationary phase. Undoubtedly this work opened the way to the introduction of the cross-linked dextrans now available as the various grades of Sephadex.[11] The use of such products, for the operations rather loosely described as "gel filtration" on "gel permeation" chromatography, has vastly extended the range of chromatographic techniques available to the protein chemist.

In retrospect, it is clear that the successful separation of amino acids and peptides by partition chromatography greatly stimulated the whole field. In the decade after 1950 chromatographic and improved electrophoretic methods of separation of proteins and automatic amino acid analysis on ion exchange columns led to an enormous expansion of available techniques. It is also apparent that the theoretical and practical contributions made by Tiselius[2,3,5,28-30,36] from 1937 until his death in 1971 contributed in a quite outstanding degree to the advancement of the field. His presence and penetrating analyses of the contributions at the meeting of the Faraday Society at Reading[36] were a particular stimulus to the exploitation of molecular-sieve and affinity effects both in chromatography and electrophoresis during later years.

I am most grateful to Professor R. L. M. Synge, FRS for advice on a number of points.[37]

REFERENCES

1. Morris, C. J. O. R. & P. Morris. 1976. Separation Methods in Biochemistry. 2nd edit. Pitman. London.
2. Tiselius, A. 1940. Arkiv Kemi. Mineral. Geol. *14B* 22.
3. Tiselius, A. 1937. Trans. Faraday Soc. *33*: 524.
4. Grabar, P. & C. A. Williams. 1953. Biochim. Biophys. Acta *10*: 193.
5. Tiselius, A. 1946. Ark. Kemi Min. Geol. *26*: 1.
6. Paleus, S. & J. B. Neilands. 1950. Acta. Chem. Scand. *4*: 1024.
7. Martin, A. J. P .M. & R. R. Porter. 1951. Biochem. J. *49*: 215.
8. Sober, H. A. & E. A. Peterson. 1954. J. Amer. Chem. Soc. *76*: 1711.
9. Porter, R. R. 1955. Biochem. J. *59*: 405.
10. Lathe, G. H. & C. R. J. Ruthven. 1956. Biochem. J. *62*: 665.
11. Porath, J. &. P. Flodin. 1959. Nature *183*: 1657.
12. Konig, P. 1937. Actae Trab 13th Congr. Sudamer. Chim. *2*: 334.
13. Svensson, H. & V. Brattsten. 1949. Ark. Kemi. *1*: 401.
14. Gordon, A. H., B. Keil & K. Sebesta. 1949. Nature *164*: 498.
15. Grassmann, W. & K. Hannig. 1950. Naturwissenschaften *37*: 397.
16. Porath, J. 1956. Biochim. Biophys. Acta *22*: 151.
17. Smithies, O. 1955. Biochem. J. *61*: 629.
18. Raymond, S. & L. Weintraub. 1959. Science *130*: 711.
19. Martin, A. J. P. & R. L. M. Synge. 1941. Biochem. J. *35*: 1358.
20. Consden, R., A. H. Gordon & A. J. P. Martin. 1944. Biochem. J. *38*: 224–232.
21. Synge, R. L. M. 1944. Biochem. J. *38*: 285.
22. Stein, W. H. & S. Moore. 1949. Cold Spring Harbor. Symp. Quant. Biol. *14*: 179.
23. Consden, R., A. H. Gordon & A. J. P. Martin. 1946. Biochem. J. *40*: 33–41.
24. Michl, H. 1952. Mh. Chem. *83*: 737.
25. Ingram, V. M. 1958. Biochim. Biophys. Acta *28*: 539.
26. Coolidge, T. B. 1939. J. Biol. Chem. *127*: 551.
27. Von Klobuzitsky, D. & P. Konig. 1939. Arch. Exp. Path. Pharmakol. *192*: 271.
28. Haglund, H. & A. Tiselius. 1950. Acta Chem. Scand. *4*: 957.
29. Tiselius, A. 1943. Kolloid Z. *105*: 101.
30. Synge, R. L. M. 1962. Arch. Biochem. Biophys. (Suppl.) *1*: 1.
31. Synge, R. L. M. 1939. Biochem. J. *33*: 1913.
32. Consden, R., A. H. Gordon, A. J. P. Martin & R. L. M. Synge. 1947. Biochem. J. *41*: 596.
33. Sanger, F. & E. O. P. Thompson. 1953. Biochem. J. *53*: 353.
34. Hirs, C. H. W., W. H. Stein & S. Moore. 1951. J. Amer. Chem. Soc. *73*: 1893.
35. Moore, S. & W. H. Stein. 1952. Ann. Rev. Biochem. *21*.
36. Tiselius, A. 1949. Discuss. Faraday Soc. (9):7.
37. Synge, R. L. M. 1970. *In* British Biochemistry Past and Present (Biochem. Soc. Sympos. No. 30) T. W. Goodwin, Ed.: 175, Academic Press. London.
38. Gordon, A. H. 1977. Trends Biochem. Sci. *2*: N243.

DISCUSSION OF THE PAPER

EDSALL: I am sure that there are questions that some people would like to raise concerning the chromatographic and electrophoretic methods. One astonishing thing is to consider the simplicity of these methods and note, nevertheless, the fact that they were discovered so relatively recently.

I remember a communication we had from Dr. A. J. P. Martin to the Survey of Sources for the History of Biochemistry. He remarked that he thought paper chromotography might have been discovered anytime after 1800 or thereabouts, but in fact of course it was not. How far is it helpful in the study of history to speculate on why certain things did not happen much earlier than they did? I do not know the answer but it is I think an interesting question.

GORDON: You probably know, Prof. Edsall, that in the dyestuffs industry in the 19th century in Germany, paper chromatograms of a fair degree of sophistication were in fact being used. However, the scientists did not pay any attention. Then there was the oil industry in this country in which chromatography was used quite a lot. And still the scientists did not pay much attention.

ROBERT BERNSTEIN (*Princeton University, Princeton, N.J.*): It is a general question of historical and philosophical interest to know why people make certain discoveries and since you are discussing, to a certain extent, some of your own work, I would like to take advantage of the opportunity to ask you a question about your agar work. I assume that there are several possible gels you could have worked with. Why did you pick agar from among these possibilities, and then did you go on to test the other possibilities that you might have thought of?

GORDON: Agar was the easiest to try but we did try gelatin as well because we felt it would be a pity to miss out on something more useful; but gelatin was very unpromising so we came straight back to agar.

KIRSCHENBAUM: I have an interest in amino acid analysis and when I go through the papers it is interesting to note that the development of amino acid analysis can be divided into two periods at least for me: before Moore and Stein and after Moore and Stein. If you go look at the amount of space in publications devoted to amino acid analysis per paper, it becomes an interesting relationship. Right after the first publications most of the papers concerned with analysis had long paragraphs on amino acid analysis using the Moore and Stein technique; every paper went into great detail. As you go through the years of the 60s, the amount of such material decreased until nowadays one usually finds two sentences: "An amino analysis was done by the Moore and Stein method." One then finds in the body of the paper the results, "The amino acid analysis can be found in table 6." So we have gone from perhaps a paragraph down to two complete sentences.

GORDON: This is quite understandable because agreement has emerged concerning the best and most convenient way to do these things. It is only repetitive to keep on quoting names after that. I believed it is no longer necessary even to say Moore and Stein; just say amino acid analysis.

KARLSON: There are two approaches and I would like to point out another approach started by Bergmann and his colleagues in devising reagents specific for one amino acid. The principle was: if you have a mixture of amino acids and you can measure by color reactions or other methods just one amino acid out of the mixture, you need not separate them. The other approach is that you have one general method of detecting amino acids, namely, the ninhydrin method, and you then have to separate them. Obviously the second approach was the more successful.

GORDON: Yes, that is right; there was also the microbiological method which had evolved where you were able to pick out amino acids from the mixture by the specificity of a microorganism.

EDSALL: Yes, the first complete amino acid analysis for a protein was done by Erwin Brand and his collaborators on β-lactoglobulin about 1945. That was done by microbiological analysis. When Moore and Stein studied the same protein a few years later, they found certain corrections had to be made in Brand's data. However, at that period the microbiological techniques were very important. It was a brief period but this method played a major part in amino acid analysis in the late 1940s.

EMIL L. SMITH received his Ph.D. at Columbia in 1936 and later worked with Keilin in Cambridge, Vickery at Yale, and Bergmann at the Rockefeller Institute. He was a member of the Biochemistry Department at the University of Utah from 1946 to 1963. Since 1963 he has been head of the department at the University of California Medical School in Los Angeles. His contributions to protein chemistry have been varied and extensive. He is a member of the National Academy of Sciences, the American Philosophical Society, and the American Academy of Arts and Sciences.

Dr. Smith, together with Abraham White and Philip Handler, is the author of the book *Principles of Biochemistry*, now in its sixth edition.

Amino Acid Sequences of Proteins—The Beginnings*

EMIL L. SMITH
Department of Biological Chemistry
UCLA School of Medicine
Los Angeles, California 90024

THE EARLY HISTORY of the study of the amino acid sequences of proteins can, perhaps, be more readily appreciated if we can recall the general state of protein chemistry immediately prior to this time. The extraordinary burst of activity in the investigation of amino acid sequences resulted from the confluence of both conceptual developments and experimental methods. Although some of these contributions will be discussed in detail by others at this meeting, it seems useful to remind ourselves of the situation at that time, although only brief attention can be devoted to some of the factors that led to the subsequent rapid developments.

Of greatest importance was the increasing evidence in the period between 1925 and 1935 that many important biological activities were due to specific proteins, including enzymes, antibodies, oxygen carriers, and some hormones. Although there was considerable information to support these views much earlier, it was not until these concepts were put on a firm experimental basis that protein chemistry aroused wide interest among biologists and chemists.

During this same period, it had been fairly well established that the major covalent linkage between amino acid residues in proteins was the peptide bond. Concurrently, it was recognized that most, if not all, of the amino acids commonly found in proteins were known,[1] with glutamine, asparagine, and threonine being the last to be firmly established as general constituents of proteins[2]; nevertheless, in no instance was there available a complete accounting of the amino acid composition of a protein.[3]

* In order to conserve space, many of the references are to comprehensive summaries rather than to individual papers. For those who wish to sample some of the flavor of the discussions of protein structure during part of the early period under review, useful sources are the Cold Spring Harbor Symposia of 1938 (Vol. 6) and 1949 (Vol. 14), and the Ciba Foundation Symposium, "The Chemical Structure of Proteins," held in December, 1952 (G. E. W. Wolstenholme & M. P. Cameron, eds., J. & A. Churchill, Ltd., London, 1953).

This work was supported by Grant No. GM 11061 of the National Institute of General Medical Sciences of the United States Public Health Service.

0077-8923/79/0325-0107 $01.75/0 © 1979, NYAS

Although some investigators continued to propose that other types of covalent linkages might exist in proteins, experimental support was lacking and these notions eventually faded from the scene.

There was one problem that was not so easily laid to rest, namely, the phenomenon of protein denaturation. Even after Hsien Wu[4] suggested that native proteins were specifically folded and held together by noncovalent forces, whereas denatured ones were randomly disorganized, there was still no clear notion as to the forces maintaining the native structure of proteins. After Mirsky and Pauling[5] proposed in 1936 that hydrogen bonds were involved in maintaining the native structure, there was still only a slow and gradual acceptance of the view that covalent bonds were not the primary forces involved. I suppose that one of the major reasons for accepting this proposal were the studies, mainly by Anson[6] and Mirsky, of the reversible denaturation of serum albumin and hemoglobin, and of trypsin, chymotrypsin, and pepsinogen by Northrop, Kunitz, and Herriott,[7] under conditions that could not readily be assumed to break or produce covalent bonds.[8] As we now know, the forces maintaining the native structure of proteins are much more complex, involving not only hydrogen bonds but also hydrophobic and ionic interactions.

It was also recognized that many proteins contain substances other than amino acids. Perhaps the best known early example of a conjugated protein was hemoglobin. It was also accepted that some proteins contain covalently linked carbohydrate. It is largely to research in the 1930s and later that we owe the discovery and characterization of most of the organic cofactors or coenzymes associated with enzymes, some of which are covalently linked. It is also noteworthy that it took some time for general acceptance of the view that the protein played an essential role and was not merely a passive inert carrier of the organic cofactor, a view strongly held by Willstätter[9] and his school, as well as by some others.

These concepts of the function and structure of proteins were of course accompanied by the development of methods that provided the experimental basis for the general acceptance of these views. Some of these will be mentioned briefly, since others will discuss more fully the history of certain of these experimental contributions. First of all, there was the growing number of well-characterized, relatively pure proteins, including some crystalline enzymes. Second, confidence in the homogeneity of these proteins was accompanied by the development of physical methods for their study. The oil-turbine and later other types of ultracentrifuges[10] aided in demonstrating that each native protein had a definite molecular weight and could be obtained in a state free from obvious impurities of different size. Of equal importance was the later development in 1937 by Tiselius[11] of an electrophoretic method of high resolving power that aided in demonstrating that preparations of proteins could be obtained that were homogeneous in charge. Thus, there developed considerable confidence in the view that once a protein was obtained in a form that was monodisperse both by ultracentrifugation and by electrophoresis at a few pH values, it was safe to assume that a high degree of purity had been achieved. That this has not always proved to be the case in later studies by employing more refined methods of chromatographic or electrophoretic analysis is beside the point. What was important at that time was that investigators could proceed with more detailed studies of composition, structure, and function.

Another development of great importance was the isolation by Northrop, Kunitz, and their coworkers[7] of relatively homogeneous proteolytic enzymes. As we all know, these studies and the earlier crystallization of urease by Sumner in 1926[12] served to establish firmly the view that enzymes were proteins, particularly since no organic cofactors were found to be associated with these enzymes. Later, Bergmann[13] and his associates devised synthetic substrates of defined structure that could be specifically hydrolyzed by the crystalline proteinases. The demonstration that such enzymes as trypsin, chymotrypsin, and pepsin could hydrolyze the specific kinds of peptide bonds that were expected to be found in the interior of the peptide chains of proteins aided in reinforcing the view that peptide bonds were present in proteins.[14] These contributions were made possible by the development of the carbobenzoxy method for the synthesis of peptides by Bergmann and Zervas.[15] This technique and later developments in methods for synthesis of peptides of almost any sequence contributed both to the characterization of proteolytic enzymes with new types of specificity as well as defining the specificity of previously isolated enzymes.

Little needs to be said here concerning the importance of the developments by Martin and Synge[16] of chromatographic and other partition methods for the study of polypeptides and proteins. The method of partition chromatography, both on columns and on paper, first used for the separation of acetylamino acids and then of amino acids and small peptides, opened the investigation of peptide and protein structure in a way that was almost unimaginable a few years earlier. In brief, the development of partition chromatography, both on columns and on paper, the related method of countercurrent distribution, and, much later, ion-exchange and molecular sieve chromatography led, on the one hand, to advances in rapid methods of qualitative and quantitative amino acid analysis, and, on the other, to the ability to separate complex mixtures of proteins or of peptides resulting from acidic or enzymic hydrolysis of proteins.

In brief, this was the situation in 1943 when Synge[17] reviewed the partial hydrolysis products that had been obtained from proteins. In retrospect, it is remarkable how few peptide sequences had been rigorously defined—approximately half a dozen dipeptides but only two tripeptides, one each from hydrolysates of gelatin and casein. Indeed, among naturally occurring peptides, only the structures of carnosine, anserine, and glutathione were known.

Within the next few years, a number of dipeptides were isolated from partial acid hydrolysates of the gramicidins. Of greatest importance were the studies characterizing gramicidin S as a cyclic pentapeptide (or as now known, a repeating decapeptide). In these studies, each of the four expected dipeptides was isolated and each was characterized by comparison with the corresponding synthetic product. Prior to 1949[18] this was the longest peptide sequence known! In this work partition chromatography was used both for the amino acid analysis and for the isolation of the dipeptides.

In the meantime, improved methods were being developed for amino acid analysis, for without a quantitative amino acid analysis of a protein, the problem of ascertaining the sequence of a protein of even modest size seemed impossible. Over the preceding years, a host of methods had been developed, many being simply qualitative, others quantitative but usually for a single amino acid. A vast array of techniques had been devised: colorimetric methods, precipitation or other isolation methods, and many procedures involving specific chemical degradation of a particular amino acid.[16] Some of the colorimetric methods proved to be reasonably reliable, notably those for amino acids with unique functional groups, e.g., tyrosine, tryptophan, and cysteine. Several quantitative methods were available for specific precipitation of individual amino acids, notably flavianic acid (1-hydroxynaphthalene 7-sulfonic acid) for arginine. In the mid-1930s Bergmann and his associates began a systematic search for specific precipitants, first using various metal complexes and later, various organic sulfonic acids. However, the most specific reagent for a given amino acid seldom gave complete precipitation, particularly from such a complex medium as a protein hydrolysate. Hence, a gravimetric method was devised by Stein and Bergmann using the solubility product of the complex of the reagent and the amino acid for a more precise estimation of a given amino acid.[19, 20] This gravimetric method did produce excellent results for several amino acids in the hands of the highly skilled developers, but it was obvious that the method was laborious and required reasonable amounts of protein.

Almost concurrently, microbiological methods[21] were being developed for analysis based on the observations that either the total growth or the rate of growth of many lactic-acid-producing bacteria was dependent on the presence of specific amino acids. Quantitative procedures were devised that for the first time permitted estimation of several amino acids, for which there had been no adequate colorimetric, chemical, or precipitation method.

In this same period, an isotope dilution method was developed for quantitative estimation of some amino acids.[22,23] These methods gave results of high precision but required a mass spectrometer for analysis of the ^{15}N-labeled amino acid and, of course, required the labeled amino acid in pure form. The method was not widely used.

With the use of colorimetric, microbiological, and some chemical methods, Erwin Brand and his coworkers[24] reported in 1945 a complete accounting of the amino acid composition of β-lactoglobulin. Although it became clear later that there were some compensating errors in obtaining this complete analysis, the importance of Brand's results should not be underestimated. This was the first time that such an analysis was reported and it had a major impact on the general thinking of the field.

With the end of the Second World War, Moore and Stein abandoned the solubility product method of amino acid analysis, and on the basis of the earlier work by Martin, Synge, and their coworkers, developed partition chromatography as a rapid, quantitative method for estimation of all of the amino acids usually found in a protein hydrolyzate.[25] It was at a symposium at The New York Academy of Sciences in 1948 that they first presented their method of using partition chromatography on starch columns to affect the quantitative estimation of all of the amino acids obtained by acid hydrolysis. This was accomplished by collecting fractions of eluate of equal volumes with the aid of an automatic fraction collector and quantitative colorimetric estimation of the content of each fraction with an improved ninhydrin reagent.

The impact of this paper and the subsequent detailed publication of these methods produced a small shock wave. Now, 30 years later, it is difficult to say which had the greater impact, the analytical method or the fraction collector. The realization that a complete amino acid analysis could be obtained with high precision in a matter of days by one competent technical assistant instead of in years by efforts requiring mastery of a variety of methods, encouraged many laboratories to enter the field of protein chemistry. Every protein that was readily available in reasonably pure form could now be analyzed at a microlevel. The triumph of this method was, however, short-lived, for the same investigators in 1951 replaced the starch column with the ion-exchange column.[26] The new method, having greater advantages in convenience, precision, and reproducibility was rapidly adopted. The further development of a completely automated rapid system of analysis[27] is now, of course, so universally known and used that it requires no further comment.

The legacy of the automatic fraction collector remains to this time. It is one of the most universal tools of the craft of biochemistry. Certainly the application of column chromatography for analysis, separation, or purification of every type of mixture of substances still demands the use of the fraction collector. Nevertheless, it is useful to recall that it was originally invented for use in amino acid analysis.

While these advances had been taking place in one part of the Rockefeller Institute, Lyman Craig[28] and his coworkers had been making important developments in another part. It was in their hands that counter-current distribution reached its sophisticated development,[29] initially for purifying various synthetic and natural products, including several peptide antibiotics and hormones, later for separating the chains of hemoglobin and, indeed, for purifying some proteins as well. An important by-product of this work was the invention of the roto-evaporator[30] for the removal of the large volumes of solvents involved in counter-current distribution, and, of course, also, in column chromatography. This initially humble tool, now available in all sorts of sophisticated modifications has been a boon not only to the dry-ice distributors and manufacturers, but to all biochemists.

Perhaps some of you may feel that I am overemphasizing the tools of the trade rather than the intellectual developments of protein chemistry. Be that as it may, this is the reality of the period. As I noted in my introductory remarks, most of the conceptual notions regarding the importance of proteins and the general view of their covalent structure were already part of our intellectual heritage. What was needed were the methods and the tools to tackle the structure of proteins in order to begin the task of trying to understand how proteins accomplished their functions.

We have now noted that by the end of the war improvements were taking place on both sides of the Atlantic in separating amino acids and peptides and in quantitative amino acid analysis. We have not yet mentioned what subsequently became one of the most important advances, namely, the development of an end-group method. In 1945 Sanger[31] published his first results using 1-fluoro-2,4-dinitrobenzene to label the amino terminal groups of peptides and proteins. The formation of an aryl derivative relatively stable to acid hydrolysis, when bound to most of the amino acids found in proteins, permitted isolation and identification by partition chromatography of the dinitrophenyl derivatives of the amino acids found at the α-amino ends of peptide chains. For the first time it became possible to determine the number of peptide chains in a protein and to determine the residues bearing free α-amino groups. Although it was demonstrated earlier that some proteins, e.g., hemoglobin, could be dissociated to smaller units, presumably without breaking peptide bonds, by using urea, acid, or alkali, these methods could not be universally applied, particularly when it was already appreciated that proteins could contain disulfide or other bonds that might cross-link peptide chains to one another.

The story of Sanger's step-by-step progress in determining the structure of insulin is too well-known to require detailed repetition here. In brief, by 1949, he had shown that insulin contains two types of chains[30] and had introduced the use of performic acid to oxidize the disulfide bonds.[32] This permitted separation of the two chains in stable form. He had also determined by partial acid hydrolysis and isolation of the dinitrophenyl peptides, the amino-terminal sequences of both chains.[33] By 1952, Sanger with his coworkers had determined the sequences of both chains,[34-36] and by 1955, the complete structure allocating the three disulfide bonds.[37] It should be emphasized that in this work the major techniques used in separating peptides and identifying the constituent amino acids involved chromatography and electrophoresis on paper.

The impact of Sanger's achievement was immediate. First of all, in addition to the relatively few laboratories already studying the sequences of proteins, many more were encouraged to work in this field. Secondly, although the three-dimensional structure of insulin remained unknown for some years, there was now hope that this problem could be approached with the aid of the knowledge of the amino acid sequence. Indeed, the structures of a number of other proteins were solved by x-ray analysis before that of insulin. Third, the success with insulin firmly established that the primary linkage between amino acid residues in proteins is the peptide bond. Further, proteins could contain one or more than one type of peptide chain. Fourth, the sequence in each chain was completely unique containing no repetitive sequences or patterns. Finally, Sanger and his associates used both partial acid hydrolysis and enzymic hydrolysis to obtain the peptides used in ascertaining the structure of insulin. An important aspect of these studies was the proof that proteolytic enzymes acted at sites in the longer peptide chains that could be predicted from the earlier knowledge of the action of these enzymes on synthetic substrates. A subsequent result is the now general practice of using the chains of insulin or other proteins of known sequence to determine the specificity of proteolytic enzymes of unknown specificity, a method first used to determine the specificity of subtilisin by Tuppy.[38]

At the present time, Sanger's dinitrophenyl method is still used, but for quantitative end group determinations the cyanate procedure of Stark and Smyth[39] is frequently superior. Most important was the development by Edman[40] of the phenylthiohydantoin method published in 1950. This technique, which permits sequential degradation from the amino-terminal ends of polypeptides and proteins, continues to be one of the major methods in current use for determining the sequences of small peptides by manual procedures, and of larger peptides and proteins by use of the automatic sequencer developed by Edman and Begg.[41]

It is interesting that no general chemical method for determining the carboxyl-terminal residues or sequences has achieved such success, although hydrazinolysis has been used for identifying carboxyl-terminal residues.[42] The crystalline carboxypeptidase A of Anson[43] has proved to be a useful tool in identifying carboxyl-terminal residues and sequences in polypeptides and proteins. Also, other carboxypeptidases have since been added to the roster of enzymes employed in sequence work. Similarly, a number of aminopeptidases[44] have been useful both for sequence work on peptides and for identifying the acid-labile amino acids, glutamine, asparagine, and tryptophan in polypeptides obtained from proteins. The list of proteolytic enzymes currently employed in structural studies on proteins is now an extensive one, as can readily be ascertained by consulting the appropriate volumes of *Methods in Enzymology*.

While Sanger and his associates were elucidating the complete structure of insulin, du Vigneaud[45, 46] and his coworkers were studying the two smaller pituitary hormones, oxytocin and vasopressin. In these investigations, countercurrent distribution gave the first pure hormone preparations, starch columns yielded quantitatively the amino acid compositions, partial acid hydrolysis produced small peptides, which were separated by partition chromatography, and both the dinitrophenyl method of Sanger and the Edman degradation were used to establish the sequences of these nonapeptides. Du Vigneaud and his associates then used the knowledge of the sequence to synthesize both of these hormones in fully active form. If there had been any doubts regarding the reliability of the aforementioned methods, almost all of which had been initially devised for studying larger molecules, such scepticism was eliminated by the unequivocal total synthesis of these hormones.

These early studies merit a few additional comments. The pioneer work of Sanger and du Vigneaud gave us the first complete sequences of a protein hormone and two polypeptide hormones. This information has only recently begun to yield information of how these hormones perform their functions. In contrast, the work that followed on enzyme structures led very rapidly to the labeling of active site residues in enzymes by use of specific reagents or inhibitors usually containing ^{14}C, ^{32}P, or ^{3}H. In short order, active site residues were identified in the partial sequences of many enzymes. The studies in some cases confirmed earlier information but most important was the discovery of entirely novel reactive residues and relationships in the active sites of enzymes. It would be invidious to select for mention only a few of the large number of investigations of this type that were published in the period beginning in the 1950s.

Among the most notable earlier investigations were those of Moore and Stein[47] and their associates in determining by 1960 the structure of pancreatic ribonuclease, the first enzyme of known sequence. Their work encompassed the use of quantitative amino acid analysis and, most important, the separation of peptides, derived entirely by enzymic hydrolysis, on low cross-linked ion-exchange columns. Their chemical and enzymic studies also served to define many of the properties of the active site of this enzyme. This work also eventually aided in the elucidation of the three-dimensional structure of this enzyme by x-ray diffraction, confirming and extending the information already gleaned from knowledge of the primary sequence and other studies by Stein and Moore and their associates and by many others.

Concurrently, Anfinsen[48] and his associates demonstrated that reduced and denatured ribonuclease would refold spontaneously and regenerate the native, active enzyme after allowing molecular oxygen to reoxidize the sulfhydryl groups to the four original disulfide bonds, reinforcing the view that the primary structure provides all the information necessary for forming the native structure. Richards and Vithayathil[49] separated the two parts of ribonuclease produced by hydrolysis with subtilisin. They were then able to demonstrate the return of activity on recombining the S-peptide and S-protein, although the hydrolyzed bond was not reformed. The "active site" of the enzyme clearly involved not a single part of the sequence but residues from both parts of the sequence. The subsequent further studies on elucidating the mechanism of action and structure of this enzyme do not require further description here.

Horse cytochrome c was the third protein for which the sequence became known.[50] Now the sequences of the homologous protein from an additional 75 species[51] have given great insight not only in structure-function relationships, but also in our understanding of some features of biochemical evolution. Such comparative studies have also been made for a considerable variety of proteins and have contributed greatly to our knowledge of evolutionary protein relationships.

Similarly, studies on myoglobin and hemoglobin helped to establish the three-dimensional structures of these proteins. For hemoglobin the sequence studies on the normal and abnormal forms together with the entire body of physico-chemical information on the oxy and deoxy forms and the detailed x-ray analyses have given us the most sophisticated information on structure-fucntion relationships as yet known for any protein.

It is noteworthy that even before the complete sequences were known that Ingram[52] was able to ascertain the sole difference in the amino acid sequences of normal and sickle cell hemoglobins. This was accomplished by using the technique of two-dimensional separation of tryptic peptides on paper by electrophoresis and chromatography. This ingenious, simple procedure of comparing peptide patterns on paper is still valuable in detecting genetic or species differences in proteins and is widely employed for these and other purposes.

There is little that remains to be said regarding this heroic early period in establishing the methods for determining the primary structures of proteins. Within a few years, partial or complete sequences of many smaller proteins were reported and now, of course, the sequences of a host of proteins of many kinds are known.

These few examples are cited only to indicate new types of studies that became possible once Sanger had demonstrated that it could be done. This is not to say that the work is easy, despite the addition of many new tools and a greater understanding of many of the pitfalls that lie in wait for the unwary. Among the latter should be mentioned the unusual lability and possible rearrangements of some peptide bonds, the presence in some proteins of novel types of modified amino acids, and the occurrence in fibrous proteins, particularly, of several unusual types of cross-linkage other than disulfide bonds. However, my task for today is not to discuss the current problems facing the investigator who undertakes the study of the structure of ever larger and more complex proteins using smaller and smaller quantities of material. This is current history.

I should like to conclude with a sentence from Emil Fischer's Nobel Lecture of 1902 quoted by Fruton in his book, *Molecules and Life*[53]: "Since the proteins participate in one way or another in all chemical processes in the living organism, one may expect highly significant information for biological chemistry from the elucidation of their structures and their transformations."

Fischer's prophetic remark seems particularly apt for those of us who witnessed or participated in the study of protein structure from the 1930s to the present.

REFERENCES

1. VICKERY, H. B. & C. L. A. SCHMIDT. 1931. Chem. Rev. *9*: 169–318.
2. VICKERY, H. B. 1972. Adv. Protein Chem. *26*: 81–171.
3. VICKERY, H. B. 1941. Ann. N.Y. Acad. Sci. *41*(2): 87–120.
4. WU, H. 1931. Chinese J. Physiol. *5*: 321–344.
5. MIRSKY, A. E. & L. PAULING. 1936. Proc. Nat. Acad. Sci. U.S.A. *22*: 439–447.
6. ANSON, M. L. 1945. Adv. Protein Chem. *2*: 361–386.
7. NORTHROP, J. H., M. KUNITZ & R. M. HERRIOTT. 1948. Crystalline Enzymes. 2nd edit. Columbia University Press. New York, N.Y.
8. NEURATH, H., J. P. GREENSTEIN, F. W. PUTMAN & J. D. ERICKSON. 1944. Chem. Rev. *34*: 158–265.
9. WILLSTÄTTER, R. 1927. Problems and Methods in Enzyme Research. Cornell University Press. Ithaca, N.Y.
10. SVEDBERG, T. & K. O. PEDERSEN. 1940. The Ultracentrifuge. Oxford University Press. Oxford, England.
11. TISELIUS, A. 1937. Trans. Faraday Soc. *33*: 524–531.
12. SUMNER, J. B. 1926. J. Biol. Chem. *69*: 435–441.
13. BERGMANN, M. & J. S. FRUTON. 1941. Adv. Enzymol. *1*: 63–98.
14. FRUTON, J. S. 1938. Cold Spring Harbor Symp. Quant. Biol. *6*: 50–57.
15. BERGMANN, M. &. L. ZERVAS. 1932. Ber. Chem. Ges. *65*: 1192–1201.
16. MARTIN, A. J. P. & R. L. M. SYNGE. 1945. Adv. Protein Chem. *2*: 1–83.
17. SYNGE, R. L. M. 1943. Chem. Rev. *32*: 135–172.
18. SYNGE, R. L. M. 1949. Quart. Rev. Chem. Soc. *3*: 245–262.
19. BERGMANN, M. & W. H. STEIN. 1939. J. Biol. Chem. *128*: 217–232.
20. MOORE, S., W. H. STEIN & M. BERGMANN. 1942. Chem. Rev. *30*: 423–432.
21. SNELL, E. E. 1945. Adv. Protein Chem. *2*: 85–118.
22. RITTENBERG, D. & G. L. FOSTER. 1940. J. Biol. Chem. *133*: 737–744.
23. FOSTER, G. L. 1945. J. Biol. Chem. *159*: 431–438.
24. BRAND, E., L. J. SAIDEL, W. H. GOLDWATER, B. KASSELL & F. J. RYAN. 1945. J. Am. Chem. Soc. *67*: 1524–1531.
25. MOORE, S. & W. H. STEIN. 1948. Ann. N.Y. Acad. Sci. *49*: 265–278.
26. MOORE, S. & W. H. STEIN. 1951. J. Biol. Chem. *192*: 663–681.
27. SPACKMAN, D. H., W. H. STEIN & S. MOORE. 1958. Anal. Chem. *30*: 1190–1206.
28. CRAIG, L. C. 1944. J. Biol. Chem. *155*: 519–534.
29. CRAIG, L. C. & D. CRAIG. 1950. Extraction and Distribution. *In* Technique of Organic Chemistry. A. Weissberger, Ed. Vol. 3 (Chap. 4): 171–311. Interscience Publishers Inc. New York, N.Y.
30. CRAIG, L. C., J. D. GREGORY & W. HAUSMANN. 1950. Anal. Chem. *22*: 1462.
31. SANGER, F. 1945. Biochem. J. *39*: 507–515.
32. SANGER, F. 1949. Biochem. J. *44*: 126–128.
33. SANGER, F. 1949. Biochem. J. *45*: 563–574.
34. SANGER, F. & H. TUPPY. 1951. Biochem. J. *49*: 463–481, 481–490.
35. SANGER, F. & E. O. P. THOMPSON. 1953. Biochem. J. *53*: 353–366, 366–374.
36. SANGER, F., E. O. P. THOMPSON & R. KITAI. 1955. Biochem. J. *59*: 509–518.
37. RYLE, A. P., F. SANGER, L. F. SMITH & R. KITAI. 1955. Biochem. J. *60*: 541–556.
38. TUPPY, H. 1953. Monatsh. Chem. *84*: 996–1010.
39. STARK, G. R. & D. G. SMYTH. 1963. J. Biol. Chem. *238*: 214–226.
40. EDMAN, P. 1950. Acta Chem. Scand. *4*: 283–293.

41. EDMAN, P. & G. BEGG. 1967. Eur. J. Biochem. *1*: 80–91.
42. AKABORI, S., K. OHNO & K. NARITA. 1952. Bull. Chem. Soc. Japan *25*: 214–218.
43. ANSON, M. L. 1937. J. Gen. Physiol. *20*: 663–669.
44. SMITH, E. L. & R. L. HILL. 1960. Leucine Aminopeptidase. *In* The Enzymes. 2nd edit. P. D. Boyer, H. Lardy & K. Myrback, Eds. Vol. 4: 37–62. Academic Press. New York.
45. DU VIGNEAUD, V. 1960. Ann. N.Y. Acad. Sci. *88*: 537–548.
46. DU VIGNEAUD, V. 1964. Nobel Lectures Chemistry: 1942–1962. 446–465. Elsevier Publishing Co. New York.
47. MOORE, S. & W. H. STEIN. 1973. Science *180*: 458–464.
48. ANFINSEN, C. B. 1973. Science *181*: 223–230.
49. RICHARDS, F. M. & P. J. VITHAYATHIL. 1959. J. Biol. Chem. *234*: 1459–1465.
50. MARGOLIASH, E., E. L. SMITH, G. KREIL & H. TUPPY. 1961. Nature *192*: 1125–1127.
51. MARGOLIASH, E., S. FERGUSON-MILLER, C. H. KANG & D. L. BRAUTIGAN. 1976. Fed. Proc. *35*: 2125–2130.
52. INGRAM, V. M. 1956. Nature *178*: 792–794.
53. FRUTON, J. S. 1972. Molecules and Life: Historical Essays on the Interplay of Chemistry and Biology. Wiley-Interscience. New York, N.Y.

DISCUSSION OF THE PAPER

HAUROWITZ: You mentioned the late discovery of nickel in urease. I remember the time when I worked in Willstätter's laboratory in 1924. He could not believe that urease protein, as such, could be an enzyme. Willstätter suspected that there was some foreign active group, whatever it might be. We also know very little about, for instance, saccharase and certain other hydrolytic enzymes. I wonder whether in these enzymes heavy metals would also be involved. I still do not know whether the role of nickel in urease is explained. I know, however, from spectrophotometry of ashes of animal organs that they contain a lot of heavy metals. You find, in the ashes of liver tissue, molybdenum, manganese, and tin. Do you think that in the future some of these metals will be found to be involved in enzymatic reactions?

SMITH: I do not know why they would not. There is no evidence, however, and the summary is correct. There was no evidence of an organic moiety in urease other than amino acids but there are atoms of nickel in the molecule. And I would suspect we are in for still more surprises, and this requires the utmost caution in making conclusions. For example, glutathione peroxidase turns out to be the first eucaryotic enzyme containing selenium, and it has now been shown by Tappel and his co-workers at Davis that the selenium is present in selenocysteine residues.

Now clearly this is a posttranslational modification, in which the HSe group replaces presumably an HS group—or perhaps it may be derived from serine or some other residue—but nevertheless, we have to look out for the surprises. The number of posttranslational modifications now is simply enormous. This does not change the fact that the protein is essential for enzyme action, and not the cofactor alone. Proteins are involved but what other residues or metals will turn up I certainly would not guess.

For example, it is now more than 40 years since it was known that boron is an essential element for the growth of plants. To my knowledge the need for boron has not been explained in terms of its site of action, or whether it is present in a protein. Nevertheless, it is essential for the growth of plants.

HAUROWITZ: You mentioned the important methods of protein structure discovery. However, you did not mention the fractional distillation of esters of amino acids. In essence it goes back to Emil Fischer who tried to separate amino acids by distillation of the esters. What is your view on the use of high-resolution mass spectrometry of amino acids or esters?

SMITH: Well, historically the Fischer ester distillation method was an extremely important method in the discovery of several amino acids, but, in even the most skilled hands such as Fischer himself, Henry Dakin, T. B. Osborne, and others, it never gave quantitative recoveries. Of course that is why the method was abandoned.

Mass spectrometry will probably have considerably more use in the study of rather small peptides. There are still problems connected with it, however. In our laboratory, at least, one of my colleagues found mass spectrometry extremely useful in determining a sequence of a peptide which had five glycine residues out of seven present, and it was utterly impossible by any of the other methods to be sure that one really had five glycine residues in a row. It was only mass spectrometry with its accurate molecular weight determination that really proved the sequence.

We have given samples of various complex peptides to our colleagues who think you can do everything with mass spectrometry, and they do not always come up with the right answers. There are still many, many problems and until they can solve some of these problems, enzymic and other methods remain the methods of choice.

DOROTHY CROWFOOT HODGKIN graduated first from Oxford in 1932 and then received her Ph.D. degree in 1936 from Cambridge, where she worked with J.D. Bernal on many compounds of biochemical interest. She first became widely known in Biochemistry when she and Bernal in 1934 were able to take detailed x-ray diffraction photographs of crystalline pepsin. Returning to Oxford in late 1934, she concentrated on solving, with many collaborators, the crystal structures and hence the chemical structures of many compounds, including cholesterol, calciferol, penicillin, and vitamin B_{12}, and eventually also solved the crystal structure of 2-zinc insulin, of which she had first taken x-ray photographs in 1935.

Crystallographic Measurements and the Structure of Protein Molecules as They Are

DOROTHY CROWFOOT HODGKIN
Chemical Crystallography Laboratory
Oxford, OX1 3PD
England

IN THE HISTORY of the study of the structure of proteins by x-ray crystallographic analysis there are three dates that stand out as of special importance:

(1) 1912, when the first x-ray diffraction patterns of any crystal, copper sulphate in fact, was obtained by Laue, Friedrich, and Knipping;

(2) 1934, when J. D. Bernal recorded the first x-ray diffraction pattern of a protein crystal, pepsin, in its mother liquor;

(3) 1953, when Max Perutz took an x-ray photograph of the mercury benzoyl derivative of horse methemoglobin and observed measurable and interpretable changes in the intensities of the x-ray spectra compared with data from the unmodified protein.

These dates have a common characteristic—they are dates of great promise on our way towards the vision of the actual arrangement of the atoms in protein molecules.

In this paper, I am principally concerned with the period between 1934 and 1953. I propose, however, to begin with a brief account of the history and prehistory of protein crystallography that determined our thinking in the thirties and forties.

PREHISTORY—EARLY OBSERVATIONS ON PROTEIN CRYSTALLOGRAPHY

During the nineteenth and early twentieth century a number of crystals were observed in animal and plant cells under the light microscope and were identified as protein in nature. The earliest of these were probably hemoglobin crystals seen by Baumgaertner in 1830.[1] Hünefeld, in 1840,[2] certainly described hemoglobin crystals growing in a drop of blood from the earthworm between glass plates (Barbara Low and I repeated his experiment in Oxford rather more than a hundred years later). K. E. Reichert in 1849[3] observed tetrahedral crystals growing in the fetal membranes of a guinea pig, six hours after its death. Detailed studies followed, notably by Preyer.[4] By 1909 a great volume of 600

0077-8923/79/0325-0121 $01.75/0 © 1979, NYAS

microphotographs of hemoglobin crystals from 106 species had been collected by E. T. Reichert and A. P. Brown.[5] A number of other protein crystals were also observed. Cohn, "Über Proteinkrystalle in den Kartoffeln," in 1860,[6] described experiments on sharp edged, cubic crystals, which split in two in the presence of water and were readily permeable to liquids and dyes. Ammonia dissolved them from outside in, acetic acid from inside out, leaving a hole in the center. The morphology of these crystals was described by A. F. W. Schimper in a summary paper in the *Zeitschrift für Kristallographie* in 1880[7] together with others, which appear later in protein crystal history, such as excelsin, for example:

1. Art.—Typus: Krystalloide der Paranuss.*

   Optisch einaxige Krystalloide von hexagonal rhomboëdrisch-hemiëdrischer Symmetrie.

   Axenverhältniss: a:c = 1:2.4

   Auftretende Formen: $R(10\bar{1}1), -\frac{1}{2}R(01\bar{1}2), oR(0001)$.

   Doppelbrechung: +, sehr schwach.

In a number of experiments Schimper demonstrated swelling and shrinking of the crystals in water, acids or alkalis. He found that with excelsin swelling does not occur normal to the principal axis, e.g.,

	vor.	nach.	auf 1 bez.
Gr. diag.	8	12	1.5
Kl. diag.	5	5	1.0

He observed that the crystals from *Musa hillii*, when dry, showed almost no birefringence between crossed nicols but brightened when brought near water, passing through the different birefringence colors, violet, orange, yellow, blue-green, green.

That the molecules in these crystals were large and that this conferred peculiar properties on them was realized. Preyer had estimated the molecular weight of hemoglobin as 13,000 in 1871 from the iron analysis (later corrected by Zinoffsky[8]). Schimper comments on variations in metal content observed in crystals of the same form: "Ist es sehr möglich dass in Molekülen, die wie die jenigen der Eiweisskörper aus vielen Hunderten von Atomen bestehen, einige Affinitäten auf sehr verschiedene Weise gesättigt werden können, ohne dass die Krystallform eine wesentlich andere werde." It was also generally the opinion in the late nineteenth and early twentieth century that the packing units in crystals were chemical molecules and the types of arrangement they might have were in process of being worked out in space group theory. But nothing more

* *Bertholletia excelsa.*

precise could be found out about protein crystals until x-ray diffraction had been discovered.

1912–34. THE DISCOVERY OF X-RAY DIFFRACTION OF CRYSTALS: FIRST APPLICATIONS TO THE STUDY OF PROTEINS

Any time after the discovery of x-rays in 1895 it would have been possible to discover x-ray diffraction: Röntgen himself tried an experiment of passing a beam of x-rays through a crystal of calcite but the beam was too weak, the crystal too thick for the observation of diffraction effects. The successful experiment in 1912[9] came about almost by accident, through a series of conversations between Paul Ewald, M. v. Laue, W. Friedrich, and P. Knipping in Munich, all physicists, and quite young, working in the laboratories of Sommerfeld and Röntgen; a full account has been given of the circumstance in *Fifty Years of X-Ray Diffraction*.[10] The diffraction pattern given when x-rays were passed through copper sulphate, the first crystal examined, was too complicated to understand and the investigators turned to zinc blende, where at least it was possible to correlate the cubic crystal symmetry with that of the diffraction pattern. Friedrich then carried out a series of experiments, passing x-rays through other crystals and through partially ordered and amorphous materials such as beeswax and paraffin.[11] Within a few months, the experiments had been repeated in England and in Japan and W. L. Bragg in England had solved the first crystal structures by measuring and comparing the diffraction effects from sodium and potassium chloride.[12]

Professor Herman Mark, who took part in the early measurements on fiber structures has written for me the following account of the next stage, the experiments in Japan and afterwards in Germany, which it seems worth quoting in some detail:

> Already in 1913—only one year after the Bragg-v. Laue discovery—Nishikawa and Ono[13] in Japan irradiated a randomly selected number of fibers with x-rays; natural silk was amongst them (and I think also hair).† They got a few broad spots and concluded that there must be some kind of molecular order in these materials. Of course, their samples were not properly purified and their x-ray tube was primitive (1913!); it was a *pioneering* effort and was not followed up. Next—quite a few years later —in 1920 Professor R. D. Herzog, Director of the Kaiser Wilhelm Institute for Faserstoffchemie in Berlin-Dahlem, decided that he would initiate a *systematic* study of fibers with x-rays in his Institute. In 1920[14]

† Silk, wood, bamboo, and asa (*Cannabis sativa*).—D.H.

he published a paper on several fibers—including silk—together with W. Jancke. As a result of better sample preparation and improved x-ray equipment they obtained "fiber diagrams" of cellulose and silk. They concluded that there is a "crystalline" component in these fibers but made no attempt at a quantitative evaluation. In 1921 another coworker of Herzog, Michael Polanyi, evaluated the fiber diagram of cellulose more numerically and determined its crystallographic basis cell.[15] This encouraged Herzog to attempt the same step with natural silk, which, of all proteinic material, gave at that time the clearest diagrams [FIGURE 1]. At his request, Polanyi asked a graduate student—Rudolf Brill—to prepare well-oriented and purified samples and to get as good x-ray photos as possible for a quantitative study. At that time (1922) I was also working, as a postgraduate, at Herzog's Institute. Rudi and I worked together in a makeshift laboratory but had—for that time—rather good x-ray tubes, which would give some 10 mA at about 40 kV for several hours. Rudi worked on silk and hair, I studied metal wires (single crystals) and cellulosics. When we got reasonable looking patterns we would show them to Polanyi and Weissenberg (both so well known and liked!) and they would explain to us what it all meant and how we should evaluate the pictures; we had a wonderful time!

After a year Rudi had finished his work and published it as his Ph.D. thesis.[16] He determined the basis cell (correctly) and concluded that there were 2 possible "structures" for silk fibroin—either a small glycyl-

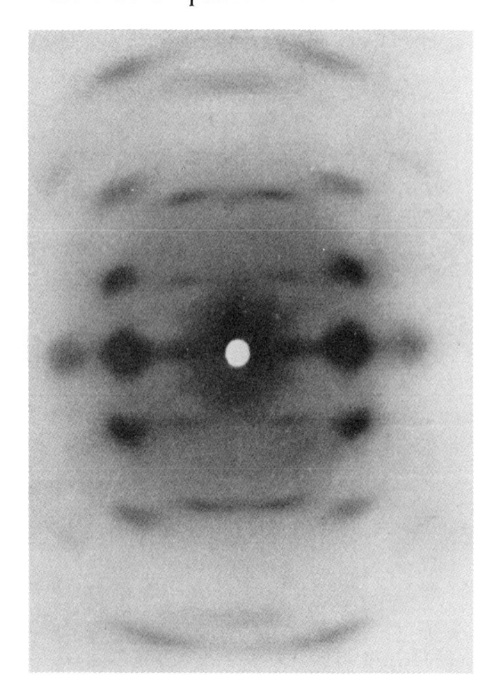

FIGURE 1. X-ray photograph of silk fibroin from *Bombyx mori* taken by Astbury.

D-alanine peptide *or* a long polypeptide chain. This *alternative* was, of course, the *last* the chemists wanted to hear because it was the time of the controversy between small aggregates and macromolecules; but, as in the case of cellulose (Polanyi) the x-ray diagrams of these days were not yet good enough for a more thorough analysis (space group, intensities, etc.).

A year later Dr. Brill took a position in the Ammoniak Laboratory of the IG Farben in Oppau where he studied mainly catalytically active inorganic materials. Again, a year later I joined the Hauptlaboratory of the IG Farben in Ludwigshafen. The Director of the laboratory was Professor K. H. Meyer, a very distinguished scientist, who felt that another, more quantitative study of the structure of fibers should be initiated. At that time Dr. G. von Susich had rather elaborate equipment —x-ray tubes, cameras, etc.—and Dr. H. Hopff helped us in the preparation of highly purified and oriented samples of "bombyx mori." Some diagrams were good enough to determine the space group and to carry out a preliminary intensity evaluation. Using the Bragg atomic radii, K. H. Meyer and I proposed a chain structure for silk fibroin, which was published in 1928[17] [FIGURE 2].

Meyer and Mark actually suggested two alternative unit cells and chain arrangements for silk fibroin consistent with their observations and different probable space groups. They did not attempt to place the atoms

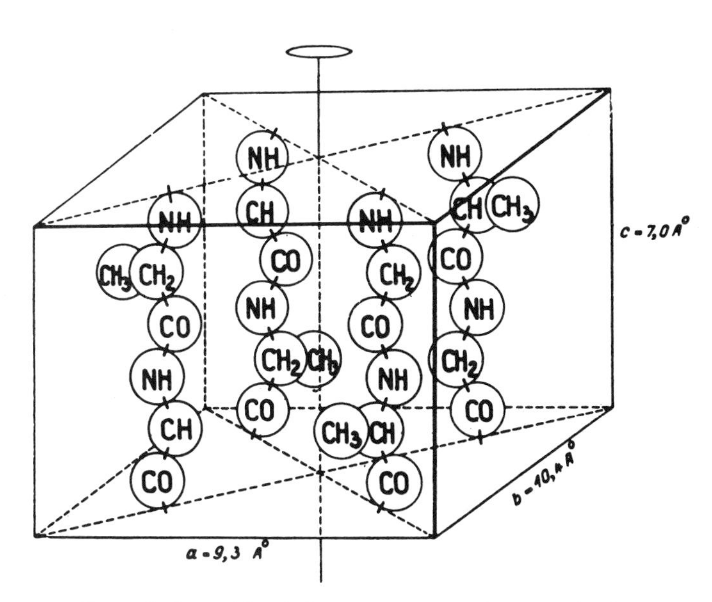

FIGURE 2. Extended peptide chain arrangement (one version) proposed by Meyer and Mark for silk fibroin in 1928. (From Meyer & Mark.[17])

precisely owing to the limited data given by the photographs. They assumed that regularity, such as it was, occurred in limited regions within micelles between which the main chain passed in a less orderly fashion. (Compare Weissenberg's theories in 1926.[18]) And among the "Bragg radii" they used were those for oxygen and nitrogen, partly derived from the first organic crystals analyzed, hexamethylene tetramine by Gonell and Mark[19] and by Dickinson and Raymond,[20] and urea by S. B. Hendricks.[21]

The observations of the Berlin school were known to W. H. Bragg. After his Christmas lectures at the Royal Institution on "Old Trades and New Knowledge" in 1924 he planned a new lecture, "The Imperfect Crystallisation of Common Things" in 1926. He asked W. T. Astbury to take some photographs for him of fibers like wool and hair (first recorded earlier by Herzog and Jancke). Astbury (as Bernal recorded) took to them from the beginning; they led to his first formal appointment as lecturer in Textile Physics at Leeds in 1928 and became his life's work. With H. J. Woods and A. Street, he discovered that wool gave two types of x-ray photograph, α when unstretched, β when stretched.[22, 23] He deduced that the α photograph, which gave only two features, an equatorial "reflection" of spacing 10 Å and a meridional "reflection" of spacing 5.1 Å, represented some type of folded peptide chain that could be reversibly stretched to a straight chain, giving the β photograph. The β photograph, which showed two equatorial reflections of 10 Å and 4.65 Å and one meridional one, 3.4 Å, clearly, to his eyes, was the same type as that of silk fibroin. As the years passed, he found the same types of photograph given by many proteins, keratin, myosin, epidermin, fibrinogen and called them the "kmef" group. Occasionally much more highly ordered structures giving many reflections, were observed, such as feather keratin, which still conformed to this type.[24] Collagen, on the other hand, gave a very different kind of photograph with a meridional reflection of 2.84 Å; Astbury, with others, speculated on possible different folds in this type of chain.[66] One small very interesting observation he made in this period with J. D. Bernal and T. C. Marwick in a quite different system, guided his thinking later. He showed that the cell wall of the single cell organism valonia ventricosa, the sea grape, gave an x-ray pattern that indicated cellulose chains crossed one another at an angle that varied over the cell surface, suggesting the chains were wound over the surface as over a ball of wool.[25]

The x-ray photographs of the natural fibers were too limited in themselves to give precise information about the atomic arrangement in fibrous proteins (FIGURE 1). J. D. Bernal began a first paper (1931)[26] on the crystal structures of the natural amino acids and related compounds with the words "a knowledge of the crystal structure of the amino acids is essential for the interpretation of the x-ray photographs of animal materials: silk fibroin, keratin, collagen, proteins, etc., which have been studied by this method for the first time in the last few years." He gave preliminary x-ray data and some guesses about the structures of the crystals of 15 substances kindly prepared for him by Dr. A. Leese at the Biochemistry Laboratory, Cambridge. He ended with a request for more crystals of related compounds, of 0.01 mm or over in size, for extended examination. In a further paper he gave some x-ray data on the very interesting crystals of cuprous glutathione prepared by N. W. Pirie in Cambridge.[27]

It was clear that detailed x-ray analysis of the kind needed was likely to be slow even for such simple crystals. But new ideas of how to proceed were being developed with the use of Fourier analysis, as Bernal mentioned. In 1928 B. Warren and W. L. Bragg published the crystal structure of the feldspar diopside,[28] a structure involving the determination of 27 parameters to place 10 atoms. They used a marvelous series of intricate arguments involving the relative scattering contributions of the different atoms to the different observed intensities of the x-ray spectra. With amplitudes derived from their measured intensities and phase constants from their found atomic positions, Bragg calculated a projection of the electron density on the main crystal plane (010). Not till very long afterwards did any one notice that most of the phase constants, 0° and 180°, were here determined by the calcium ion contribution alone. The history of the introduction of the direct approach to crystal solving through the use of Fourier series in electron density calculations, which W. L. Bragg and West discussed in relation to this structure, is a little odd. W. H. Bragg (W. L. Bragg's father) first suggested the application of Fourier methods in 1915.[29] In a theoretical paper by Duane in 1925,[30] followed by Havighurst's practical demonstration on sodium chloride,[31] it was pointed out that in the alkali halide structures, the chlorine ion contributions determined the phase constants necessary for the calculation, here signs, plus or minus; in 1927 Cork[32] demonstrated with the alums that the same information could be derived through the study of isomorphous crystals. Yet, as first, the electron density calculation was used mainly in the representation and refinement of crystal structures as with diopside or the first organic crystal seen in an electron density map, hexachlorbenzene (Lonsdale[33]). W. L. Bragg used to quote the structure of copper sulphate found by Beevers and Lipson in 1934 as one of the first crystal structures actually solved by the "direct" Fourier approach.[34] ‡

‡ In this analysis, Beevers and Lipson also devised "strips" we all used in our first calculations of Patterson and electron density maps for proteins.

As for the protein crystals themselves, every one who worked seriously on proteins in this period knew they existed. Their number had been added to very dramatically in the twenties by the first crystallization of several enzymes by Sumner, Northrop, and Kunitz; pepsin, trypsin, and chymotrypsin and others, and of insulin by J. J. Abel. There were many attempts by crystallographers to obtain x-ray photographs of protein crystals in this period, e.g., by W. H. George[35] and by Clark and Corrigan[36] on insulin (Clark and Corrigan recorded some difficult-to-interpret reflections with long spacings), and A. L. Patterson on hemoglobin. All observed almost nothing but vague blurs for reasons that became obvious in 1934.

1934–53. THE FIRST X-RAY CRYSTALLOGRAPHIC MEASUREMENTS ON SINGLE PROTEIN CRYSTALS AND RELATED STUDIES

As with the discovery of x-ray diffraction itself, there were both accidents and purposes contributing to the taking of the first x-ray photographs of protein crystals by J. D. Bernal in the spring of 1934. John Philpot was working at that time on the purification of pepsin at Uppsala; a preparation he left standing in a fridge while he was away skiing produced very large—2 mm long—and beautiful crystals of the hexagonal bipyramidal form obtained earlier by Northrop and Kunitz. They were seen by Glen Millikan, a passing visitor and friend of Bernal's, who knew of his interest in proteins and his appeal for crystals. John Philpot gave Millikan a tube of crystals in their mother liquor to take back to Cambridge.

Bernal had an old fashioned classical Cambridge education in mineralogy and a very large petrographic microscope§ with rotating nicols, which made it easy for him to see the crystals in the tube in which they grew and that they were moderately birefringent. As Schimper had done for the seed globulins before him, he recorded the axial ratio, $c/a = 2.3 \pm 0.1$, and the birefringence, positive, uniaxial. He first took a crystal out of its mother liquor, noticing as he did that the birefringence fell. He mounted it and took an x-ray photograph, which showed nothing but vague blackening. He then thought of drawing crystals in their mother liquor into thin walled capillary tubes of Lindemann glass—fortunately available in the laboratory since he and Helen Megaw had been growing ice crystals in them to study the expansion of ice. The following photographs, taken on a 3 cm radius cylindrical camera in a series of 5° oscillations, showed hundreds of x-ray reflections, rather large, corresponding

§ Now on the working bench of Max Perutz.

with the size of the crystals, and noticeably even in intensity distribution.[37] The *a* axis was easily measured as ~67 Å but the reflections defining the long *c* axis were not completely resolved owing to the large size of the crystals, the experimental conditions used, and some disorder that introduced smearing. The *c* axis was recorded as $n \times 154$ Å—a value derived from the axial ratio. Much later experiments by Max Perutz (1949)[38] and very recently by Tom Blundell at Birkbeck have provided accurate data for the pepsin crystal lattice constants, $a = 67.9$ Å, $c = 292$ Å, space group $C6_122$, $n = 12$. Tom Blundell's photographs are shown in FIGURE 3. Since no one can now find them, it seems almost certain that the original pepsin photographs were destroyed in the bombing of Birkbeck College during the war.

The large lattice constants observed for the pepsin crystals were in general agreement with the magnitude of the molecular weight of the pepsin molecule, about 40,000, derived from ultracentrifuge measurements. The unit cell dimensions suggested a packing of oblate spheroidal molecules, 25 Å × 35 Å. The communication to *Nature* comments on the intensity distribution, "From the intensity of the spots near the centre we can infer that the protein molecules are relatively dense globular bodies, perhaps joined together by valency bridges but in any event separated by relatively large spaces which contain water. From the intensity of the more distant spots it can be inferred that the arrangement of atoms inside the protein molecule is also of a perfectly definite kind, though without the periodicities characterising the fibrous proteins." The latter remark very much worried Astbury who had obtained diffuse reflections at $4\frac{1}{2}$ Å and 10 Å spacing, β keratin-like, from dried crystalline pepsin powder[39] given him by Northrop. For a moment it suggested very radical changes might occur on drying a protein crystal, polymerization between the molecules to form chains.

The proper exploitation of the x-ray photography of pepsin was too large an undertaking for Bernal's very small research group, heavily otherwise engaged in the summer of 1934. But the observations led immediately to various events important in protein crystallography. Professor Herman Mark visited Bernal's laboratory and saw the photographs. He was so excited he forgot to arrange, as intended, for his research student, Max Perutz, to work with Hopkins and arranged instead for him to work with Bernal. Bernal lectured in Manchester that summer about his results and I. Fankuchen heard him and asked after the lecture if he might work with him. And after I had returned to a research fellowship in Oxford in the autumn, Professor Robert Robinson put a sample of insulin crystals he had been given into my hands.

It is interesting that the observations on pepsin crystals were not

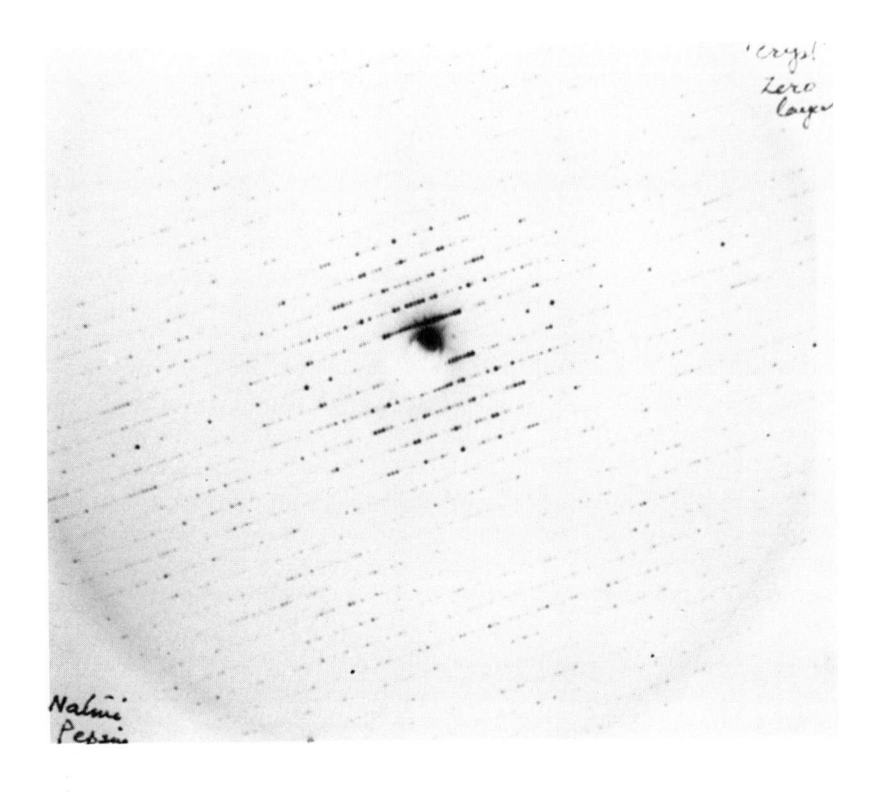

FIGURE 3. (Top) Precession photograph of pepsin taken in 1978. (Bottom) Crystals of pepsin, 1978. (From T. L. Blundell.)

immediately taken up in other countries although at the time there was considerable interest both in proteins and in x-ray diffraction. The research workers most immediately implicated all turned their efforts in other directions. A. L. Patterson had earlier (1930) considered working on protein crystals but decided the methods available for the interpretation of x-ray diffraction phenomena were inadequate. He left the Johnson Foundation where he was working in 1931 and took a year off on his own resources—which extended to two—to investigate at M.I.T. the properties and uses of Fourier series. Out of this came the understanding of Fourier series calculated with the squared amplitudes of the x-ray spectra as coefficients. These series, which we all call by Patterson's name, define distributions in which the density corresponds with interatomic vectors in the crystal.[40] No protein crystal structures could have been solved without their use. R. W. G. Wyckoff had been introduced to x-ray analysis and space group theory by Nishikawa in 1918 and had later worked specifically on crystals of biological interest, on the crystal structures of urea and thiourea. In 1936 with R. B. Corey he took x-ray powder photographs of a number of protein crystals kept in their mother liquor[41] and also of Stanley's preparation of "crystalline" tobacco mosaic proteins.[42] The x-ray photographs showed sharp lines of long spacing but clearly could be of no immediate use in the detailed study of proteins. Wyckoff turned to the development first of small ultracentrifuges and then of electron microscopes—with marvelous effect on the seperation of individual proteins and observation of their external shapes. Corey was invited to Pasadena by Linus Pauling, to work on the exact crystal structures of the amino acid constituents of proteins, more immediately amenable to x-ray analytical methods than the protein crystals themselves. His work and that of E. W. Hughes and their students established the structures of diketopiperazine, glycine, alanine, and diglycylglycine, which provided evidence extremely helpful in the eventual understanding of protein structures.

Bernal and Fankuchen were themselves diverted from the main line of protein crystallography by the preparation of the tobacco mosaic virus given to them by Bawden and Pirie (1936).[43] They developed techniques for producing very fine intense monochromatic x-ray beams and studied the diffraction effects obtained from the virus preparations under a variety of conditions, in solutions of varying concentration, salt concentration, and pH, still or flowing. The strong lines in their patterns were similar to those found by Wyckoff and Corey. They showed it was possible to separate a wide angle pattern, which stayed unchanged through a number of concentration variations, from small angle patterns, which varied with concentration. The wide angle pattern appeared very like fiber patterns of the complexity of that obtained from feather keratin; it clearly indicated structure within the virus particles themselves. From the small angle pattern and the relations between the different liquid crystal states it was possible to derive the shape and approximate size of the virus particles; they were clearly rod shaped and 150 Å across; rough estimates of their lengths as about ten times their width agreed well with the value 1500 Å shown in the first electron microscope pictures. From the wide angle pattern the authors concluded that the particle had a structure not unlike that of a small protein crystal with a spiral close packed arrangement of subunits 11 × 11 × 11 Å, rather smaller than the units they thought existed in protein molecules, "in some ways simpler." The paper in which they recorded their many fascinating observations on the organization of the virus particles in different liquid crystalline states, positive and negative tactoids, isotropic and anisotropic gels, was written just before the war and published in the *Journal of General Physiology* in 1941,[44] after I. Fankuchen had returned to the United States. (Bernal passed over the final preparation of the text to me and the drawing of some of the illustrations.)

Before war broke out in 1939 single crystal x-ray photographs were taken of five different protein crystals, insulin (1935, 1938, 1939)[45-47]; excelsin (1936)[48]; lactoglobulin (1938, two forms)[49]; hemoglobin (1938)[50]; and chymotrypsin (1938).[50]

The insulin crystals first photographed were rhombohedral in form, similar to the first variety obtained by J. J. Abel in 1926. They were grown from 10 mg of a microcrystalline sample of Boots insulin, following D. A. Scott's prescription (1934) from a phosphate buffer with added zinc. To grow the crystals large enough, the solution was warmed to 60°, brought to a pH of 6.2, and cooled slowly over three days. Since the original air-dried crystals were brightly birefringent, the crystals were dried and photographed. Though this was formally a step back, it did make it easier to attempt to measure the molecular weight of the protein; it also established that it was possible to get interpretable, if limited, x-ray data from dry crystals (59 reflections in fact), provided the crystals were large enough and the drying slow. The first measurements posed problems that have continued with protein x-ray molecular weights. The unit cell weight of 39,700, with rather wide limits of error on account of the limited dry crystal x-ray data, should crystallographically contain $3n$ protein units and water. The first measured water content, 5.35%, was too low and hence led to rather too high a figure (37,600) for protein in the unit cell compared with 35,100 measured by Svedberg and Sjögren in solution. The correction was made as protein chemical analyses themselves improved. It was Du Vigneaud who first observed during the sulphur analysis of insulin that the water content of crystalline insulin tended to be measured too low unless special precautions were taken: Chibnall provided an accurate measurement, 10.1% of the insulin "dry" crystal water content in 1942, which enabled the protein molecular weight to be corrected to $35,330/n$. By this time it seemed fairly clear from osmotic pressure measurements that n was 3, and the true molecular weight of insulin was ~12,000. But there were indications from the very beginning of a lower figure, about 6000—crystallographically measured as 5,888—though this was not established until Sanger's work on the sequence many years later. A further worry in the early 1940's was that the preferred ultracentrifuge overall molecular weight, from sedimentation-diffusion measurements, was higher than the old Svedberg sedimentation equilibrium value. So that a new discrepancy appeared between the crystallographic and centrifuge figures, in the opposite direction from the old one. This applied also to some other proteins, such as lactoglobulin. (In retrospect, the original crystallographic measurement with insulin can be seen as fitting with the sequence value, $5,776 \times 6$, or 34,656 within the estimated limits of error.)

The next crystal to be x-ray photographed, excelsin, also suffered from drying, at least in part, and this Astbury always thought was very lucky.[48] Kenneth Bailey prepared, by different methods, a number of excelsin crystals, rhombohedral in general habit. One large one was balanced on the tip of a glass capillary tube, partly immersed in water On the oscillation photograph taken with x-rays along the threefold axis. a pattern of sharp crystalline reflections appeared, together with a strong fiber pattern showing reflections at 11.4 and 4.55 Å, repeating around the three-fold axis. This pattern occurred whether the crystal was moving or not. It seemed to Astbury that the peptide chains in the protein molecules must collapse on crystal drying into an arrangement very much like that suggested for fibrous proteins such as silk fibroin.

Lactoglobulin, given to Crowfoot and Riley by R. A. Kekwick, was the first protein to be photographed both wet and dry and in two different crystalline modifications.[49] ¶ Horse methemoglobin, prepared by Adair, and chymotrypsin, prepared by Northrop, were photographed soon after by Bernal, Fankuchen, and Perutz[50]; the three proteins were published together with a little additional note by Crowfoot and Fankuchen[51] on the possibility of interpreting photographs of air-dried cubic crystals of a tobacco seed globulin (contrary to our views of what ought to be done, but we could not get large wet crystals). Zinc insulin crystals were prepared and photographed wet a little later (Crowfoot and Riley,

¶ There is a small error in the first measurements, corrected later.[53]

1939)[44] and added to our list. Very slowly during the war and at a gradually increasing pace afterwards x-ray data on new protein crystals were added: ribonuclease 1941,[52] ferritin and apoferritin 1943,[53] the tobacco necrosis virus derivative 1945,[54] lysozyme,[55] tomato bushy stunt virus,[56] myoglobin,[57] lactoglobulin (new measurements), all 1948,[58] turnip yellow virus with and without nucleic acid (1948),[59] and six new varieties of hemoglobin crystals between 1946–48.[60] John Kendrew's list of 1953[61] includes x-ray data on 55 protein crystals, but one should add perhaps that twelve of these are myoglobins of different species measured by John Kendrew and seventeen are different varieties of hemoglobin, all but two of which[62,63] were measured by Max Perutz and his students.

From the taking of the first pepsin photographs the problem before us was clear. How could the thousands of observable x-ray spectra be used in practice to give us a view of the electron density in protein crystals? And from the beginning this question was posed in the form, how could the appropriate phase constants be directly determined? In Bernal's laboratory in 1934 there were a number of much smaller structures being investigated, particularly the sterols, where exactly the same questions were being asked. In 1932, Bernal obtained crystals of thallium sodium tartrate, isomorphous with Rochelle salt, with a view to getting one of us students to practice the structure determination of isomorphous crystals; the project eventually passed over to Beevers. When Bernal received the news of the x-ray photography of the insulin crystals, he looked up D. A. Scott's papers himself and wrote a quick letter to me (dated 1935 by the address), noting that cadmium as well as zinc could be used in the crystallization of insulin and promising to get cadmium insulin from one of his Cambridge friends (Chibnall?), which he did. The calculations I made on the difference in the scattering contributions of zinc and cadmium and one abortive trial on the photography of cadmium insulin made me pessimistic about the use of this particular pair as isomorphous derivatives.** But I did spend some time trying to obtain iodine-substituted insulin crystals, thinking a heavier atom might be easier to detect, particularly from Dr. L. Reiner of Burroughs Welcome who had made iodo-benzene-azo insulin. None of the iodine-containing crystals proved useful—too little iodine was present distributed over too

** In the abortive trial of cadmium and zinc insulin I compared powder photographs of wet crystals (which I realized were inadequate) because I found it impossible for a time to grow large enough single crystals. My first letter to Bernal about my difficulties in growing cadmium insulin crystals is dated March 15, 1935. Probably the preparations were not pure enough. Much later, in 1942, Chibnall offered me new cadmium insulin, but I was beginning work on penicillin and I put it aside for very much later use, as things turned out.

many different sites and the crystals were very small. In the meantime the use of isomorphous derivatives to solve a number of very much simpler structures was investigated; as a result, J. M. Robertson pointed out in 1939 that mercury in an insulin crystal could be as effective in phase determining as nickel in phthalocyanin.[64] My own trials with cholesteryl chloride and bromide were less promising since the crystals proved not sufficiently isomorphous. I tried out the calculations while photographing insulin at the Royal Institution in 1935.

In 1938 a possible new approach to the phase problem was raised by further theoretical ideas about the treatment of x-ray diffraction and the actual properties, as observed, of the wet protein crystals. Formally the scattering factor of a single molecule is a continuous function in which the succeeding changes of amplitude and phase might visually be traced, could it be observed alone. In practice, the study of x-ray diffraction effects from crystals confined the observation of the transform to intervals in diffraction space defined by the crystal lattice constants. Bernal, Fankuchen, and Perutz in their paper about hemoglobin and chymotrypsin in 1938[50] said, "As can be seen from Figure 2 [of Reference 50], the dried crystals of chymotrypsin show not only alterations of spacing but also of relative intensities of reflections. If we assume that drying takes place by the removal of water from between protein molecules, studies of these changes provide an opportunity of separating the effects of inter- and intra-molecular scattering. This may make possible the direct Fourier analysis of the molecular structure, once complete sets of reflections are available in different states of hydration."

The idea that in protein crystals the molecules were rigid and remained internally unchanged, simply moving relatively when the water content of the crystal changed (a necessary condition for the application of this method), was encouraged by the calculation of Patterson vector distributions for wet and dry insulin crystals and for hemoglobin crystals in different shrinkage stages. The maps showed very complicated distributions of peaks that included strong concentrations at 4.5–5 A and 11 Å (FIGURE 4), suggesting some form of chain packing within the molecules as Astbury hoped. but impossible to interpret in detail[53] (though a number of attempts were made). The peak pattern relative to the origin, however, remained unchanged on drying or shrinking the crystals and encouraged the attempt to trace the molecular contribution as such to the x-ray scattering.

It was Max Perutz's major work for the next ten years of his life to test these ideas within the framework of the crystal structure of horse methemoglobin, and particularly in relation to the centrosymmetrical projection based on the h01 reflections where the phase constants re-

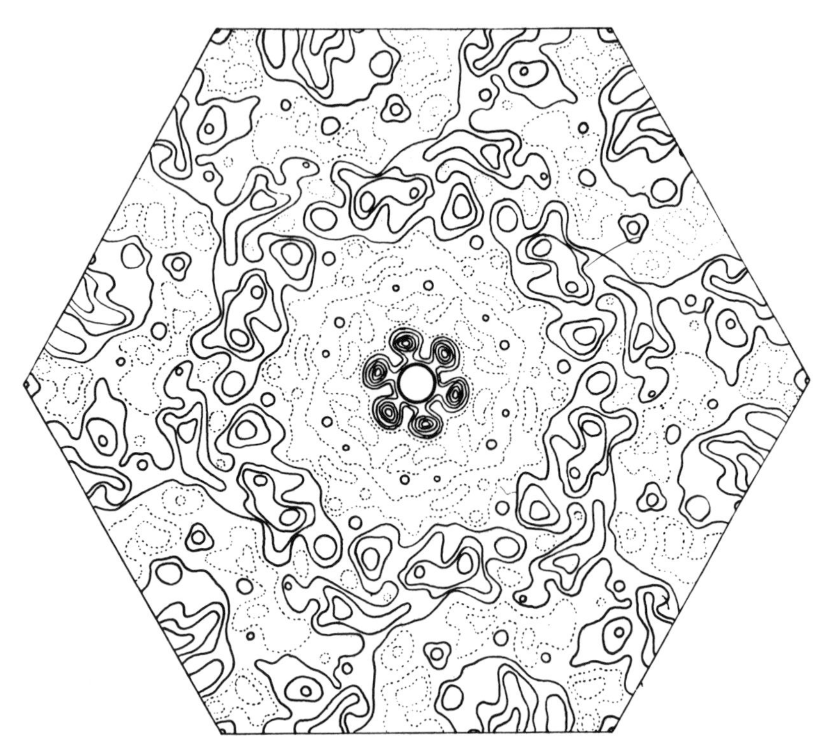

FIGURE 4. Section in three-dimensional Patterson distribution for wet insulin at ~ 3.0-Å resolution, showing 5-Å vectors (calculated in 1940 by D. Crowfoot).

duced to signs.[65] He was interrupted by internment at the beginning of the war, and later by work on Habbakuk in Canada,†† but by 1942 he had obtained a series of shrinkage states of the crystals that made it possible to narrow down the solution of the sign contributions for the 001 reflections of hemoglobin from 64 to 8. Further studies of salt-free crystals make it possible to limit the alternatives to two, one of which gave a very improbable solution. So he calculated a first one dimensional projection of all the electron density in the hemoglobin molecule. I found a letter from W. L. Bragg in my files, dated 3rd August 1942, trying to stimulate me into further action on insulin in which he says, "I have been very interested in Perutz's latest work, and light seems to be beginning to break in the case of hemoglobin." In the following years Perutz was able to trace the relative sign relations along parallel lines of

†† Habbakuk was a gigantic aircraft carrier, made of ice.

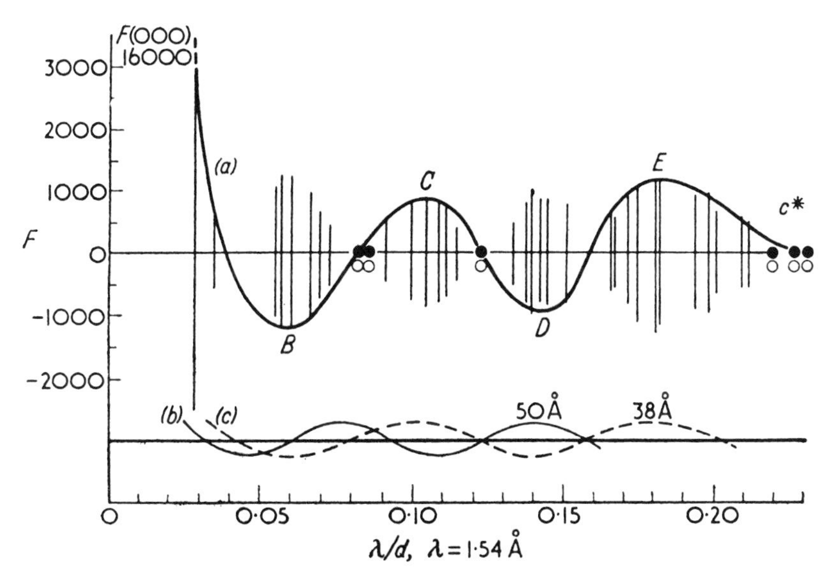

FIGURE 5. The tracing of the molecular transform of hemoglobin in one dimension from the measurement of the (001) reflections (vertical lines) at different shrinkage stages of horse methemoglobin. (After Bragg and Perutz.[65])

h01 reflections but the crystal shrinkage was in one direction only (see excelsin), which made it extremely difficult to interrelate the phase relations of reflections in different row lines.[61] In retrospect, the researches are of great interest both in the light they throw on the behavior of water between protein molecules in various states of humidity and on the possibility of limiting from the observations the overall shape of hemoglobin.[77] But their major importance historically is that they required for their achievement the measurement of the intensities of the x-ray intensities with the highest possible accuracy, preferably on an absolute scale (FIGURES 5–7).

Throughout the 1930s and 1940s there were many attempts to devise model structures for protein molecules with varying degrees of detail and precision as knowledge of the structures of amino acids and peptides increased.[66, 67] In the early years speculation was very free, even the presence of peptide chains was questioned by some. The most complete theory proposed, the cyclol theory of Dorothy Wrinch.[68] in which peptide chains were further condensed to form fabrics had for a time a stimulating effect on both protein and crystallographic research though it soon appeared to be untenable in detail. On the other hand it was also

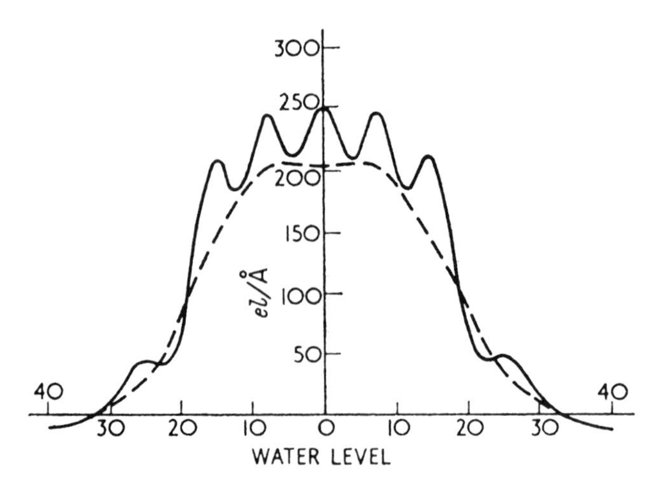

FIGURE 6. Electron density in hemoglobin projected on $\underline{c}^*$ axis. (After Bragg & Perutz.[65])

shown from the general characteristics of the Patterson distributions that protein molecules could not contain even approximately identical peptide chains in simple close packing. A serious reexamination of the probable geometry of peptide chain folding and hydrogen bond formation occurred in a number of laboratories immediately after the war. It led Pauling, Branson, and Corey, using all the accurate x-ray analyses of the Pasadena school to propose precise pleated sheet models for the atomic arrangement in β folded chains and a wholly new helical model, the α helix, for α folded chains in protein molecules.[69] It led also, gradually, to the solution of a number of other fiber structures, collagen among proteins, and DNA.

There were a number of other historically interesting developments in the immediate postwar period. A quite new approach to the problem of phase determination was raised by observations by D. Harker and J. Kasper[70] on mathematical equations relating the magnitudes and phases of the structure amplitudes. Though almost immediately E. W. Hughes[71] (compare Wilson[72]) showed that with many atom molecules as complicated as proteins these relations were unlikely to be useful, they focused everyone's mind again on absolute and accurate intensity measurements and new, more rapid methods of accurate measurement were developed: film scanning and counter diffractometers. There were also very great developments in electronic computing, which made it possible by the early 1950s to handle the enormous calculations required by protein x-ray analysis. And Bokhoven, Schoone, and Bijvoet[73] showed that

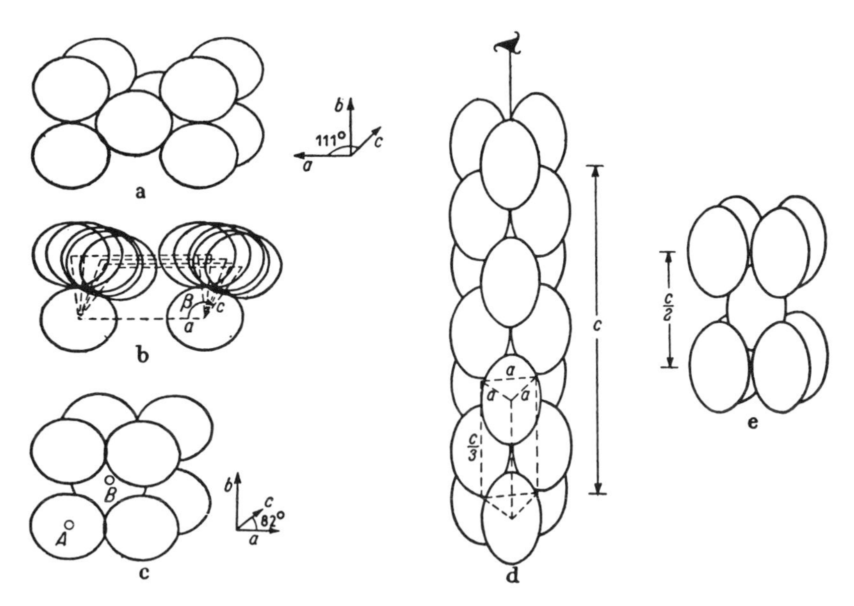

FIGURE 7. Attempts to derive shapes for the hemoglobin molecule by considering packing in different crystals (After Bragg & Perutz.[65])

by accurate measurements on more than two isomorphous derivatives of noncentrosymmetric crystals such as strychnine, general phase angles could be calculated directly from the x-ray data to show atoms in their full three-dimensional relationship.

John Kendrew's review of "Crystalline Proteins; Recent X-Ray Studies and Structural Hypotheses," published early in 1954,[61] gives a picture of the state of our knowledge at the end of this period—measurements, calculations, speculations, good ideas, no certainty, no details. All that was to change as the result of Perutz's experiment in 1953. I give here an account he wrote for me about the sequence of events.

> I had been doubtful if any heavy atom would change the intensities of hemoglobin successfully until I asked Bill Cochran if he would let me use his counter spectrometer–consisting of a Unicam oscillation camera with a geiger counter on one arm–to measure the absolute intensity of the hemoglobin reflections. I was surprised how small the absolute Fs were and did some simple calculations which showed that a heavy atom would produce changes that should be easily measured. The original purpose of this experiment had been quite a different one: to put the molecular transform derived from Bragg and my salt-water Fourier on an absolute scale. This happened in 1951 or 1952. [Compare FIGURE 5.]
>
> At that stage I had no idea how I might attach a heavy atom to hemo-

globin. As a side line I had done some work on the crystal structure of sickle cell hemoglobin. One day I received a set of reprints from the *Journal of General Physiology*, a journal that I would not normally have looked at, from an unknown man at Harvard called Austin Riggs. He had wondered whether hemoglobin A and hemoglobin S differed in the number of reactive SH groups and had titrated them with paramercuribenzoate. He also examined the effect of PMB on the oxygen equilibrium curve and found that heme-heme interaction was largely preserved.

I got very excited by this observation, because it suggested that you can attach molecules of PMB to hemoglobin without changing its structure significantly. I discussed the crystallization of PMB-hemoglobin with Vernon Ingram who had used this reagent before and kindly made the compound for me.

When I developed the first precession picture of PMB-hemoglobin and compared it with that of native hemoglobin, I saw that the two crystals were isomorphous and that the intensity changes were just of the magnitude that my measurements of the absolute intensities had led me to expect. Madly excited, I rushed up to Bragg's room and fetched him down to the basement dark room. Looking at the two pictures in the viewing screen, we were confident that the phase problem was solved. [Compare FIGURE 8.]

Two weeks later, the first two-dimensional electron density projection of hemoglobin was calculated, and six *years* later—strictly out of our period—the first protein, myoglobin, was seen in a three-dimensional electron density distribution at sufficient resolution (2.0 Å) to show clearly the arrangement of the atoms in the molecule.

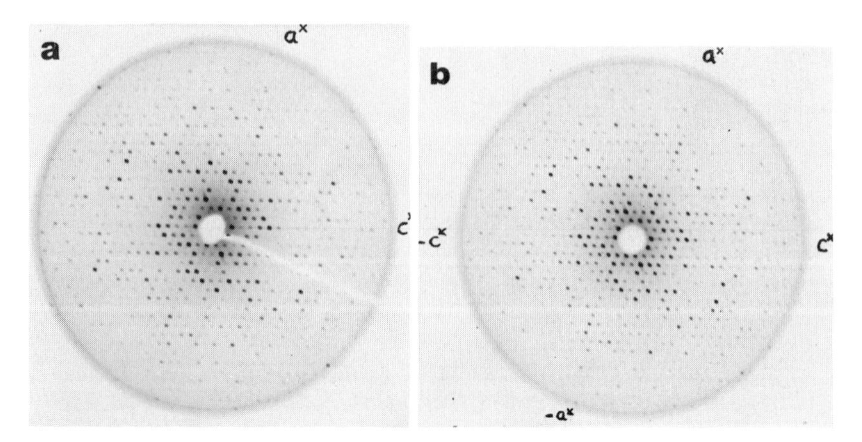

FIGURE 8. X-ray photographs of (h01) reflections of unsubstituted (a) and *p*-mercuribenzoyl (b) horse methemoglobin crystals taken by M. F. Perutz.

Today, through the x-ray analysis of isomorphous crystals,[74, 75] we know the atomic arrangement in detail of some fifty protein molecules (I cannot risk an exact figure, the number seems to increase so rapidly).

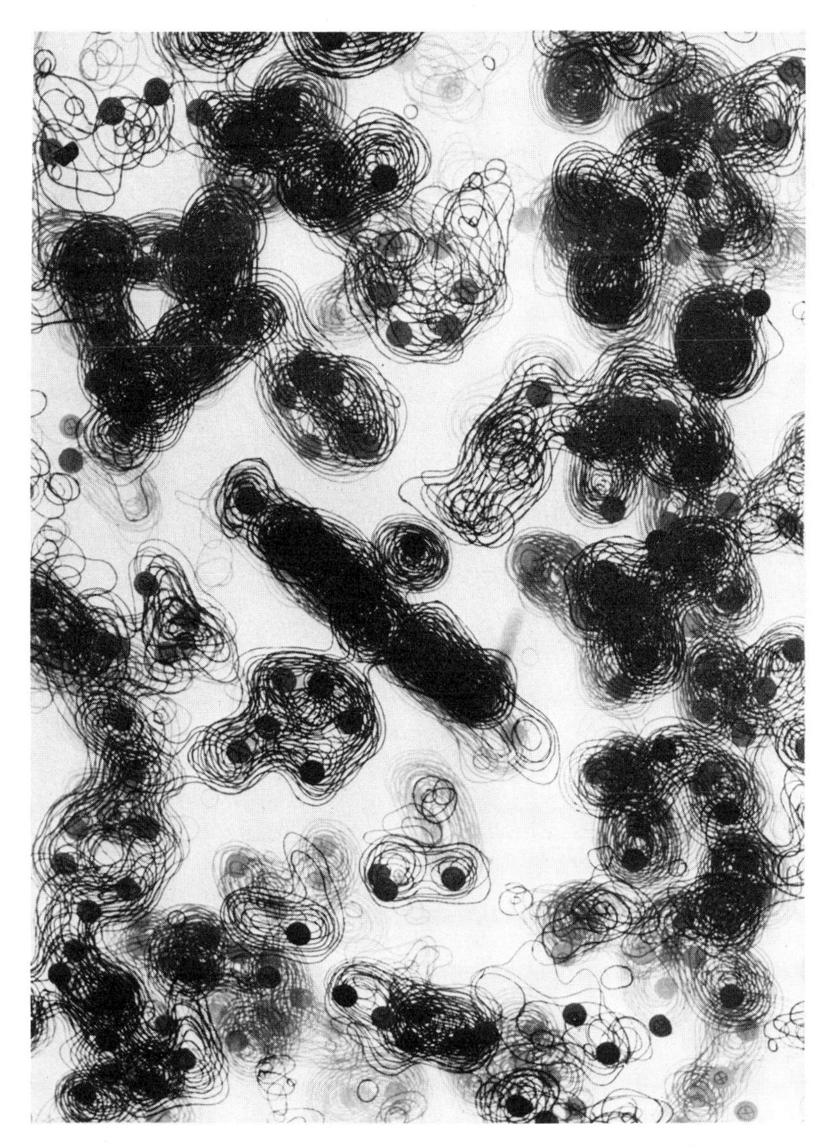

FIGURE 9. Part of the electron density map of myoglobin at 1.4-Å resolution. (Center) The heme group on edge. (Top left) An α helix, end on. (Right) An α helix, on its side. (After J. C. Kendrew, Les Prix Nobel, 1962.)

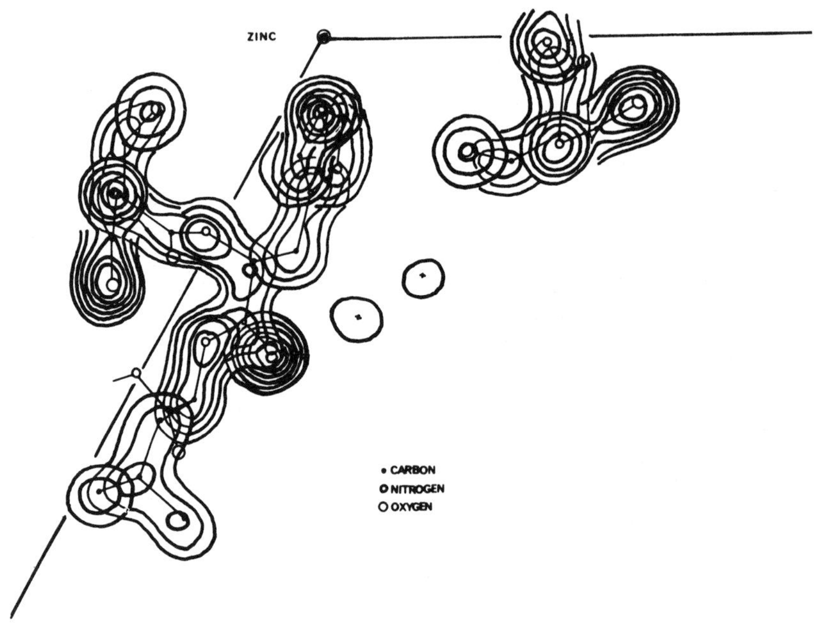

FIGURE 10. The electron density over the residues B9–B11 in 2-zinc insulin crystals showing individual definition of the atoms in serine, histidine, and leucine residues and in the peptide chain (1.5-Å resolution).

In many three-dimensional electron density maps we can observe individually the atoms in the peptide chains and residues (FIGURES 9 & 10). We see in these maps the answers to many of our old controversies.

Though the structure of the pepsin crystal first photographed has not yet been solved, we know that it is very likely to be closely similar to that of the acid protease shown in FIGURE 11. Within this molecule there is a single peptide chain but its course is at first sight rambling; there is very little specific α helical structure and the β pleated sheet structure that occurs is far from regular. It is not surprising that we wondered briefly long ago whether chains existed at all within it; it is reasonable to us now that the β chain character should become rather clearer as Astbury observed when the crystals dry. In protein molecules in general the proportions of α helices and β structures vary widely; both tend to be more distorted than we expected; β sheets are usually much twisted. And within the more irregular strands of chains that run between the specific structures we can trace small stretches of chains in other folds, postulated in various papers and notebooks of those who thought about proteins in the 1930s and 40s. Many protein molecules we find are very

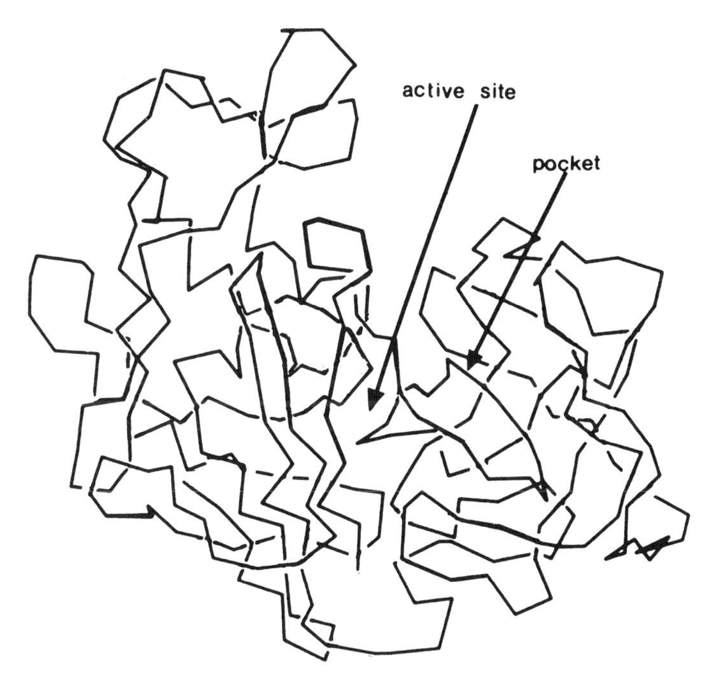

FIGURE 11. The course of the peptide chain in the acid protease; only α carbon positions are marked. (After T. L. Blundell.)

irregular in shape, and many aggregate as we suspected long ago: insulin into hexamers, lactoglobulin into dimers, hemoglobin into tetramers. We are far still from understanding all the details of the structures we find: protein molecules are marvelously various.

REFERENCES

1. BAUMGÄRTNER, K. H. 1830. Beobachtungen über die Nerven und das Blut. :25 (Table V). Freiburg.
2. HÜNEFELD. 1840. Die Chemismus in der thierische Organisation: 160. Leipzig.
3. REICHERT, K. E. 1849. Arch. P. Anat. u. Physiol. :198.
4. PREYER, W. 1871. Die Blutkrystalle. Jena.
5. REICHERT, E. T. & A. P. BROWN. 1909. The differentiation and specificity of corresponding proteins and other vital substances in relation to biological classification and organic evolution. Carnegie Institute of Washington.
6. COHN, F. 1860. J. Pr. Chem. *80*: 129.
7. SCHIMPER, A. F. W. 1880. Z. Krist. *5*: 131.
8. ZINOFFSKY. 1886. Z. Physiol. Chem. *10*: 16.
9. FRIEDRICH, W., P. KNIPPING & M. v. LAUE. 1912. Sitz. Ber. Math. Phys. Klasse. Bayer. Akad. Wiss. München. :303.

10. Ewald, P. P. Ed. 1962. Fifty Years of X-Ray Diffraction. International Union of Crystallography. Oosthoek, Utrecht.
11. Friedrich, W. 1913. Phys. Z. *14*: 317.
12. Bragg, W. L. 1913. Proc. R. Soc. A *89*: 248.
13. Nishikawa, S. & S. Ono. 1913. Proc. Math. Phys. Soc. Tokyo 7: 113.
14. Herzog, R. D. & W. Janke. 1920. Berichte *53*: 2062.
15. Polanyi, M. 1921. Naturwissenschaften *9*: 288.
16. Brill, R. 1923. Annalen *434*: 204.
17. Meyer, K. H. & H. Mark. 1928. Berichte *6*: 1932.
18. Weissenberg, K. 1926. Berichte *59*: 1535.
19. Gonell, H. W. & H. Mark. 1923. Z. Phys. Chem. *107*: 181.
20. Dickinson, R. G. & A. L. Raymond. 1923. J. Am. Chem. Soc. *45*: 22.
21. Hendricks, S. B. 1928. J. Am. Chem. Soc. *50*: 2455.
22. Astbury, W. T. & H. J. Woods. 1930. Nature. *126*: 913.
23. Astbury, W. T. & A. Street. 1931. Phil. Trans. R. Soc. *230*: 75.
24. Astbury, W. T. & T. C. Marwick. 1932. Nature *130*: 309.
25. Astbury, W. T., T. C. Marwick & J. D. Bernal. 1932. Proc. R. Soc. B. *109*: 443.
26. Bernal, J. D. 1931. Z. Krist. *78*: 363.
27. Bernal, J. D. 1932. Biochem. J. *26*: 75.
28. Warren, B. & W. L. Bragg. 1928. Z. Krist. *69*: 467.
29. Bragg, W. H. 1915. Phil. Trans. R. Soc. *215*: 253.
30. Duane, W. 1925. Proc. Nat. Acad. Sci. U.S.A. *11*: 489.
31. Havighurst, R. J. 1925. Proc. Nat. Acad. Sci. U.S.A. *11*: 502.
32. Cork, J. M. 1927. Phil. Mag. *4*: 688.
33. Lonsdale, K. 1931. Proc. R. Soc. *133*A: 536.
34. Beevers, C. A. & H. Lipson. 1934. Proc. R. Soc. *146*A: 570.
35. George, W. H. 1929. Proc. Leeds Phil. Lit. Soc. *1*: 412.
36. Clark, G. L. & K. E. Corrigan. 1932. Phys. Rev. *40*(ii): 639.
37. Bernal, J. D. & D. Crowfoot. 1934. Nature. *133*: 794.
38. Perutz, M. F. 1949. Research *2*: 52.
39. Astbury, W. T. & R. Lomax. 1934. Nature *133*: 795.
40. Patterson, A. L. 1935. Z. Krist. *90*: 517, 543.
41. Corey, R. B. & R. W. G. Wyckoff. 1936. J. Biol. Chem. *114*: 407.
42. Wyckoff, R. W. G. & R. B. Corey. 1936. J. Biol. Chem. *116*: 51.
43. Bawden, F. C.. N. W. Pirie, J. D. Bernal & I. Fankuchen. 1936. Nature *138*: 1051.
44. Bernal, J. D. & I. Fankuchen. 1941. J. Gen. Physiol. *25*: 111.
45. Crowfoot, D. 1935. Nature *135*: 591.
46. Crowfoot, D. 1938. Proc. R. Soc. A *164*: 580.
47. Crowfoot, D. & D. P. Riley. 1939. Nature *144*: 1011.
48. Astbury, W. T., S. Dickinson & K. Bailey. 1935. Biochem. J. *29*: 2351.
49. Crowfoot, D. & D. P. Riley. 1938. Nature *141*: 521.
50. Bernal, J. D., I. Fankuchen & M. Perutz. 1938. *Ibid.* 523.
51. Crowfoot, D. & I. Fankuchen. 1938. *Ibid.* 522.
52. Fankuchen, I. 1943. J. Biol. Chem. *150*: 57.
53. Crowfoot Hodgkin, D. 1950. Cold Spring Harbor Symp. Quant. Biol. *14*: 65.
54. Crowfoot, D. & G. M. J. Schmidt. 1945. Nature *155*: 504.
55. Palmer, K. J., M. Ballantyne & J. A. Galvin. 1948. J. Am. Chem. Soc. *70*: 906.
56. Carlisle, C. H. & K. Dornberger. 1948. Acta Crystallogr. *1*: 194.

57. Kendrew, J. C. 1948. Acta Crystallogr. *1*: 366.
58. Senti, F. R. & R. C. Warner. 1948. J. Am. Chem. Soc. *70*: 3319.
59. Bernal, J. D. & C. H. Carlisle. 1948. Nature *162*: 139.
60. Kendrew, J. C. & M. F. Perutz. 1948. Proc. R. Soc. A *194*: 375.
61. Kendrew, J. C. 1954. Prog. Biophys. Biophys. Chem. *4*: 244.
62. Zinsser, H. H. & Y. C. Tang. 1951. Arch. Biochem. *34*: 81.
63. Tang, Y. C. 1951. Acta Crystallogr. *4*: 564.
64. Robertson, J. M. 1939. Nature *143*: 75.
65. Bragg, W. L. & M. F. Perutz. 1952. Proc. R. Soc. A *213*: 425.
66. Astbury, W. T. 1939. Ann. Rev. Biochem. *8*: 113.
67. Huggins, M. L. 1942. Ann. Rev. Biochem. *11*: 27.
68. Wrinch, D. 1936. Nature *137*: 411.
69. Pauling, L., R. B. Corey & H. R. Branson. 1951. Proc. Nat. Acad. Sci. U.S.A. *37*: 235.
70. Harker, D. & J. S. Kasper. 1948. Acta Crystallogr. *1*: 70.
71. Hughes, E. W. 1949. Acta Crystallogr. *2*: 34.
72. Wilson, A. J. C. 1942. Nature *150*: 152.
73. Bokhoven, C., J. C. Schoone & J. M. Bijvoet. 1951. Acta Crystallogr. *4*: 275.
74. Green, D. W., V. M. Ingram & M. F. Perutz. 1954. Proc. R. Soc. A *225*: 287.
75. Kendrew, J. C., R. E. Dickerson, B. E. Strandberg, R. G. Hart, D. R. Davies, D. C. Phillips & V. C. Shore. 1960. Nature *185*: 422.
76. Jenkins, J., I. Tickle, T. Sewell, L. Ungaretti, A. Wollmer & T. Blundell. 1977. Acid Proteases. :43. Plenum Publishing Corporation. New York, N.Y.
77. Bragg, W. L. & M. F. Perutz. 1952. Acta Kristallogr. *5*: 323.

DISCUSSION OF THE PAPER

Gordon: I got the impression that the pepsin crystal when it was wet was so very much better than when it was dry as tried first. Now is it a fortunate thing that Bernal happened to find such an enormous difference in the sharpness of the spots for that one? Or, if he had used another crystalline protein and got some sort of pattern from the dried crystal might he not have bothered to go on to the wet one?

Hodgkin: I am always glad that Bernal took the first x-ray photographs of pepsin himself a little hurriedly, and saw almost no diffraction effects as others had observed before him, who all had tried taking x-ray photographs of dried crystals. We now know that it is possible to obtain extensive x-ray reflections from most wet protein crystals and more limited data from most dried protein crystals, including pepsin, provided they are allowed to dry sufficiently slowly. In the wet crystals the molecules are in regular contact with one another, with water filling the spaces between them. If you let the water out, the molecules sag irregularly, order diminishes, and most of the x-ray reflections fade or become diffuse. I think that Bernal even if he had obtained such a limited diffraction pattern would have realized it could not correspond with the beautiful appearance of the wet pepsin crystals.

W. DUAX (*Medical Foundation of Buffalo, Buffalo, N.Y.*): I found a letter from Max Perutz about his excitement when he realized that he was on his way to the answer concerning the hemoglobin structure. Do you have any similar feeling in regard to either insulin or vitamin B_{12}? Was there a time when you suddenly realized it was finally going to be downhill the rest of the way, or was it just a sort of excitement of discovery?

HODGKIN: Yes, there are always these moments of discovery. I should also add that I was so deeply involved in the protein story, that I shared Max's excitement. I remember how Max rang me up the day he got the first projection of the hemoglobin structure drawn out and I just got in the car and drove straight over to Cambridge to see it, and I knew I would see essentially nothing because the projection was not interpretable down 63 angstroms. But I could not help going to look at it.

E. PATTERSON (*Institute for Cancer Research, Philadelphia, Pa.*): My husband would have been so excited had he lived to know what the method did. I think the reason he felt that it never would be so useful was because he could not conceive of the development of computers, which it seems to me have been exceedingly helpful in solving protein structures. Again the intellectual development and what you might call a mechanical development have gone hand in hand with the solution of the structures.

HODGKIN: Of course, the very interesting thing is that the first steps, the calculation of the one-dimensional projection, the calculation of the two-dimensional projections, and the first Patterson projections, were done by hand computation on adding machines. However, the first three-dimensional map at 2 angstroms that showed where the atoms really were, the map of myoglobin, depended upon computers.

KARLSON: I have a question regarding model building, we all know that the detection of the alpha helix and also of the pleated sheet relied very much from the first on Corey's measurement of peptides and, secondly, on the model building with the exact angles and distances of the atoms in Pauling's laboratory. What was the attitude in Astburys's laboratory and in Cambridge to this kind of model building? Did Astbury ever try such a thing and what did they do at Cambridge?

HODGKIN: Yes, actually there was a history of this going on all the time. The actual determination of crystal structures gave more and more exact ideas of the size of atoms and how they were arranged. The very first step, the old silk fibroin model, really depended upon structure analyses of urea by Hendricks of hexamethylenetetramine in Pasadena and in

Berlin. Then later more structures showed that you not only had to have the sizes right but that you had to have the angles and conformations right. Just before the helix model was produced there was a very long model-building paper by Bragg, Perutz, and Kendrew in the *Proceedings of the Royal Society*, in which a variety of structures were proposed for peptide chains. Astbury certainly also took new evidence into account and revised earlier proposals in his later model building. The alpha helix was just a stroke of genius on the part of Linus Pauling himself. Again I know because I was there. He was having a cold or flu in Oxford the year he was there and was rolling bits of paper around his fingers, which took him away from the crystallographic repeats and allowed the helix to go its own way. Before that, structures very close to the alpha helix had been built by several people, particularly H. S. Taylor and by Maurice Huggins, following the idea that a helix was a likely kind of repeating pattern; but it was the nonintegral character of the alpha helix repeat that really produced a change of outlook, and that was due to Pauling.

One of the stories is that when John Kendrew's group calculated the first three-dimensional electron density map of myoglobin in Cambridge, at 2.5 angstroms with all the terms included, all of the group stayed up until the alpha helix came through. When they could see that there were definitely alpha helices in the structure most of them went home to bed.

T. SEJNOWSKI (*Princeton University, Princeton, N.J.*): I have two questions. The first is a matter of strategy. I wonder why the decision was made to put all the effort into hemoglobin first, rather than, for example, myoglobin, which is closely related but much simpler and easy to work with.

And the second question concerns remarkable persistence over many years. There must have been moments when the group was discouraged, or at least it may not have been clear that the end was in sight. And I wonder what was the driving force that kept you on that particular problem rather than going on to some simpler one first?

HODGKIN: If you look back at the whole history of protein crystals, which goes back to the 1830s, hemoglobin is the first protein observed to crystallize. Different hemoglobins give very beautiful crystals and Max Perutz was keen on working on hemoglobin as a problem in biochemistry before he even came to Cambridge in 1936. Myoglobin was not isolated until sometime later, I think during the war or soon after. When its characteristics were realized, Bragg and Perutz decided to concentrate work on myoglobin with a new very good research student, John Kendrew, who came to work with them directly after the war. However, Max Perutz still had his old love of hemoglobin and had already put in an enormous amount of his own time measuring hemoglobin reflections.

So the work on myoglobin ran ahead once three dimensional calculations became possible, first at low resolution, the 6-angstrom map, and then at high resolution, the 2-angstrom map—because, as you say, fewer x-ray reflections had to be measured to get the answers. But hemoglobin, partly for the reasons given in Max Perutz's letter, partly because he had measured hemoglobin crystals so often already, was the first protein for which heavy atom derivatives that were interpretable were obtained. These led to making similar derivatives for myoglobin.

Actually I myself did put protein x-ray analysis largely aside and worked on simpler compounds for many years before beginning to concentrate again on insulin. And even Max Perutz did some pieces of research over the years, not connected with hemoglobin, to sustain his morale.

SMITH: Myoglobin was actually first isolated from horse heart in the 1930s. It was very difficult to obtain it in pure form and it was very difficult to obtain in any quantity. Kendrew's x-ray work was done on whale myoglobin; that became available in large quantities immediately after the war because people thought that whale meat was going to be a good substitute for the meat shortage in Britain.

HODGKIN: Thank you very much, yes. I should say that the immediate next step, which I skipped over, was that after the set of photographs of protein crystals that were measured up to 1950, there was a fantastic thrust in which a large number of crystals were studied, but nearly all of these are different species of myoglobin and different species of hemoglobin.

PNINA ABIR-AM (*Université de Montréal, Montréal, Québec*): It was mentioned that a gentlemen's agreement existed between Bernal and Astbury with regard to crystalline versus fibrous material for x-ray work. Bernal and Astbury agreed that Bernal would work on crystalline material and Astbury on fibers. Bernal said later that this division, by a gentlemen's agreement, led to his having reached DNA later on. I wonder what your opinion on this is.

HODGKIN: I do not think they took this agreement quite that seriously. Bernal certainly became most interested in solving the DNA structure just after the war and his line of study was the same as that adopted for the α helix. That is to say, he planned to have studies made of the structures of crystals of the nucleotides and nucleosides of which DNA is composed and one such study was done at Birkbeck by S. Fürberg. Fürberg certainly built what he hoped might be models of DNA as a consequence of his work. Of course, they had only enough diffraction data to do rough calculations. He just did not happen to find the right structure.

Dialogue: A Discussion among Historians of Science and Scientists

FREDERIC L. HOLMES, *Chair*

Department of History of Medicine and Science
University of Western Ontario
London, Ontario
Canada

HOLMES: We are going to have a round table discussion among historians of science and scientists, and so that we will not take that theme too seriously I should point out that there is no round table and that I think there is some question whether we can really be divided into historians of science and scientists since most of the history of science at this session has been provided by scientists and some of them can regard themselves as historians of science.

We are to be entirely informal and yet not totally undirected. I would like to open with the very brief remark that ever since the history of science has become visible as a professional activity the question has been asked: Is it part of science or is it part of history? The problem is that if you ask historians they say it is part of science and the scientists say it is part of history. It is always thrown in the opposite camp.

I found in my personal experience that while historians collectively accept in principle that the history of science is part of history, individually they tend to avoid involvement with the subject. Scientists collectively reject the history of science; they say it is not part of science. Yet individually they are willing to become involved. This conference manifests the fact that there are quite a few scientists who are willing to become involved in the history of science.

I would like to ask those of you who are scientists, who have become sufficiently involved to come here and perhaps contribute, why you feel that it is worthwhile to look back? Those of us here who are historians of science look back because that is our profession, but those of you who are principally scientists, who most of the time think of today and tomorrow, have taken special time out to look back, and I would like to ask you: What have you found valuable about it? What among the things that have been said today in the various papers have you found particularly revealing or unexpected or illuminating?

0077-8923/79/0325-0149 $01.75/0 © 1979, NYAS

PAUL C. ZAMECNIK (*Harvard University, Boston, Mass.*): There is pressure for space in all the journals, particularly some which have a limit such as *Proceedings of the National Academy of Sciences* of five pages, and I think some of that pressure is expressed by shortening the bibliography. There is a tendency to omit anything older than five years, and not to include the primary sources. That is a tendency not only for one journal but for many, and this conference is in a way an effort to try to set the record straight and to have some of the primary people give a chronological story of events as they see them. My feeling about it is that the all of the explosive growth of science during the period from say the dawn of NIH, high visibility, and the large budgets in the early 1950s has resulted in an enormous growth in science and an inadequate history of that time.

PIRIE: Giving references is fine, but I think the writers ought to show evidence that they have in fact read the papers cited. There is a tendency to have a lot of references that are included because it is traditional to put them in. Needham in his vast survey of the history of China rather interestingly put an asterisk on those things he quoted but had not in fact read. Now that, I think, is a discipline that other people ought to adopt too. You see the most shocking misstatement of what so-and-so said because people just quote from reviews or literature surveys, and they have not read the papers. So if you are quoting a paper I think the editor ought to insist on some evidence that you have read it.

JOHN PARASCANDOLA (*University of Wisconsin, Madison, Wisc.*): Perhaps we can go beyond just the question of why people came here or what they got out of the sessions today. I would like to ask the scientists here if they think that it is important that biochemists today and those being trained in biochemistry have a feel for the history of their discipline? If so, why is it significant that they have some understanding of how everything came about?

KARLSON: I would say it is certainly important for any scientist to know what has developed in his or her field in the past 3, 4, or 5 years and what the evidence is for the current theories. That is one point. Another question is if an enzymologist of today should know, say, the theories of enzyme action around 1850, or the reason why Wilhelm Kühne called them enzymes and not ferments, as was formerly done. I think it is a reward even for scientists today to look back at least for the last 50 years or so. We have a column in the journal *Trends in Biochemical Sciences* with a heading "50 Years Ago," which I always read with keen interest.

I may mention another point which was quite shocking to me. Three or four years ago the Deutsche Gesellschaft der Naturforscher und Ärzte celebrated its 150th birthday and a historian of medicine was invited to write about its history. He replied saying he was prepared to do so but he would stop at 1910 because historians do not regard the last 50 years as appropriate for historical studies. As a scientist I cannot appreciate this view. The really important part of the story lies in the last 50 years or so. I find the history of biochemistry, say, from 1920 to 1970 most rewarding for an actively working scientist.

GORDON: I wanted to suggest a rather different, perhaps more important reason for an emphasis on the history of science now. The image of science now is not the shining image we used to think it was when some of us came in. There is a great deal of criticism from the public in general, at least in Britain, that science has gotten the world into a pretty awful mess. Now I do not think we believe that, but it is a point of view which has to be met, and I think by analysis of how it has all come about people will be in a better position to do that job as it comes along.

Of course one of the problems for the younger generation is deciding whether they become scientists or not. Perhaps it will help them to make the right decision about that if there is available much more material on the history of science.

SCHACHMAN: I would like to perhaps be the devil's advocate because I think a lot of what I hear is not what my students or any of the students I see are feeling. I shall start with a contrast between the scientists who are presenting background information on the development of the field and their apparent age versus those that I gather to be the historians of science in the room, who are significantly younger.

It is my general impression that the interest in the history of science is a personal matter. I disagree completely with Karlson's statement, that you cannot be a good scientist without knowing the history of science. The young kids who are doing molecular biology today, who are making a living out of a bottle of cesium chloride, have never heard of Svedberg, they probably could not care less about the ultracentrifuge or the theory of sedimentation equilibrium. They do not understand the experimental or theoretical foundation for the use of the reagents that they are using. And they are doing beautiful work. The development in them of an interest in history of their field will come about when they acquire that personal longing and the leisure of no longer competing at the ferocious pace under which they now operate.

But I do not think that we ought to kid ourselves into thinking that the young people, at least as I see them, have a great interest. When Lipmann or Krebs comes to Berkeley to give a lecture, the attitude of the graduate students is, "So what. What have they done for us lately?" The students in my classes do not seem to have a compelling feeling that the only way they can be good in biochemistry and molecular biology is to understand the history of science.

I am interested in it because I think it is a marvelous way to teach. I teach via the vehicle of showing the evolution of ideas, but I do not kid myself into thinking that without that interest I also could not do research. And when the bibliographies that Zamecnik is talking about are written by the molecular biologists, their history is not 50 years, it is 2 years, and anything that happened earlier, so far as they are concerned, was never discovered.

HOLMES: If you think that the younger biochemists are not necessarily interested in this nor need it, do you think that the older biochemists are interested in history because some of their own contributions have become historical?

SCHACHMAN: Well, to some extent that is true. It is a question really of what one does with one's life, and the nature of our science is such that as one gets older one recognizes that there are a lot of new techniques, and a lot of new phenomena being discovered, and the younger people have a training that is no longer available for their elderly professors. The professors spend their time doing creative things in other directions. One of the things, just one of them, is history. It would be interesting for me to learn what you think about the students who are studying history. Is this as exciting a field for graduate students in the history of science as it is to study molecular biology or biochemistry when the field is in a state of ferment, when everything that happens yesterday or tomorrow is exciting?

N. H. HOROWITZ (*California Institute of Technology, Pasadena, Calif.*): I agree with Schachman about the use history as a pedagogical tool. I find when I teach that it is impossible not to teach in a more or less chronological sequence of whatever subject it happens to be. To me that is the only way to really understand the subject.

Now many of my young colleagues do not adopt a chronological view at all. They adopt a logical view of the subject and they can give the same material in a totally logical way without ever touching on the history of it, without ever mentioning the mistakes that were made, the great models of discovery, and how they contrasted with the great errors. These younger colleagues do very well even though their bibliographies do not go past the Watson and Crick paper of 1953. They do marvelous science, as Schachman said, and their students do good science. But to me it is absolutely essential to know the history of this thing I am teaching.

As for this meeting here in New York this week, I have never before been involved in this sort of institutionalized history of science, and I was a little surprised to get the invitation. I wondered, when people begin thinking about the history of a branch science, it seems to imply that maybe the subject is over. After all we do know that the funding of science is declining relative to real dollars, relative to the real cost of science.

On the other hand, what also may be involved here are personal interests of men like Edsall and Fruton. And there is a great interest in the history of science elsewhere. The Genetics Society has a committee on the history of science and Tracy Sonneborn is always peppering us with reminders to save our letters, lecture notes, and so on.

Whether this interest in history is related in some way to the difficulty science is having on the national scene I will leave to the historians to decide. It seems to me that it is a coincidence that is almost unexplainable otherwise.

DAVID BEARMAN (*American Philosophical Society, Philadelphia, Pa.*): There are a number of different uses for the history of science and different people have different uses for them. The testimony that today's graduate student perhaps does not need the history of science, yet some people wish to teach with the history of science, that is, use the history of science as a pedogogical tool, suggests that the history of science is sometimes used as part of a person's socialization in a disciplinary specialty.

One of the things that has interested me about this particular conference is that here we have a history of science being used, in a sense, to legitimate the discoveries in the field, to document the existence of a specialty which, at least in many of its stages from much of the testimony that we have heard this morning, was rather amorphous and considerably less defined in the minds of the investigators and in the minds of their audiences at the time that this was going on than it is now.

I would like to suggest some of the uses as a historian of science because there are ongoing questions in our own field as well. One of those questions is about the socialization of people in this field. Clearly one of the things that you learn as you enter a specialty is something about its history, maybe not going back 50 years but its history in the immediate past. And one of the things that has been stated throughout this conference this morning is that the question of protein structure was an important problem, almost as if it were recognized universally as an important problem from 1900 on. I would like to ask, which audiences thought it was an important problem? To whom was it an important problem? What problems did it solve for other specialties in the scientific community? Were immunologists always as convinced that the structure of proteins was an important problem as they became in late 1930s? Did nutrition researchers have a special interest in protein structure?

These questions are very much related to the question of what does the history of a science do for the people who are active researchers. How does your view of the development of a field help to shape the problem that you ask? This is the question that a historian asks as he looks back on the development of a field. But in a way it is also a question about what the utility of the history of science is to scientists.

FRUTON: I would venture a few thoughts about some of the ideas that were offered. First, as regards the question of the hotshot molecular biologist, or for that matter any biochemist who is busily engaged in some exciting endeavor, is it important for him to know what ridiculous ideas may have been entertained 50 years ago about some aspects of the field he is now engaged in? Important in terms of the solutions to the problems that he is working on? I think the answer is clearly no. It may in fact be a distraction from the real business of solving the problem in hand to know too much about the mistakes of the past, perhaps because one learns better by making the mistakes all over again. So I do not sympathize with the view that a knowledge of the history of science is something needed for current research.

As regards the teaching of science, certainly using historical examples can occasionally add entertainment, and indeed even enlightenment, to the understanding of a subject. But, one of the things that has clearly happened in the teaching of any advanced branch of science that you care to mention is that in order to compress a body of knowledge into some sort of reasonable time it is necessary to impose on it a logical structure which obscures the chronological development. Who would imagine teaching an introductory course in organic chemistry today on a historical basis? We must recognize therefore that there are limits to the historical approach as a tool in teaching. I certainly sympathize with the view that it helps to lighten the subject when it becomes a little on the dull side, although occasionally I do get worried about such entertainment when it is offered because too often it consists of mythology perpetuated by us in recounting our accomplishments.

My own feeling is that history is a part of culture just as science is, and the history of science is therefore a part of culture. Those of us who like to feel that we are at points in time with much that has come before us, and like to get some comfort out of this sense of belonging to a historical stream, may occasionally get some pleasure out of trying to understand how our present state of thinking came to be what it is. Whether it will help in any way to catch a glimpse of the future is another matter.

And then of course, so far as a conference such as this is concerned, there is a certain element among the older people here of a search for immortality, and we would like our achievements to be recorded for posterity. I should warn my friends in the historical community that we are offering our documents for them to judge and to compare with other documents and by careful study perhaps to glean a little bit of what actually happened.

BERNSTEIN: Like many of the historian scientists in the room I got my original degree in science. I think that is a general trend in the history of science. Thus many of us have had some experience in being taught or actually doing research in the sciences we are studying. Out of my own experience I have concluded that there are three things that the history of science can do for those actually doing science.

The first thing has to do with understanding the trends in a particular field, the type of experiments that are being done. I have a particular interest in immunology, in current immunology and in historical immunology, and it was not until I compared the two that I realized that in today's immunology there are three or four basic types of experiments that everybody does. If you look at a textbook, you always find these types of experiments done. Then I compared this with historical examples and found that people like Arrhenius and Ehrlich had done all sorts of experiments that were completely different. They have information in those old books that you cannot find in any textbook today. Now maybe the people who are doing active research know about this, but the student does not. I learned just as much from reading the old texts as I did from reading the current ones. To me that was extremely exciting; it showed me that there are other ways of doing immunology that people may not be thinking about now because the trends, techniques, and so on have developed in a particular path for so many years. You need some way of getting back to what was done. There are leads that have been dropped in the past because they did not pan out then; maybe they will pan out now.

The second example has to do with actual loss of information. I was looking at some modern information concerning DNA and the double helical structure and I went around asking some of my genetics friends, this included both professors and graduate students across the country, about five of them. I asked them what the proofs were that DNA was actually antiparallel, because one usually sees only one or two experiments in the textbook. Theoretically it is possible that they could be parallel, or they could be mixed. But the point is there are other ways you could have the DNA. I began to wonder if it had ever been questioned; did anybody think of that or did everyone just accept Crick and Watson's theory? And I asked my genetics friends; I asked my biochemistry friends; nobody knew. They just said that nobody doubted it, it was so obvious. Then I read Olby's book and there it was: yes, it had been questioned. But that is a piece of information that genetics professors, that people being trained to be professors, do not know now; they do not know that it was ever questioned; they do not know that it is theoretically possible or how to prove it.

The third thing that I think the history of science can do is to help scientists to know how to pose problems. I had one professor who is outstanding in my mind during my undergraduate career, and what was outstanding about him was that he was able to use a historical perspective and present how problems came up. A problem would arise; it was solved; but solving it raised some other problems. He was the only professor who taught me how to look at scientific literature and understand that there are things left to do, that an answer is not just an answer but is in itself a set of questions. He showed this through his historical techniques and no other professor that I had was able to do that. Now whether that is directly related to the historical techniques or not, I do not know, but it can be used that way.

HOTCHKISS: I would like to add one more caution to the self-examination that we scientists have been making. I have said a good many times through the years that many people do not seem to have quite enough knowledge of the history of the subject they are working on. Sometimes I will be with another teacher or scientist and he will say the same thing. You would think we were agreeing on the same point until we start to look closer. I mention that I started working in the 1930s, and that the history that the students do not seem to know is the history of the 1920s, the history that fascinated me and got me involved; and somebody who started in, say, 1955 is telling us that the students do not know the history of the 40s and 50s.

So I think you will find that you have to be a little on guard against us in that what we are fascinated by is that part of the history we have lived through or just had to accommodate to in our student years. What is much more genuinely important is that we should have the interest, as I guess most of us here have exhibited, to listen to the history of other subjects than our own particular one. When you find a scientist who has the patience and concern to watch a neighboring field, then you will find that he is probably becoming a historian like you historians are, and I suppose that is one of the mechanisms by which those science students who went into science history moved out into history, because they began to be fascinated by the whole march, the whole parallel stream of all the things going on.

I had a sort of object lesson today. I used to say pretty often that I would like to talk about the past for what teaching I do, which is little, to show how people think, how they went about solving problems of the past, in other words, the triumphs of the past, how they were reached, and how we can face new challenges. And then I began to look around and see who would I be telling that to? That would be the young people.

Well the young people by and large are not really listening to this kind of thing except when we have a captive audience in a teaching course. But there is another kind of lesson too, and that is the mistakes.

Today I was reminded of a sort of corollary of that when Pirie asked what really is in the supernatants from which people get the crystals? Is there something else there? Perhaps there is a little more to the story than just that particular protein that you get out. And somebody asked pretty soon are there families of proteins? Maybe. And I had a chance to think. I remember the feeling that that was possibly one of the big things that was being overlooked. Maybe it is quite true Sanger had crystalline insulin and worked out the structure of that one, but what was in the mother liquor? Was there another set of insulins?

And I would see today, with the history or background I have been lucky enough to have, why I thought that kind of thing. One of my particular interests was the antibiotics, the gramicidins, the tyrocidines, which are made by a mechanism different from that for proteins. Amino acids can be dropped out and others put in according to the convenience of the organism; just adding different amino acids to the environment of the mold will cause it to incorporate those different ones. When you have a nonribosomal system you have that kind of mechanism. So it does not break the rule and at the same time it tells us why we could have had these questions in the past. That is what I was getting out of this kind of rethinking: a question that was moot once and open once may have to be reopened. And that was one of the values in going through our history and seeing again, are we satisfied with the evidence that has been accumulated and do we want to reopen any of these other questions?

Robert Olby (*University of Leeds, Leeds, England*): I was very interested in Prof. Karlson's unfortunate experience in Germany of historians not being interested in research into the history of science that was more recent than 50 years ago, and I think that there are countries outside Germany where this attitude is not so common. I hope that Prof. Karlson's experience is the exception rather than the rule for the history of science today.

There is of course a natural objection to treating very recent material simply because of the limitations upon the availability of written documents. And the attempt to write history on the basis of purely oral historical interviewing and very limited availability of letters, and so on, is of course a risk that many historians feel they do not wish to take, partly because as you so often find, orally recalled memories do not agree with the documents when the documents can be found, and on the whole one tends to find that the documents are a surer guide than the oral recollection.

I have been doing some work recently in Britain where I find it difficult because of the 30 year rule. We do not have the Freedom of Information Act in Britain so we do not have the same opportunity to get hold of recent documents that you have in the United States. I hope that some scientists among us will accept that there are these practical problems about doing recent history, and the reticence about doing such things is not necessarily without some good grounds.

The other point I wanted to comment on was Prof. Schachman's interesting remarks about the attitude of the professional research scientists who are making their careers in biochemical research, molecular biology, and so on. It seems to me that there is a clear distinction between the rigorous professional development of a scientist at this stage in his doctoral and postdoctoral career, and the belonging to a community of scientific research, which is considered not only to be a profession where you get a good salary, or not a good salary as the case may be, and so on, but is part of the culture of the intellectual world. And it is, I suggest in a very crude way, this distinction between belonging to culture, as opposed to more narrowly pursuing a professional goal of advancement, which perhaps does make a very practical difference in one's attitudes and sympathy towards the history of one's subject.

I suppose that the didactic use of history of science in teaching is always going to continue to be of use in certain areas of science, and perhaps not in other areas as Prof. Fruton pointed out. And the image of science has now become a more sensitive and visible matter of concern and thought in the scientific community than it hitherto used to be. Here is the sort of question that could be tackled or is tackled by historians of science and which might be of value for the image aspect of the scientific community.

This is not necessarily what you might call "sign post" history information. The sort of question, for instance, which I have had put to me several times is, what about war? What influence have the last two wars had upon development of molecular biology, or upon the development of protein studies? It has been claimed to me, and I do not know whether many people would accept this claim, that the development of information theory in the Second World War was of particular importance for the development of the information side to molecular biology. This is something which I do not myself think is a very strong point at all, but there surely is a very close connection between the availability of tracers for biochemical research and the building of reactors which would quickly give a wide range of suitable isotopes.

These are the sort of questions which I think are involved when it is said that science students are not particularly interested in history; the cause may be that the historical questions being expounded to them are more in the sign post tradition and less in the science tradition. Perhaps we need to look again at this and think of other approaches to the sort of history of biochemistry which would be of relevance and of interest to present day science students.

Lastly, I think that even if we do not agree about the sort of aims that we have in writing the history of a subject, I am sure we are agreed over the need for support and the need to preserve material. And despite what has been said earlier, at least somebody within the historical community should be brave and attempt to write up some of the developing fields not too long after they have emerged. I must recall here my experience of it being a tremendous advantage to be able to send drafts of chapters to people who were involved in the work to get their comments back. And it is this cooperation and coordination between historians and scientists which not only should benefit the subject but also should be an enrichment to our mutual relations.

Of course this can work the other way. I have to confess that I wrote a chapter about the development of the phage group in a book which was published a few years ago. As a result of a number of comments made, I started off with a rather critical chapter, very critical of the sort of historical account that had been written in the Delbrück Festschrift of 1968, and some of these rather critical and perhaps almost harsh comments about something that was said in that publication were, as a result of correspondence, removed. I am not sure whether years afterwards I would not have been happy if they had stayed in.

In other words, one is subject to the influences of one's time. It depends on how thick-skinned one is and how sure one is of one's judgment as to whether in the long run you find that you have sifted out and benefited from the cooperation and coordination of the scientists in the field, or whether you in fact have not benefited. This is I think one of the main worries that anybody has who is writing about recent history, that they are too much caught within the framework of the concepts of the time to be really confident that they are able to arrive at an objective judgement on the development of the field.

Think for instance of the early successes of molecular biologists, and the view expressed that here we have one base change in the polynucleotide chain and this will make the difference between a sickle cell or a non-sickle cell. Here we have a functional expression of one base change. I do not think that many people today would take the view that it is likely to be a general rule that one base change would have such an effect. This is the sort of thing where you could get caught up with the enthusiasm of the simple picture of the new development, and the historical account would suffer from this.

DIANA LONG HALL (*Boston University, Boston, Mass.*): I was just trying to think about the possible uses of the history of science, about what it is that really captures the attention of students in the biology department to whom I teach the history of science. It seems to me that there are two aspects of it: One is that upper level undergraduates and beginning graduate students do find the stories of how people solve problems exciting. There are problems which perhaps initially they thought always had been solved, and they find it exciting to know how each solution actually took place, that this took place within a certain period of time, and it was done by people who were concerned about certain intellectual problems and had certain tools to work with. That is something that catches the attention of science students and it seems relevant to them and even inspiring to them.

The other aspect, however, is in areas in which, so to speak, they feel the shoe pinching themselves. If it is pinching intellectually because it is an unsolved problem, then it is in some sense reassuring to them to know that it was a difficult problem that has challenged and in fact has defeated people in the past. There are stories of the history of science with which they can identify in terms of their own intellectual experience. That is one part of the pinching shoe which I find that the students respond to, and these are usually graduate students who have begun to work on a research problem and to discover how difficult life really can be.

Another is one that refers back to an early remark by Dr. Gordon, and that is that there are almost no undergraduate or graduate students now working in science, certainly in this country, who are not aware of the cultural uneasiness about the role of science and especially the technology that has been associated with recent developments in biological sciences. They are uneasy about their own professional role as scientists and are for that reason eager to turn to the history of science to find out as you put it how we came to have this large enterprise that seems to be playing this ambiguous social role. It seems that students will respond to that and look to the history in some detail, whereas they might not in other areas. They will take the time out of their professional training within science to actually work at it, as I said, with some attention to the actual circumstances under which the research was done, social circumstances as well as the intellectual circumstances.

HAUROWITZ: I would like to briefly say that in the Chemistry Department at Indiana University we have an undergraduate course for the seniors, which we call undergraduate research. It is one hour a week and it is not obligatory. We ask each of them to present a seminar on one scientist whom they may select from where they want, a biochemist, organic chemist, physical chemist, and so on. We had about 10 or 15 students, just once a week for a semester, and we find some of them get enthusiastic about the assignment. We also find in this way the students who are interested in the history of science.

On the graduate level we did the same thing and though it was I think a very interesting experiment, it was a failure. Some of my young colleagues got together and got the idea that the department buy one of these films which produce lessons and they selected a film given by a famous Dutch physical chemist.

I attended one of these performances; it was just about intramolecular forces and noncovalent bonds, and so forth. It was a brilliant film, a pleasure to listen to. But after the film, which continued through the whole semester twice a week, nothing happened; the students and one of the three teachers who was present left the room. I later came to him and said, "Why don't you discuss this film at least?" He got very embarrassed. He said, "Yes, we should probably do that." But when nothing happened later I asked the other two teachers why they did not discuss it. And they admitted the whole thing was a failure. They admitted it was a brilliant film, but the nomenclature used and the language used was a language of about 20 years ago. They did not know the terms. They missed the present language used in biophysical chemistry, and they were lost. They felt they could not explain; they could not give answers. And they could not cancel it because the film had been bought for the whole semester. I was very depressed and very sad, but I do not know what could be done about it. In 20 years the language changes.

KARLSON: I was very glad to learn that in Great Britain, historians of science are obviously more interested in the recent development than is true for Germany, and I also think that feedback between the actively working scientists or the man who is still alive and can tell about the achievements done say 30 years ago or so and the historians writing about it is very important.

I would like to pose now a question to the historians. As a man working in science for a long time and interested in the history but not a professional historian, how should we interpret papers 50 years old or 100 years old? I have the impression that normally the tendency is to interpret them in a way bringing out ideas which seem to be true, which are true, but which may not be the ideas the man had in mind when he wrote the paper.

Let me give an example. The Willstätter idea of having a colloidal carrier and a small molecular group acting chemically was a brilliant idea for his time in providing a general theory of enzyme catalysis. It was then stated, with the advent of the co-enzymes, especially nicotinamide-adeninedinucleotide or diphosphopyridine nucleotide which it was called at that time, that this was the chemically active group. It was not. Obviously it is a co-substrate. It is not the group Willstätter was looking for.

The group Willstätter was looking for is the thing that we call now the active center. But this is also not identical because Willstätter denied that the enzymes were proteins, and therefore I hesitate to give any definite meaning to this idea put forward by Willstätter from our present knowledge. What would a historian do in such a case?

PARASCANDOLA: Obviously you have hit on a very central question and a very difficult one that historians face or that anyone faces who really likes to look back and interpret something. Whether you are trained as a historian or not, if you go back and actually wish to try to interpret what Willstätter or someone else was saying 50 years ago, you are becoming in a sense a historian in attempting to understand this historically. This is one of the central problems that faces people who do look back. And historians are as prone as scientists are to interpret something in the light of our current knowledge. When we go back and trace the history of particular concepts or theories we may tend to focus on things which became eventually the mainstream and are now strong. For example, if we should look back at the history of atoms, we would find in the culture of ancient Greece, the Democritean atoms. But there is a danger of confusing this with Dalton's atom, which is very different, or with Bohr's atom, which is also quite different, just because similar kinds of terms are used. There is also a danger of focusing on that theme and neglecting the fact that perhaps in Greece, for example, the atomic view was not particularly important. Or if we go back to look at the period we have been talking about today, the 1920s and 30s, we often see such terms as the "dark age of biocolloidology" where people will look back at the colloidal theory and somehow denigrate it, saying that this view was not upheld, that it seems ridiculous in a sense now. Or else you see terms like this used and it is ahistorical. We then lose sight of the fact that this view may have been very influential at the time. A historian going back is interested in understanding what was the understanding of the nature of proteins or whatever the question was at that time. He is not so much interested in whether or not the view was right or wrong by current standards, but in trying to understand how people interpreted these problems and dealt with them.

If historians have any advantage in this area it is simply a matter of having learned through their training to be alert to this procedure of trying to go back and put yourself in this time period and framework, and second of all of having the luxury of being able to devote their full time to this.

If a scientist is writing a review article and wants to put in a historical introduction, he is obviously not going to spend years on the historical part. A historian writing a book, say, Olby's book on the road to the double helix, may spend years looking at that one problem and thus has the opportunity to go back and try to put himself in the proper framework. But obviously we have scientists like Edsall and Fruton who have done the same thing. It is just a matter of taking the time really to go back and read these papers, not simply pick out one phrase or one quotation and then attempt to interpret that in light of our modern views because a particular term is used that is similar.

On my own I have done some work in the history of pharmacology and looked at papers by people like Erhlich. We can take a term like receptor or chemoreceptor which was used byEhrlich, but we find it is very different from our modern concept of receptors. But if you simply go back and pull something out of Ehrlich's papers, you can give it a very modern flavor and modern sense by simply abstracting a quotation or statement and interpreting it in terms of current theories. There is no simple answer to it.

Barbara Brodsky (*Rutgers University, Piscataway, N.J.*): It seems that one of the functions of the history of science is to instill a feeling of community. Much of what scientists do, meetings, belonging to societies, and so forth, create a sense of community and the history is part of it. Maybe this is similar to the *Roots* phenomenon in a way. For example, today's meeting of a community seems to be looking back at this whole sensation of belonging to something. I agree with what a previous speaker said, that perhaps one of the reasons graduate students are not so enthusiastic is that they are not so sure that it is such a prestigious thing to be part of this community. They are not sure they want to join. They are not so enthusiastic about the history because they are not sure they are part of it.

Also, it struck me today how recent many of the major discoveries had been and how the researchers must have realized that they were making historical things. Many of the developments were really major historical developments like solving the first protein structure, or doing the first sequence. And I wonder if people working on protein structure have that same feeling today. It seems to me that these feelings are in people that are doing things like genetic engineering and looking at the DNA sequences and finding these pieces that are not translated. The feeling of making history is another part of science today.

Smith: I should like to take up a little of what Prof. Fruton said, that part of what we are talking about in the history of science is intellectual history. The history of science is a history of an important intellectual endeavor. If this were a meeting, for example, on the history of how one determined the structure of one protein, one would regard it as a rather trivial exercise, all of which can be gleaned from the literature at no great cost. I think the reason why this meeting is of particular interest is because proteins have occupied an important intellectual place in the history of science and particularly in the history of biology. One can do this for a certain number of phases of the activity of biochemistry; one cannot do it for all. DNA is obviously another. How did we get the information that this is genetic material? This is an important aspect of intellectual history.

Experimental history is something entirely different. How did one come to the ideas? And how did one find that the ideas were proved? We talk now about the peptide bonds as representing one of the major factors in protein structure. In actual fact the peptide bond was first proposed a long time before Emil Fischer by Berthelot, who pointed out that amino acids had to be hooked together in peptide bonds because there was no other way of putting amino acids together. Nobody took it very seriously at the time. Nobody knew most of the amino acids in proteins, and there was nothing that one could do about it at that time. It was only when proteins began to become recognized as having all of the important biological activities that they possess that this became an important problem. And Fischer's contribution started when he, and a large number of other people, became convinced that proteins were central to a large number of different biological activities. That is way I quoted that remark of Fischer's at the conclusion of my talk this afternoon.

I do not think the history of science has to be justified anymore than the history of literary criticism. A man can write a novel without knowing all of English history or any other kind of literary history. He writes a novel. It is the job of a historian or literary critic to put that novel in a certain context, in a certain kind of historical perspective. This is equally true of sculpture, painting, or any other type of intellectual history. Science is part of our intellectual endeavor, and the history of science deals with it.

HOLMES: There is a theme which came up repeatedly in one form or another in the papers during the day, that generally there is an assumption that the further one goes back in biochemistry or its precursors the simpler things were and that the closer one comes to the present the more complex and more sophisticated they are. I would like to challenge that assumption and suggest that it seems to me that, for an individual, both the intellectual complexity that he had to handle and perhaps the complexity of methods that he had to employ have been relatively constant so long as these areas have been organized professional activities and that for every example of something which has become more complex over any given period we find other factors that have become simplified.

KARLSON: I think this is entirely true and that for the individual, for the scientist working at the frontiers, the complexity is about the same. Think about the very complex method of determining pH in the days of Michaelis with the hydrogen electrode and how simply this is done today. There are other similar things in experimentation. As far as concepts are concerned we have of course certain knowledge of our concepts, but in 20 or 30 years from now these concepts may be outdated and replaced by other concepts which are simpler perhaps, or more complicated, or at least different. I believe that for the actively working scientist this situation remains about the same.

EDSALL: I think the view that things are getting more complicated than they used to be is partly a matter of one's personal perspective. We learn things when we are young and we master certain techniques, certain intellectual frameworks, certain approaches to things. As we get older I know I find that now I see all sorts of new techniques springing up which I am not competent in. They tend to look very complicated and therefore I think science is getting more complicated. But from the point of view of the younger people who are actually doing the work, it may be just about the same to them as it was for me at a comparable age.

FRUTON: Of course the question you raised is a very intriguing one. But I think the question needs clarification because some of the references to the growing complexity I think were in relation to the accumulation of detailed knowledge of the more intimate structure of a piece of matter, whether it is a cell or part of a cell or an enzyme or a protein. In the sense of the accumulation of knowledge, the fact that we have more detailed information leads us to recognize that our present knowledge has more bits of information in it than our previous knowledge and to that extent the structure of this piece of matter is more complex than we thought it was yesterday.

But if that is what you mean, then I suppose to that extent there is greater complexity in our appreciation of certain parts of the natural world. But in terms of the consideration of the solution of a problem, or discussion of a problem, there is one thing that I think perhaps might have been emphasized a little more, and that is the element of uncertainty and controversy.

Dr. Pirie said that in the Cambridge lab around 1930 everybody believed that enzymes were proteins. Now that suggests that there were different clusters of individuals in various laboratories in the world interested in enzymes and proteins who had different views about this sort of thing. Is that an element of complexity? Certainly in the minds of individuals who were worrying about this problem the situation seemed very complex because there was evidence offered by Willstätter claiming that enzymes are not proteins; there was the conviction, on what basis was not made clear, that they were. Is that the kind of complexity that you mean? It would be very interesting to have the question clarified from that point of view, as the question appears to historians.

HOLMES: I suppose I was deliberately vague. I was responding to the simple use of these terms as they came up repeatedly during the day, and I was not sure precisely what was meant by them. But it was I think a feeling that things have gone from simple to complex.

I might respond to one aspect of your question. I think that even in terms of the amount of knowledge or information a person has to deal with that there are compensating simplifications. For example, there are all those complex subtle distinctions between peptones and albuminoses of the late 19th century defined by Kühne. It seems to me that that was a very complex kind of information which it is no longer necessary to worry about. Although there is an accumulation of information which grows more complex, there is also a packing down and a sifting out and a letting go of thins, complexities which previous generations worried about.

MERRILEY BORELL (*University of Edinburgh, Edinburgh, Scotland*): It seems to me from listening to the papers that were presented today that the peak of complexity in the development of protein biochemistry was in the 1930s when there was the greatest number of competing theories for protein structure. A couple of the papers remarked that after the problem of protein purity had been clearly defined with the development of new techniques, ultracentrifugation, electrophoresis, and chromatography, that the problems of sequencing and analysis were really considered to be quite straightforward, and, particularly after Sanger's work, that the work for biochemistry is quite clear. So it seems from the historical evidence that you have presented today we have had a considerable decrease in complexity in the last 30 years.

DUAX: I would just respond to that that there is plenty of complexity left around, such as the questions of how do the proteins function? Why do they function the way they do? Are they flexible? Are they different in solution than they are in the crystal? There are many questions as to what is going on and there are strong controversies, what is the nature of protein-protein interaction? How do proteins fold up? How do things get to be the way they are and then how do they go on to the next stage?

And there is the steroid hormone action: the steroid binds to a protein in the cytoplasm, goes into the nucleus, interacts with the chromatin and causes different kinds of interactions that are related to the protein structure. We now have to figure out how is that structure controlling the function, which I think is a far more exciting problem. I think it is the function of protein that is interesting not the structure. The structure controls the function but I think that if you know all about the structure that you wish to and, if that does not help you understand how it functions and how it works the way it does, then I do not see what you have learned. You may have increased the body of knowledge but if we are concerned with using that knowledge for improved health or controlling what we are doing to the environment, I think you have to know why they function the way they do.

HOTCHKISS: Well, I am from New England where they say the person answers a question by asking another. Complex *for whom* is the question I wanted to ask. Let's refer to Schachman's case of a student who with a bottle of cesium chloride can get a degree or two. I know he said that playfully, but it points up what I have always assumed, that there are periods of exploitation and consolidation in fields. I saw it happen for vitamins and for antibiotics and various things and now the sequences. But there will always be people whose task it is to add it all up and say, what does it all mean? That is what the last speaker was trying to ask. There is always the new complexity and so I suppose it is arbitrary. It is obviously *a priori* that if people of approximately equal ability through the years have been operating more or less at the limits of their ability they would do equally difficult things at all those difficult times.

I had another point I would like to make. It seemed to me that when I was starting out in biological science we were obliged to speak to our peers within the very same field. I think something has happened; I cannot tell whether it is something that has happened to me. One is always having trouble with his own history versus the history of science. But it seems to me that nowadays we speak much more to people in other disciplines than we did when I was growing up.

If that is true then the writing, the communications, the documents will all seem different, and I would especially like to see whether the historians or anyone else has an opinion about that. It seems to me that nowadays, although to get a paper in a technical journal may take just as much supplication to the specialized experts, one is talking much more to people in other fields. Back then we had biology, microbiology, chemistry, a few fields; we did not have all the in-between fields. Now we are always talking to everybody in a sense.

BEARMAN: I should like to respond to Dr. Hotchkiss's point because I think it is an interesting one and one which historians are concerned with. Obviously there are more scientists; there are more historians too; the whole enterprise has grown. This may be one aspect of the complexity perhaps. There certainly has been interspecialty collaboration in this work that has been talked about today, physicists inventing new techniques, chemists coming up with new methods, people coming from different fields traveling about; these are questions that I think interest historians very much and they point to an aspect of the history of the field which has not been discussed a great deal this evening, which is that the history of the field does consist of an intellectual history. This is one of the tensions in the history of science, as the image of science or the intellectual history of science. But I think they are both part of culture, which has produced science, and that the history of science also involves its institutional history in its place within society.

One of the questions that I thought was intriguing was a question that was not addressed directly by the scientists but a question which historians are interested in: Where was all this work being done? There are new settings. Astbury's setting is a very peculiar, very interesting cultural phenomenon. The creation of industrial research organizations which see science as fitting into their purpose, an association of companies interested in propagating scientific research for a particular purpose, that is something that is very peculiar to our culture that certainly has been created in the West since say 1900 or 1870 or whenever but was not part of culture earlier. And that has changed the meaning of science and the way in which science has been done.

I think that those kinds of questions are equally interesting to historians who are coming here to listen to these firsthand accounts of what actually went on. One of the points that was raised earlier by a number of speakers is that they did not have as much material, they did not have as much money in those days. That is of course true and that is a limiting factor in the research. But it is interesting to a historian to ask, who was patronizing research? Who was interested in the research? When did pharmaceutical companies become interested in protein structure? What did they do about it? What was the role of the foundations? There were large private foundations that were giving money for, among other things, much of the traveling that spread the techniques and spread the ideas from one place to another. Why were they interested?

So these are other problems which historians are dealing with and go back to. It is very difficult to ask those kinds of questions from the published scientific record, that is, the reports of the actual conclusions that people have reached in their research. For those kinds of questions, we would very much have to wait until we have other kinds of documentation: institutional records, the records of government and foundations, the records of individuals who built research laboratories, who invited people to work with them, who shared different kinds of perspectives and ideas. These are the kinds of questions that often are in absence for the immediate past but which can be asked by historians and fit into more general problems in the development of culture 30 and 40 years back.

FRUTON: I agree entirely with the importance of doing what might loosely be called the sociology of the development of biochemistry. But I think it is worth reiterating that a good many of the people at this meeting and a good many of the people who contributed to the development of what we now call biochemistry were not identified as biochemists. And therefore, a judgment is made by someone, in this case a historian interested in the sociology of science, about the significance of a particular scientific contribution to the development of a field regardless of whether the person had been labeled a biochemist during the 1920s or not. He might have been labeled a histologist or an x-ray crystallographer. In order to do a study of the social relationships, institutional, financial, and the rest, a judgement has been made, or will have been made about the relative importance of various individuals and the institutions in which they worked in order to decide which ones to examine. This implies that a prior study will also have been made that will have led to a basis for the choice; a value judgement will have been made about who is important and who is not important.

So let us not separate too sharply the historical study of the intellectual side of the scientific effort, the values that are attached to its results, and the examination of the sociology that attends it.

FREDERIC L. HOLMES received his Ph.D. in the History of Science at Harvard in 1962. He then joined the Department of the History of Science and Medicine at Yale University. In 1972 he became Professor of the History of Medicine and Science at the University of Western Ontario in London, Canada. In July 1979 he returns to Yale to be chairman of a newly established section on the History of Medicine.

Dr. Holmes has published studies on a wide range of historical topics covering a period from 1750 to the recent past. In 1974 he published a book, *Claude Bernard and Animal Chemistry*, which received the Pfizer Award of the History of Science Society and the Welch Medal of the American Association of the History of Medicine.

Early Theories of Protein Metabolism

FREDERIC L. HOLMES

Department of History of Medicine and Science
University of Western Ontario
London, Ontario
Canada

SYSTEMATIC EXPERIMENTAL INVESTIGATIONS of protein metabolism began before the words "protein" and "metabolism" were invented. The characterization of a special class of "animal matters" emerged so gradually during the eighteenth and early nineteenth century that it would be artificial to say that at some particular time proteins were first recognized. Theories regarding the physiological role of these substances were so interwoven with the slowly evolving knowledge of their chemical nature that it is equally difficult to fix the beginnings of conceptions corresponding to the modern knowledge of protein metabolism. For my purpose, however, it will be convenient to start with a memorable investigation completed in 1826 by Friedrich Tiedemann and Leopold Gmelin. Combining their respective knowledges of anatomy and of chemistry, these two scientists for the first time applied chemical analyses extensively and effectively to try to trace the changes that various "simple" foods undergo within animals. At varied intervals after feeding dogs and other animals, they killed the animals and analyzed samples of the contents of the stomach and small intestine, as well as the lacteal fluid and blood. Of the nutrients they tested, albumin, fibrin, gelatin, casein, and gluten were regarded as very similar "animal" or "albuminous" substances. All contained nitrogen and gave characteristic reactions with salts such as lead acetate and potassium ferrocyanate. They were distinguished from one another by their different solubilities in water, alcohol, acids, and bases, as well as by whether they coagulated when heated. Tiedemann and Gmelin applied such tests to identify the materials they removed from the digestive tracts of animals. They were not able to define clearly any specific chemical changes that these "animal" substances as a whole underwent, although they did believe that they had found that when fibrin is digested it is converted to albumin. More important than their own results was the general way to investigate nutrition that Tiedemann and Gmelin introduced. The term "nutrition" then encompassed the successive stages through which food was believed to be converted to chyme, chyle, and blood, and finally assimi-

0077-8923/79/0325-0171 $01.75/0

lated to the body tissues. The method of Tiedemann and Gmelin was to trace the chemical changes by comparing the composition of the substances at different points along the anatomical pathway they follow into the organism. The impressive study of Tiedemann and Gmelin influenced other physiologists and chemists to follow the same general approach.

During the 1840s this approach was overshadowed by concepts based on quite different types of evidence. G. J. Mulder had established by combustion analyses that albumin, fibrin, and casein have the same elementary composition. Each of them, he thought, contained the same unit, $C_{40}H_{62}N_{10}O_{12}$, which he named the "proteine" radical. Finding an albumin in wheat, Mulder inferred that the principal constituents of animals are produced in plants. The eminent German organic chemist Justus Liebig pursued these investigations further. He identified a plant substance equivalent to each of the animal substances albumin, fibrin, and casein. These agreed not only in their elementary compositions, but in their solubility properties and their reactions with acids, alkalies, and salts. From these results Liebig elaborated the generalization that animals receive the chief components of their blood and tissues pre-formed in their food. To be assimilated to the animal body, he believed, these proteins required only minor alterations in form, not changes in their composition. In France, Jean-Baptiste Dumas independently reached similar conclusions, and went even further than Liebig in asserting that animals cannot synthesize any substances. They secure all of their chemical constituents from plants, and can only decompose them, through successive oxidations, which release heat and mechanical work.

For Liebig and Dumas and their followers, therefore, the tracing of the changes that nutrient materials undergo as they enter the organism, in the manner of Tiedemann and Gmelin's research, was relatively unimportant, for they assumed that these changes were minor. In their view the central problem was the decomposition reactions which the protein constituents of the tissues underwent as a consequence of the vital functions of the body. Liebig called this process the *Stoffwechsel*, or in English, metamorphosis of the tissues. Liebig maintained that by measuring the formation of urea, the end-product of this process, one could directly determine the rate of this tissue *Stoffwechsel*. In a hypothetical equation Dumas represented the conversion of albumin into urea, carbonic acid, and water as an oxidative decomposition reaction.

Liebig assumed that the decomposition of protein produced, in addition to urea, a nonnitrogenous residue that was then oxidized in the blood to produce animal heat. Carbohydrates and fats he and Dumas both believed to be directly oxidized in the blood. Liebig pointed out that by supplying some of the heat required, nonnitrogenous nutrients could reduce the "wasting of the tissues," that is, could "spare" protein from unnecessary consumption.*

These views were less arbitrary than they later appeared to be. The assumption that no major chemical changes are necessary to convert nutrient proteins into the nitrogenous constituents of tissues was a natural reaction to the striking new discovery that all of this class of substances, in both animals and plants, had nearly identical compositions. It seemed to chemists like Liebig and Dumas that to search for chemical transformations during the process of digestion and assimilation was merely to create unnecessary complications in the face of the great simplifying generalization that the substances involved were all alike.

Liebig's theoretical conception of "animal chemistry" formed the starting point for a field of metabolic investigation, based on feeding experiments with intact animals, which flourished over much of the next half century. His most important disciple, Carl Voit, painstakingly developed the techniques for precisely measuring the contents of the food and excretions that enabled him and others to test with rigorous control the effects of dietary variations upon the formation of urea—therefore upon the rate of the nitrogenous *Stoffwechsel*. Central to all subsequent research was Voit's discovery that an animal on a stable, adequate diet enters a condition in which the nitrogen excreted, mainly as urea, equals the amount it ingests in its food. The maintenance of this "nitrogen equilibrium" became the primary criterion of adequate nourishment. Voit at first followed Liebig in believing that all nutrient protein must become part of organized tissue before it can decompose and form urea. Others, including Theodor Frerichs, Friedrich Bidder, and Carl Schmidt, contended that dietary protein in excess of a minimum daily requirement was oxidized directly in the blood, a process they termed *Luxusconsumtion*. After 1866 Voit advocated a third alternative, that there are two conditions of the protein in the body: "organized" and "circulating" protein. The former is incorporated into the tissues, and is stable, whereas the latter is more rapidly decomposed.[2, 3] Whatever position they took over these controversial issues, however, each school retained as a general premise that animals acquired their body proteins in their food, and that the significant chemical transformations in which the proteins took part all belonged to the stage of "regressive metamorphosis," or oxidative degradation, which serves as the prelude to their elimination from the body. Voit habitually fed his animals meat

* The subjects of this and the preceding paragraphs are discussed more fully in Reference 1, where references to the relevant primary literature may be found.

from which the fat had been carefully removed, and he treated this "flesh" as virtually identical with the protein constituents of the whole animal. He did not consider that the compositions of the proteins of different kinds of tissue might differ. This point of view prevailed until near the end of the century. In 1873, for example, the influential physiological chemist Felix Hoppe-Seyler treated the problem of *Stoffwechsel* investigations as synonymous with the question of "where and how the decomposition processes of the proteins and other nutrients occur."[4]

Over the same decades other physiologists and physiological chemists were continuing the investigation of digestion within the framework set by the pioneering work of Tiedemann and Gmelin. Theodor Schwann's brilliant characterization of the gastric digestive ferment pepsin, in 1836, permitted later investigators to study digestion readily outside the stomach as well. In 1846 Louis Mialhe gave a new impetus to the question of what changes proteins undergo in digestion, by asserting that fibrin, casein, and albumin are all reduced to an identical substance, which he named "albuminose." Carl Lehmann, author of the most authoritative physiological chemistry text of the middle nineteenth century, believed that the products of the digestion of different proteins were not quite identical, and he named them peptones.[5] During the 1850s and 1860s a number of people attempted to clarify the nature of peptones and the conditions under which they formed. The main distinguishing feature of peptones was that they were not coagulable and were not removed from solution by most of the salts that precipitated proteins. Such characteristics led some physiological chemists, including Mulder, to regard peptones as proteins modified to make them more soluble, so that they could be absorbed into the circulation. There, they assumed, the albumin and fibrin of the blood were regenerated from the peptones. The weakness in this conception, as Ernst Brücke pointed out in 1859, was that peptones were defined almost entirely by negative properties.[6]

The uncertainty concerning the chemical nature of peptones, including the continuing question of whether there really were more than one of them, produced correspondingly indefinite conceptions of the nature of the transformation that converted protein to peptone. Because under some conditions substances that seemed intermediate between proteins and peptone were produced, the process appeared to be not a discrete chemical reaction, but a loss of the characteristic properties of proteins, one after the other, in a gradual process. Some referred to it as a decomposition, but Brücke thought that the change involved merely a separation of proteins into smaller particles rather than a cleavage of their molecules.[7] Combustion analyses showed that the elementary composition of peptone was very similar to those of proteins. Hoppe-Seyler pointed out that the process did not result either in the separation of carbonic acid or ammonia, or in the consumption of oxygen. He regarded peptone as hydrated protein.[8] Because the nature of these chemical changes involved in digestion was still so obscure, these investigations did not seem to challenge seriously the generalization that animals assimilate their proteins from their food with little change. Thus in 1881 Hoppe-Seyler could readily accommodate the digestive conversion of protein to peptone (and other less-modified substances) within the prevailing framework[9]:

> Nearly everyone accepts that the proteins are formed only in plants, that in the digestive canal they are converted more or less to peptone, or acidalbumin, or globulin, and are absorbed into the blood; that they remain in the blood or lymph, or are transported through these fluids into the organs, and here either decomposed or incorporated into protoplasm. . . .

By 1860 the characterization of the composition of proteins by their stark empirical elementary formulas was already being supplemented by efforts to identify partial decomposition products. Among the many substances that appeared when proteins were hydrolyzed in acids or alkalis, three compounds had come to appear particularly significant. They were glycocoll (or glycine), leucine, and tyrosine. These compounds had been isolated in relatively pure, crystalline form, their elementary compositions had been established, and the first two had been synthesized. They were recognized to have both acidic and basic properties, and in 1858 Auguste Cahours generalized that glycine and leucine (along with alanine) were members of a fatty acid series containing an amino group, analogous to a known series of aminobenzoic acids. Two of these "amido acids," leucine and tyrosine, had been obtained from a wide variety of protein materials.[10] Physiological chemists began to consider that these well-defined compounds might be produced during intermediate stages of the "regressive metamorphosis" by which proteins that had ceased to constitute organized tissues were supposed to decompose to form urea. If this were so, then one would expect that by depriving an animal of oxygen one could inhibit this oxidative process and cause leucine, glycine, and tyrosine to accumulate in the body fluids or organs. A number of investigators during the 1870's attempted this, by placing animals in atmospheres without oxygen, by restricting their breathing, or by poisoning them with phosphorus to remove the oxygen from their blood. Some obtained positive results, but they were not decisive.

Hoppe-Seyler could not detect any of these compounds in the urine of animals poisoned by phosphorus, and he regarded the view that they are precursors of urea as likely but unproven.[11]

The isolation of amino acids from proteins soon influenced the continuing investigation of digestive processes. Despite Claude Bernard's discovery of the special action of pancreatic fluid on fats, in 1848, most studies of digestion during the next two decades still focused on the action of the gastric ferment pepsin. In 1867, however, Willy Kühne completed an important investigation of the pancreatic digestion of protein. He tested the action on fibrin of pancreatic tissue infusions and found that this fluid converted the protein to peptone far more rapidly than pepsin did. The peptone produced a distinct Millon reaction, a test which was by then regarded as the most general indicator for proteins, but it was so much more diffusible than proteins that it would pass through parchment membranes. Out of the 95% alcohol solution from which he had precipated the peptone derived from fibrin, Kühne was able to crystallize tyrosine and leucine. He concluded that these amino acids arise by a "deeper decomposition" of the peptone.[12]

Kühne was able to interpret his discovery within the framework of current nutritional concepts. If during digestion protein is not only converted to the more diffusible modification, peptone, but the latter is "further decomposed into substances, which up until now one had been accustomed to assign to the so-called regressive metamorphosis," the phenomenon could be explained by assuming that the direct decomposition of excess protein nutrients, the *Luxusconsumtion* recently advocated by Frerichs and Bidder and Schmidt, really did occur. The process did not take place in the blood, as they had supposed, but in the small intestine. "It is possible," he added, "yes, very probable, that the greater part of the highly diffusible peptone escapes the further decomposing action of pancreatic fluid through absorption" into the circulation.[13] Thus Kühne saw no basic contradiction between his dramatic new finding and the established view that animals obtain their proteins directly from their food. The outcome of his experiments, however, reverberated through the following decades independently of his own conclusion, stimulating further inquiry and deeper questions about the conventional assumptions.

Two other people showed soon afterward that pancreatic fluid also produces tyrosine and leucine from proteins other than fibrin, and Kühne himself confirmed his original discovery nine years later by producing peptone, tyrosine, and leucine from fibrin injected into an isolated section of small intestine in a living animal. During the 1870s and 1880s, however, the discovery of the digestive decomposition of peptone to amino acids, which some people attributed to bacterial action, attracted less attention than the question of the physiological fate of that portion of the peptone that Kühne and others assumed to be absorbed from the intestine. Peptone could not be detected in the blood, except in very small amounts immediately after proteins had been digested. To find out what happened to it, Adolf Schmidt-Mühlheim, a pupil of Ludwig, in 1879 injected peptone into the jugular vein of a dog. Within 15 minutes it had disappeared from the blood, and Schmidt-Mühlheim inferred that it had undergone a rapid chemical transformation.[14,15] In 1881 Franz Hofmeister pursued this problem further. He used the characteristic "biuret" color reaction for proteins, and measurements of optical rotation, which together enabled him to estimate the concentration of peptone in a solution, after he had precipitated the unaltered proteins out of it. Hofmeister found that most of the peptone he injected into the blood was eliminated in the urine. The outcome was the same when he injected the peptone into subcutaneous tissue so that it would enter the blood no more rapidly than the peptone produced by digestion. These results proved, he thought, "that the theatre for the suggested 'transformation' cannot be displaced into the bloodstream."[16]

It seemed absurd to Hofmeister to accept that the peptone absorbed from the intestine would be eliminated from the blood in the same way as the injected peptone, for the conversion of protein to peptone would then be a wasteful, purposeless process. He assumed instead that the peptone must be further changed before it enters the bloodstream, consequently within the intestinal wall. He could not specify what kind of change this might be, but believed that the peptone must enter some sort of combination that renders it nontoxic and protects it from being excreted, yet "without obliterating its characteristic properties." He hypothesized that this combination took place within the white blood cells, which are abundant in the mucous membrane of the intestine.[17]

To test his view further, Hofmeister searched for peptone in the blood and several tissues of animals killed at different time intervals after they had been fed. He used the intensity of the biuret reaction in tissue extracts, from which he had precipitated the proteins, as a measure of the peptone present. Of all the organs he tested, only the intestinal wall yielded peptone regularly, a result that he considered supportive of his theory. Yet he did sometimes find small amounts of peptone in the blood and in such organs as the spleen. Moreover, the quantity of peptone in the small intestine and intestinal lining reached a peak after the same length of time at which the excretion of urine was known to approach a maximum during digestion. This parallelism, he acknowledged, favored the assumption that part of the peptone taken up in the body is "quickly decomposed into its end products." Altogether, he was forced to conclude, this type of investigation could not directly answer the question "in what way and in what places does the absorbed peptone undergo those changes" through which it disappears as a detectable substance. Trying another approach, he measured the peptone in the mucous membrane of a stomach removed from a dog that had been digesting meat. Half of the lining he heated to a temperature that he assumed would destroy the activity of its ferments. The half that was not heated contained large quantities of peptone. Stomachs removed a longer time after the digestion had taken place yielded less peptone. These comparisons favored his interpretation that peptone is transformed within the lining of the intestinal tract rather than in blood; but he had become even less certain about the nature of that change. The question of whether "a reformation of protein, or a cleavage process takes place," he said, cannot be decisively answered by this investigation. He remained confident that the answer was accessible to further research.[18]

Meanwhile other workers performing similar experiments were turning out results no more decisive than Hofmeister's, and sometimes inconsistent with each other. In 1888 Kühne encouraged one of his associates, Richard Neumeister, to reexamine the whole problem. For years Neumeister had worked with Kühne on the development of methods to sort out the various products of the action of the digestive enzymes pepsin and trypsin on proteins. By means of tedious fractional precipitations from solutions of salts in various concentrations, at differing temperatures and degrees of acidity, they distinguished a series of very similar substances, which they regarded as intermediate stages in the conversion of proteins to peptone.[19] They attempted to give a more precise definition of peptone itself, as that fraction that remained in solution in saturated ammonium sulfate and gave a positive biuret reaction. Neumeister hoped that with this more rigorous test he could clear up the difficulties that had prevented his predecessors from ascertaining the fate of peptone in the body. He first injected peptone directly into the intestines of rabbits, by means of a canula, and found afterward "never even the slightest trace of peptone or albuminose in the blood." Neumeister confirmed Hofmeister's observation that peptone injected into the blood appeared in the urine, but he found in addition that the peptone disappeared from the blood even when he had tied off the ureters to prevent urine from being excreted. That result was more difficult to explain, but Neumeister nevertheless believed that he had proven, from comparisons of the effects of these injections with injections of various proteins and albuminoses, that peptone was a "foreign substance" in the blood. The view that "peptones may serve directly as the building blocks [*Bausteine*], out of which the various organs construct their different proteins," was, he thought, no longer tenable. Like Hofmeister, he inferred that peptone must be transformed in the wall of the digestive tract, and he extended Hofmeister's experiments on the mucous membrane of the stomach to the lining of the small intestine, showing that it too can cause peptone to disappear.[20, 21]

Neumeister struggled at length with the question which he, like his predecessors, now confronted: What is the nature of the conversion that removes this peptone? Ruling out Hofmeister's hypothesis that it forms some unknown combination in the white blood cells, Neumeister posed the alternatives clearly: the disappearance of the peptone is caused "either by a transformation of these bodies back into protein, or a further splitting of them into smaller molecules." He reviewed the evidence that others had presented in favor of the regeneration of protein and found it unconvincing because those who had done the experiments had not used the latest precipitation methods to insure that what they identified as peptone included no protein. On the other hand, the decomposition of peptones to amino acids, observed in experiments such as those of Kühne, might represent merely secondary reactions, and it was difficult to assure that microorganisms had not produced them. In the fluid surrounding the pieces of intestinal membrane that caused peptone to disappear, Neumeister himself tried to find leucine, tyrosine, and a substance that he had earlier identified as a decomposition product of proteins and named tryptophane. He detected in some cases a little tryptophane by its peculiar color reaction, and a very small amount of tyrosine; but the results were irregular, and he could not be sure that bacteria had not also interfered in his experiments. He concluded that his investigation could "not definitively decide for one view or the other." [22]

Despite the efforts of many people, applying increasingly sophisticated analytical methods, the central question remained unsolved. After Neumeister's strenuous effort no one ventured a major experimental investigation of the problem for over a decade. His final ambivalence, perhaps also the diffuse style of his articles, allowed the conservative impression to prevail, that he had verified the older view—peptone is a diffusible form of protein, which regenerates albumin and fibrin as it enters the blood and is afterward incorporated into cellular protein.[23] This position was represented most strongly in the widely known physiology textbook of Gustav von Bunge, five editions of which appeared between 1887 and 1901. In the last edition he still maintained that the production of tyrosine and leucine from proteins mixed with pancreatic juice was probably not due to the action of the digestive enzyme. Rather it was a putrefaction caused by microorganisms, which readily develop in pancreatic fluids under alkaline conditions. Bunge wrote:

> That under normal conditions the quantity of amino acids formed in the intestine should be considerable must be doubted *a priori* on teleological grounds. It would be a squandering of chemical potential energy, which is converted by the cleavage to kinetic energy without purpose, and a re-unification beyond the intestine of the products of such an extensive decomposition is very improbable.

Bunge also thought that the chemical relationships between proteins and peptone were not clearly enough defined to demonstrate that peptones were products of the cleavage of protein molecules. The former might arise, he thought, by a "rearrangement of the atoms without a change in the size of the molecules." [24]

From our perspective it is tempting to view Bunge as taking advantage of every loophole in the current experimental situation to defend the long-familiar conception that animals must obtain their constituent proteins in their food. The fact that his arguments were destined to become obsolete almost immediately, however, should not lead us to consider them unreasonable when he wrote them. Within the context of the chemical methods then in use and the assumptions about metabolic processes, which had guided investigations for nearly 50 years, his position was still as plausible as the alternative.

Sometime around 1900 Otto Cohnheim decided to continue the investigation of the fate of peptone from the point at which Neumeister had left it. He began by repeating Neumeister's observation that peptone disappears in a solution containing pieces of intestinal lining. Then, carrying the problem one crucial step beyond, he asked the question whether there was an increase in the amount of protein present in the pieces of lining itself—as there ought to be if the peptone were reconverted to protein within the intestinal wall. After numerous experiments, during which he failed to identify any protein that could have formed in this way, he was "led to the correct trail" by investigating the solution that had originally contained the peptone. This fluid no longer gave a biuret reaction; yet, by means of the recently developed Kjeldahl method for estimating nitrogen, he could show that the solution retained approximately the same quantity of nitrogen that the peptone had contained. This simple, elegant demonstration proved, he concluded, that "The peptone is not restored [to protein], but on the contrary it is further decomposed by the mucous membrane, and is converted to crystalline cleavage products." [25]

Having satisfied himself that the peptone is further decomposed, Cohnheim inquired whether the process required the "living" intestinal membrane or was an enzyme action. With an aqueous extract of the tissue he was able to reproduce the same reaction, indicating that a new digestive enzyme was present. Under these conditions he was also able to prove more directly that a decomposition reaction takes place, by obtaining pure crystals of leucine and tyrosine. With the help of Kühne's separation methods he was able partially to purify the enzyme, and to show that it differed clearly from the pancreatic enzyme trypsin. Among the distinguishing properties of the new enzyme, which he named eripsin, was that it acted on peptone and the intermediary albuminoses, but not on unmodified proteins.[26]

With these simple but decisive steps, Cohnheim had resolved the issue with which others had contended inconclusively for over 40 years. In some ways his investigation was a logical extension of the experiments Kühne had performed in 1867, utilizing the more advanced analytical methods now available in order to establish unequivocably what Kühne's results could only suggest. Cohnheim also interpreted the biological meaning of his investigation rather similarly to Kühne. In conjunction with some recent feeding experiments showing that protein consumption is closely related to energy requirements his results implied, he believed, that instead of being synthesized into body protein, which is again decomposed, nutrient protein is "broken into easily combustible pieces before being absorbed, and only these pieces are transported to the organs." Cohnheim assumed, as Kühne had not, that in the cells proteins may be synthesized from some of these pieces; but the principal implication of his work, he believed, was that "at least the main mass of the ingested protein serves simply as a source of energy." [27] Cohnheim did not explicitly associate his conception with the older theories of *Luxusconsumtion*, but his view in fact was a modified descendant of that position.

Cohnheim's investigation was a continuation of the long tradition of experiments on digestion growing directly out of the early 19th century achievements of Tiedemann and Gmelin, and Theodor Schwann. At very nearly the same time a complementary and equally nodal investigation came out of the line of feeding experiments traceable to the inspiration of Liebig and the labors of Carl Voit. In 1901 Otto Loewi set out directly to test the question of whether animals "can synthesize proteins independently," or whether they require protein in their diets. To do this he fed dogs the product of the pancreatic digestion of casein. Believing that the biuret reaction provided the most reliable indicator of undecomposed protein, he used as the criterion for judging his digested mixture protein-free, that it react negatively to this test. He also believed that the crucial determinant would be whether he could maintain an animal in nitrogen equilibrium; for even over a short time it must lose some of its protein nitrogen if no replacement protein is forming from its food. Loewi was not the first person to try such experiments; previous attempts had not been successful, but Loewi believed the difficulty arose from general digestive upsets caused by unfamiliar diets. Patiently trying different dogs and different combinations of nonnitrogenous nutrients with the digested protein, he finally succeeded in May, 1902. The animal did better than remain in equilibrium—it actually accumulated nitrogen. The data "spoke for itself," according to Loewi. "An animal can synthesize protein, and is not dependent upon nutritional protein." [28]

The conclusion Loewi drew was well supported by his experiments, but his investigation itself was most likely guided by the preconception that an animal must be able to synthesize protein. Although he did not know, when he began, of Cohnheim's demonstration of the complete disappearance of peptones during intestinal digestion experiments, Loewi was already convinced from earlier investigations by others that at least some of the proteins are "deeply split" into crystalline decomposition products. In his view, "The assumption that the decomposed portion is worthless for the replacement of protein in the body is from the start improbable." Loewi thus took teleological arguments as seriously as Bunge did. In fact, he took the trouble to answer Bunge's teleological argument that the decomposition and resynthesis of protein would squander energy. Max Rubner determined for him the heat of combustion of the biuret-free digestion product he used in his feeding experiments, and found it to be about ten percent less than for undigested meat. This moderate energy loss Loewi thought might be balanced by other processes; but even if it were not, it "remained an open question whether this loss is a 'purposeless waste,' or a great advantage for the organism." [29]

The mutually reinforcing publications by Cohnheim and Loewi stimulated a wave of further feeding experiments, which gained momentum year by year. Some of the earlier results were ambiguous enough so that it was possible to give them the interpretation that the products of proteolysis act as "sparers of protein," in preference to "the more radical hypothesis of protein synthesis." [30, 31] Such attempts to remain within the bounds of the older nutritional framework, however, were soon swept side. Within a year the immensely clarifying theories of protein composition resulting chiefly from the work of Emil Fischer[32] were making it easier both to conceive of the reversible decomposition of proteins into amino acids; and to perform feeding experiments with control over the nature of the products of protein hydrolysis. Fischer's student Emil Abderhalden began feeding experiments with "hydrolyzed protein products of known composition" in 1904. At first he succeeded in maintaining nitrogen equilibrium only with products containing combinations of amino acids with more complicated "polypeptides;" but by 1907 he had attained positive results with meat that he considered "almost completely decomposed into the simplest *Bausteine*." [33-36]

The principal author of the *Bausteine*, or "building-block" concept, which Abderhalden here invoked, was Albrecht Kossel. Kossel developed the view that the large chemical aggregates—fats, carbohydrates, proteins, and nucleic acids—are formed by the union of smaller units "according to a definite plan or architectural idea." In 1911 Kossel outlined the broad physiological implications of this concept, particularly for protein metabolism. The specificity of each of the many proteins that comprise an organism is determined by the types, the quantitative proportions, and the order of the *Bausteine* that make them up. Consequently, when nutrient proteins are decomposed by digestion, "the individuality of the proteins is completely lost." Afterward, "out of these same *Bausteine* entirely new structures may be built up in other parts of the organism." Since there are nearly as many amino acids forming the *Bausteine* of proteins as there are letters of the alphabet, "vast numbers of the properties of the organism" may be stored in the potential rearrangements. Animals therefore do not obtain their proteins from their food in the direct manner that had been supposed until recently. Rather the food proteins serve only as the source of the *Bausteine* from which animals construct their own proteins.[37, 38]

The change that led from the 19th century ideas about *Stoffwechsel* to the viewpoint that Kossel expressed, was one of the more radical revisions of biological concepts that have taken place within the past hundred years. Not only had the belief in "direct nutrition," which had been embedded in experimental investigations of metabolism for more than a half century, been overthrown; the level at which the central processes of intermediary metabolism were thought to occur had shifted from the very large, complex, still highly mysterious proteins as a whole, to the small simple amino acid molecules, which could join together to form them.† A large field had opened up for examining the physiological consequences of the variations in the amino acid content of different proteins. The new view also marked one of the major phases within a broader and longer progression, away from an earlier stress on the contrasts between the processes plants and animals carry out, and toward a focus on the fundamental processes common to all organisms. Consider-

† See the opening paragraph in McCance,[39] reviewing this transition.

ing the depth of the reorientations involved, the switch was extraordinarily rapid. Within a decade the position Bunge could still support in 1900 was nearly irrelevant. To a large extent this departure was due to the emergence of more refined analytical methods and the experimental resourcefulness that enabled Cohnheim, Loewi, and others to solve a specific research problem; yet it was also deeply conditioned by broader trends in related areas. By 1900 proteins had come to appear much more complex than they had seemed in 1840, and there were many more of them. The similarity between the elementary compositions of the few known "albuminous substances" that had so impressed the generation of Liebig and Dumas was now overshadowed by this ever growing evidence of multiplicity and specificity.[40] The evidence from combustion analyses that the conversion of one type of protein to another involved no major chemical changes was compelling in 1840, but so superficial by the end of the century, that the belief that animals obtain their proteins in their food was more a residue of habit than a reasoned conclusion. By that time the belief that an animal could procure all of the many constituent proteins that define its identity from the substance of a different organism was almost *a priori* less plausible than the conclusion that it might have greater synthetic powers than previously imagined. The new peptide theory of protein composition not only made protein synthesis seem less mysterious, but placed the process within the class of mild hydrolytic condensation reactions, which, it had already been admitted, animals can carry out. The argument that it would be wasteful of energy for an animal to decompose and recompose proteins was dissolved by the knowledge that the formation and breaking of peptide linkages involved very small energy changes.[41] Thus by the time that the new feeding experiments provided the direct evidence that animals can synthesize proteins, there were many concurring reasons to adopt that point of view. Most of the reasons for opposing it had already eroded away. The mental metamorphosis involved therefore encountered relatively little effective resistance. Nevertheless, physiologists had become so accustomed to the older concepts that for some time it still seemed rather astonishing that simple compounds like amino acids could really serve as the sole nitrogenous food.[42]

REFERENCES

1. Holmes, F. L. 1974. Claude Bernard and Animal Chemistry. Chap. 1 & 7. Harvard Press. Cambridge, Mass.
2. Holmes, F. L. 1964. Introduction. *In* Justus Liebig, Animal Chemistry: xc–cix. Johnson Reprint Corp. New York, N.Y.
3. Holmes, F. L. 1976. *s.v.* Voit, Carl von. *In* Dictionary of Scientific Biography. C. C. Gillispie, Ed. Vol. 14:63–66. Charles Scribner's Sons. New York, N.Y.

4. HOPPE-SEYLER, F. 1873. Pflüger's Archiv. *7*: 399.
5. HOLMES. Claude Bernard and Animal Chemistry. :301–303.
6. BRÜCKE, E. 1859. Sitzungsberichte der Wiener Akademie der Wissenschaften-Mathematisch-Naturwissenschaftliche Klasse. *37*: 169–184.
7. *Ibid.* :183–184.
8. HOPPE-SEYLER, F. 1881. Physiologische Chemie :225–228. Hirschwald. Berlin.
9. *Ibid.* :998.
10. VICKERY, H. B. & C. L. A. SCHMIDT. 1931. Chemical Reviews *9*: 188–197, 208–210.
11. HOPPE-SEYLER. Physiologische :987–993, 1000.
12. KÜHNE, W. 1867. Arch. für Path. Anat. *39*: 130–174.
13. *Ibid.* :168–170.
14. HOPPE-SEYLER. Physiol. Chem. :265.,
15. BUNGE, G. V. 1901. Lehrbuch der Physiologie des Menschen. :204–205. Vogel, Leipzig.
16. HOFMEISTER, F. 1881. Z. Physiol. Chem. *5*: 126–151.
17. *Ibid.* :148–149.
18. HOFMEISTER, F. 1882. Z. Physiol. Chem. *6*: 51–68, 69–73.
19. FRUTON, J. 1972. Molecules and Life. :114–115. Wiley-Interscience. New York, N.Y.
20. NEUMEISTER, R. 1888. Z. Biologie *24*: 272–292.
21. *Ibid.* 1890. *27*: 309–373.
22. *Ibid.* :337–346.
23. COHNHEIM, O. 1901. Hoppe-Seyler's Z. Physiol. Chem. *33*: 452.
24. BUNGE. Lehrbuch. :203–208.
25. COHNHEIM. 1901. Hoppe-Seyler's Z. Physiol. Chem. *33*: 451–456.
26. *Ibid.* :456–463.
27. *Ibid.* :464–465.
28. LOEWI, O. 1902. Arch. für Exper. Path. Pharm. *48*: 303–330.
29. *Ibid.* :304, 325, 328–329.
30. HENDERSON, Y. & A. L. DEAN. 1903. Am. J. Physiol. *9*: 390.
31. LESSER, E. J. 1904. Z. Biol. *45*: 497–510.
32. FRUTON. Molecules. :112–120.
33. ABDERHALDEN, E. & P. RONA. 1904. Hoppe-Seyler's Z. Physiol. Chem. *42*: 528.
34. *Ibid.* 1905. *44*: 198–205.
35. *Ibid.* 1906. *47*: 359–365.
36. *Ibid.* 1907. *52*: 507–514.
37. KOSSEL, A. 1911. The Harvey Lectures. :33–51.
38. KOSSEL, A. 1912. Bull. Johns Hopkins Hosp. *23*: 64–76.
39. MCCANCE, R. A. 1930. Physiol. Rev. *10*: 1.
40. FRUTON. Molecules. :120–131.
41. KOSSEL. Harvey Lectures. :37.
42. CATHCART, E. P. 1921. The Physiology of Protein Metabolism. :40–41. Longmans, Green. New York, N.Y.

DISCUSSION OF THE PAPER

DIANA LONG HALL (*Boston University, Boston, Mass.*): One of the things that you make clear is the unresolved nature of these problems over a long period of time. There were repeated attempts to come to some clear understanding about the nature of animal protein in animal metabolism. Were there any points at which, in spite of this tension and lack of clarity, these researchers—and Liebig comes immediately to mind—nevertheless felt called upon to make clear statements in relationship perhaps to doing applied research in nutrition? I am thinking of Liebig's involvement with his meat extracts and with practical issues of nutrition. Was this typical, or was this unusual? Did they try to develop applications?

HOLMES: In terms of applied nutrition, the question of what happens in an intermediary sense is not so important. That is, much of the goal of the whole school of Voit was to try to determine the minimum quantitative requirements for proteins, carbohydrates, and fats, and it really did not matter what their various theories and speculations about the intermediary processes were in terms of this practical goal of establishing daily nutritional requirements. So they could make clear statements at that level in spite of the controversial aspects of their views about what happened in between.

HALL: They were, though, involved in those discussions at the simpler level of nutritional requirements?

HOLMES: Yes.

CARTER: Voit did indeed detect a difference in nutritive values of certain proteins, did he not?

HOLMES: Yes, certainly, later on. In the formative work in the 1860s and 70s he simply used meat. Later on he began to become much more interested in the qualitative question himself, so I do not want to draw a strict contrast between Voit and the later generation.

DAVID SHEMIN (*Northwestern University, Evanston, Ill.*): Yesterday some young person asked Dr. Hotchkiss about the excitement of doing research. You mentioned Loewi. I just want to recall one story he told about the experiment in which he fed the hydrolysate to a dog. He was so excited he woke up a very staid German professor at 6 in the morning to announce the news to him. He knocked on his door and then opened the window and reported his results to him.

KIRSHENBAUM: You mentioned the techniques of detection and you mentioned the Millon and the Kjeldahl method, but I have not been able to find out whose name should be attached to the biuret technique. I am not sure who the person is who introduced it and how come it got to be called the biuret test and not, let us say, von Schmelling's technique.

Also, I wonder if you have noticed how long it takes between the introduction of a technique and its utilization in a procedure? That is, the Kjeldahl method was introduced for finding nitrogen in foods. How long was it before that technique was introduced in metabolic studies? Do you by any chance have any idea about that?

HOLMES: I cannot answer in a general way because I just happened to trace it in this particular set of investigations, and it could well have been utilized earlier on other problems. I do not think terribly long, in that case. On the biuret reaction I might refer it to Dr. Fruton.

FRUTON: I am not sure of the exact dates and names, but it is somewhere in my history book, and, as I remember it, the reaction of alkaline copper sulphate with this curious derivative of urea was discovered sometime around 1840. Then it became recognized that the color reaction was given by a variety of amides, but not by simple amides, and then sometime around 1850 it was applied to materials derived from albuminoid substances. I think it may have been Lehmann around 1850 who demonstrated the reaction with peptone, or what he called peptones. I am not certain of the exact date but it is somewhere around there.

KIRSHENBAUM: I am still wondering how come we have here a test with no name. It may be, as I say, a facetious question, but every test that I have come across has someone's name attached to it. I wonder what is so special about the biuret reaction.

PIRIE: It is not quite true that there is always a name attached to a reaction because about the same time we had the xanthoproteic reaction, which does not have a name. On your question about Kjeldahl, Vickery somewhere about 20 years ago wrote quite a long essay on its history. Vickery quoted Arnold as having brought it to the U.S.A., and he claimed that Haldane's father, J. S. Haldane, introduced the Kjeldahl method from Denmark into Britain as a result of the same visit to Denmark during which he brought one of the English ballgames into Denmark. The King of Denmark saw him hitting a ball around on the seashore and wanted to know what on earth an adult was doing playing around like that. And then the King got hooked on the game as well.

HOLMES: I just want to make one general comment on name reactions. People do not usually name reactions after themselves. It is someone else deciding that person has made a distinctive enough contribution to warrant naming a reaction after him. Maybe no one stood out sufficiently in the development of that test.

SARAH RATNER received her Ph.D. at Columbia in 1937. She then became an associate of Rudolph Schoenheimer and pioneered the use of ^{15}N for the study of protein metabolism. In 1945 she joined the Department of Biochemistry at New York University and in 1954 transferred to the staff of the Public Health Research Institute in New York, where she is now a Member Emerita. In addition to work on protein metabolism, Dr. Ratner has contributed to the elucidation of the pathways of urea synthesis by the discovery of argininosuccinic acid, the demonstration of the role of aspartic acid in the formation of urea, and a clarification of the relationship between the two Krebs cycles and elucidation of the reaction mechanisms.

The Dynamic State of Body Proteins

SARAH RATNER

Department of Biochemistry
The Public Health Research Institute of the City of New York, Inc.
New York, New York 10016

MY TITLE "The Dynamic State of Body Proteins" would have been chosen by Rudolph Schoenheimer. It is closely linked to his famous book *The Dynamic State of Body Constituents*[1] where he presented, with remarkable clarity, his new concepts on the dynamic state of the body fats, sterols, and proteins. This volume comprised the three Dunham Lectures given at Harvard in 1941. Up to that time the components of body tissues were regarded as being structural in function and metabolically inert. Schoenheimer applied the term "dynamic state" to some of the most complex constituents of the cell about which very little was known.

The new findings and novel concepts described in those lectures were the results of experiments conducted over six or seven years in which the stable isotopes deuterium and ^{15}N were applied to the study of intermediary metabolism by ingenious labeling of fatty acids, cholesterol, or amino acids. By chemical means the isotope was introduced into selected, chemically stable, locations of the molecule. This made it possible to follow the fate of a suspected metabolite in the living animal and to detect its participation in a multiplicity of hitherto unrecognized reactions. I was associated mainly with the studies on amino acid and protein metabolism. The material I plan to present might equally well bear the title "The Discovery of Protein Turnover" or in current parlance "The Molecular Basis for Protein Turnover."

The application of isotopes to the study of intermediary metabolism was an entirely novel undertaking and required a new methodology. It is appropriate to acquaint you with the circumstances under which the first tracer experiments of this kind began.

Following on Harold Urey's discovery in 1932 of the stable isotope deuterium, the Rockefeller Foundation, through Urey's enthusiasm, undertook to support the exploitation of deuterium for biological studies. For this purpose, Rittenberg, who had been Urey's student at Columbia and was trained by him in isotope techniques, came to the Department of Biochemistry at the Columbia College of Physicians and Surgeons (P & S) in 1934 to see what interest he could find. Hans T. Clarke, a highly gifted organic chemist and an influential proponent of the ap-

0077-8923/79/0325-0189 $01.75/0

proach of organic chemistry to the development of biochemistry, was the eminent head of that department.

The year before, Schoenheimer left Freiburg to escape political oppression and came to join the Biochemistry Department at P & S at Clarke's invitation to continue his studies in sterol metabolism. Out of this fortunate meeting with Rittenberg there developed the basic ideas and techniques for applying deuterium, and later ^{15}N to the study of intermediary metabolism. Schoenheimer's background was highly propitious for this undertaking. He received his medical degree in Berlin in 1922. In the interlude before his move to Aschoff's Pathological Institute at the University of Freiburg as Chemist, he had spent three years acquiring broad training in biochemistry and organic chemistry under Karl Thomas in Leipzig.

In 1937, Harold Urey and his collaborators succeeded in concentrating ^{15}N, the stable isotope of nitrogen.[2] In that same year, as soon as his fractionating columns began to produce ammonium sulfate of increased ^{15}N abundance, he made these precious samples available to us; and continued to do so. The analysis of ^{15}N, unlike that of deuterium, requires the use of a mass spectrometer and Rittenberg had for the purpose, undertaken to construct one. Although of known design, the project was, nevertheless, a major undertaking. I had completed my thesis work some months before, under Clarke's direction, and having acquired experience in the chemistry of amino acids, was asked to participate in the nitrogen metabolism studies.

The group of interrelated papers on protein metabolism represented the close collaborative efforts of Schoenheimer, Rittenberg, and myself, under Schoenheimer's leadership. Talk was open and frequent and the atmosphere informal; we were nevertheless highly conscious of the rarity of the material we were working with. Each of us performed a different role in the work; my memory is taxed therefore with a triple burden.

A second reason for my choice is the relatively brief discussion alloted to this subject among the wealth of material covered in the Dunham Lectures. A long series of papers under the general title "Protein Metabolism," 16 in number,[3-18] came to a close with two papers on antibody turnover.[19,20] Some results for the first of these were at hand, but work on the second had only been started shortly before Schoenheimer's death in 1941. As the antibody results were published separately in 1942,[18-20] this may explain why they have to some extent been overlooked.

Schoenheimer combined the gifts of a highly imaginative biochemist with the approach of organic chemistry. We planned, from the outset, to feed isotopic amino acids labeled in the α-amino groups with ^{15}N to intact rats and to test the Folin hypothesis. Although we were extremely uncertain as to the outcome, and even as to the feasibility of such experiments, nothing else was left to chance. Owing to the limited supplies of ^{15}N, and the range we hoped to encompass, careful and thorough planning characterized our work. Months before the first animal experiment, I carried out "cold" runs to prepare for each forward stage. The methods of synthesis[5] were selected to conserve ^{15}N, and then modified further. Methods for the isolation of purified amino acids from protein hydrolysates had to be worked out[7, 9, 12] on a scale suitable for all our analytical needs. It was necessary to establish that the C-N bond of the amino group was stable under conditions of protein hydrolysis and amino acid isolation, to insure that any transfer of ^{15}N-amino nitrogen, after ingestion of the compound, could be attributed to a metabolic reaction.[6] The analytical methodology developed by Rittenberg for converting Kjeldahl nitrogen to gaseous N_2 for injection into the mass spectrometer represented one of his many essential contributions.[3,4]

An experiment in which glycine labeled with ^{15}N of low abundance was added to the diet of a rat along with some benzoic acid, was in the nature of a trial[8] to determine the origin of the glycine used in the formation of excreted hippuric acid, also the extent of isotope dilution, and whether or not an intact animal could tolerate an enrichment of ^{15}N or discriminate between ^{14}N and ^{15}N.

Our plans to feed ^{15}N-labeled amino acids would enable us to test the Folin hypothesis. His generally accepted theory made a clear distinction between *exogenous* nitrogen metabolism, which largely originated from the nitrogen of the diet, and *endogenous* nitrogen, which arose from a small amount of "wear and tear." Borsook and Keighley[21] had also questioned it and had proposed, as a result of sulfur balance studies, the idea of a continuing nitrogen metabolism. However, the Folin hypothesis had never been subjected to direct proof.

With all deliberate speed we were finally ready to go ahead. A small amount of isotopic DL-tyrosine was added for 10 days to the stock diet of an adult rat in nitrogen and caloric equilibrium. If the Folin hypothesis was correct, most of the ingested ^{15}N would be found in excreted urea (upper half of FIGURE 1). The results were totally unexpected. As shown in the lower half of FIGURE 1, only 50% of the ^{15}N was excreted, while all the rest had been retained in the body tissues.[9] With a natural isomer the amount retained was appreciably larger. Almost all of the retained ^{15}N was present in the proteins of body tissues. Several amino acids were isolated after hydrolysis of liver and muscle proteins. The ^{15}N was present not only in tyrosine, but also, in lower ^{15}N concentrations, in aspartic and glutamic acids, arginine, and histidine. If anything, we had expected

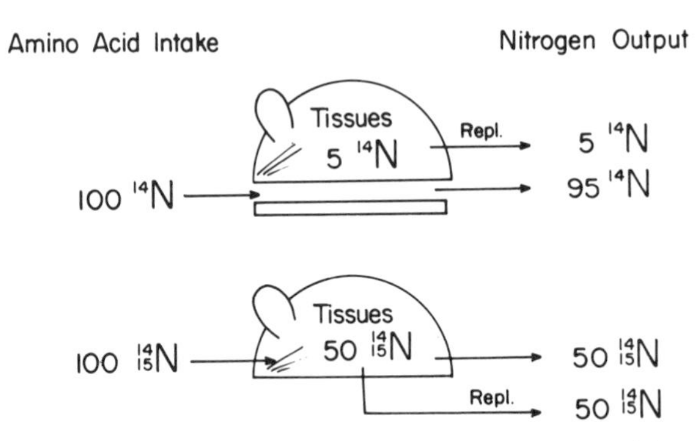

FIGURE 1. Changes in concepts of nitrogen metabolism without and with ^{15}N as tracer. Upper portion: Folin's hypothesis viewed ingested nitrogen as being largely excreted (*exogenous* metabolism, 95%), 5% was retained to replace wear and tear (*endogenous* metabolism). Lower portion: observed distribution of ^{15}N-labeled dietary nitrogen (DL-tyrosine), 50% in tissues and 50% excreted. (After Schoenheimer *et al.*[9])

only the administered amino acid, ^{15}N-labeled tyrosine, would become incorporated into tissue proteins. It will scarcely be possible for most of you to appreciate how little was known 40 years ago about amino acid and protein metabolism. The exciting results of our first experiment with one animal presented us with any number of complex problems to unravel. Amino acids were interacting with each other and also with tissue proteins, which until then were thought to be metabolically inert.

From there our reasoning went: if incorporation of the isotopically labeled amino acids into proteins is a replacement process, this would involve the displacement of preexisting unlabeled amino acids and would require, for each amino acid, the opening, and then the closing, of two peptide bonds. Was the shift of amino groups associated with total amino acid synthesis? How was this shifting about related to replacement of amino acids in proteins? The answer to that quandary led to the formulation of the idea of the metabolic pool. If the shift of amino groups occurred while the amino acids were in the free state, in a metabolic pool of free amino acids, incorporation of amino acids into proteins by direct replacement could be understood. One more important question: was the labeled amino group incorporated while still attached to the original carbon chain? What if there was some unique way by which only amino groups were accepted into the protein. To prove that the intact amino acid molecule was incorporated by direct replacement, and to answer other questions, we turned to more elaborate experiments re-

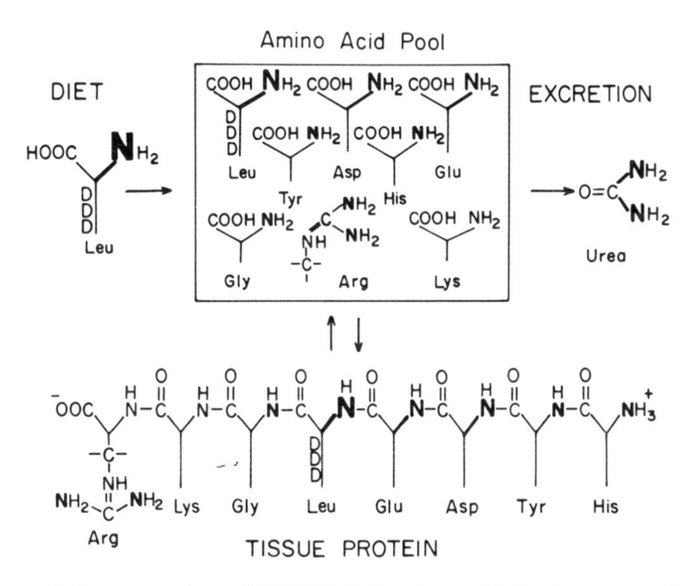

FIGURE 2. Incorporation of ^{15}N-labeled amino acids in tissue proteins after ingestion of doubly labeled [D, ^{15}N]-L-leucine. Concentration of ^{15}N is suggested by letter thickness; ^{15}N was present in all amino acids except lysine and the α-amino N of arginine. The ^{15}N concentration shown in the "metabolic pool of amino acids" is assumed to be the same as in the polypeptide. (After Schoenheimer *et al.*[12])

quiring an amino acid labeled with two isotopes. For this purpose, I synthesized the indispensable amino acid leucine with deuterium in stable linkage distributed along the carbon chain, and with ^{15}N in the α-amino group.[12] The amino acid was fed in the form of the natural, optically active, isomer for three days.

Only 30% of the ^{15}N in the diet was excreted, and about 60% had replaced unlabeled amino acids in body proteins; some 12 organs and tissues were analyzed. Eight or nine amino acids were isolated from the proteins of the intestinal mucosa, the liver, and muscle. Except for lysine, all amino acids contained ^{15}N in the same relative amounts shown in FIGURE 2. Leucine always had the highest amount of ^{15}N, compared to other amino acids, and had become incorporated together with the deuterium label (FIGURE 2). The ratio of deuterium to ^{15}N changed in the direction of dilution of ^{15}N indicating a partial replacement of labeled by unlabeled amino groups. We calculated that at least 30 percent of the dietary leucine introduced into the body proteins was incorporated without detachment of the amino group; the detached 70 percent was found in other amino acids of the body proteins. For each tissue the ^{15}N

concentration in glutamic and aspartic acids was always the highest, after leucine, and glycine, tyrosine, and histidine, and the amidine group of arginine also contained ^{15}N regardless of the protein source. As to our conclusions, I quote from the original paper.[12]

> It is scarcely possible to reconcile our findings with any theory which requires a distinction between these two types of nitrogen. It has been shown that nitrogenous groupings of tissue proteins are constantly involved in chemical reactions; peptide linkages open, the amino acids liberated mix with others of the same species of whatever source, diet, or tissue. This mixture of amino acid molecules, while in the free state, takes part in a variety of chemical reactions: some reenter directly into vacant positions left open by the rupture of peptide linkages; others transfer their nitrogen to deaminated molecules to form new amino acids. These in turn continuously enter the same chemical cycles which render the source of the nitrogen indistinguishable. Some body constituents like glutamic and aspartic acids and some proteins like those of liver, serum, and other organs are more actively involved than others in this general metabolic mixing process. The excreted nitrogen may be considered as a part of the metabolic pool originating from interaction of dietary nitrogen with the relatively large quantities of reactive tissue nitrogen.

At the time *in vitro* studies with amino acids were just beginning. Braunstein and Kritzman[23] found in homogenates of muscle tissue that the amino groups of glutamic and aspartic acid were transferred to α-keto acids to form new amino acids reversibly. Our high ^{15}N values for glutamic and aspartic acids showed that these amino acids played a central role in the shift of amino groups from one amino acid to another in all the tissues of the body. The work of Krebs and Henseleit[24] on urea formation with liver tissue slices was confirmed by our finding that ^{15}N was present in the amidine group of arginine isolated from the liver protein.

It has been known for some time that a small amount of amino acid nitrogen is present in cells and plasma. The concept of an intracellular pool of metabolically reactive amino acids derived from dietary and body proteins was arrived at intuitively. It explained our observed isotope distributions and by extension included almost all amino acids. This concept and its relation to wandering amino groups removed obstacles in the way of firm proof that leucine and other amino acids are incorporated in proteins by direct replacement.

Experiments with ^{15}N-glycine[16] as the dietary supplement gave analogous results. The pattern of isotope distribution among amino acids isolated from body proteins and evidence of the occurrence of rapid shifts of amino groups among amino acids were the same.

It was not possible to resist comment on the question of mechanism of amino acid incorporation. To quote from the leucine paper[12] in this connection:

> There are two general reactions possible which might lead to amino acid replacement: (1) complete breakdown of the proteins into its units followed by resynthesis or (2) only partial replacement of units. Metabolic studies with isotopes indicated only end-results but not intermediate steps of a reaction. We have no indication as to what had happened to the protein molecule in the animals. Both reactions are conceivable. The second type, replacement of units, has been shown by Bergmann and collaborators[25] to occur *in vitro* under the action of proteolytic enzymes on polypeptides and the occurrence *in vivo* of these reactions has been postulated.

The citation[25] referred to a new enzymatic approach attempting to meet the problem of protein synthesis in 1938. Clearly preference as to mechanism was deliberately avoided since the nature and the design of our *in vivo* experiments could not lend support to any formulation of mechanism. This reluctance, repeated again elsewhere, was the expression of a keen awareness of the complexities required for protein synthesis.

In 1935 Schoenheimer and Rittenberg began to publish the results of their studies with deuterium under the general title "Deuterium as an Indicator in the Study of Intermediary Metabolism." Several of these studies were concerned with body fats. Paper VI published in 1936 bore the subtitle "Synthesis and Destruction of Fatty Acids in the Organism."[26] The introduction reads:

> In previous publications we have shown that deuterium can be used to label foodstuffs so that their transportation and conversion into other substances could be followed. We present in this paper a general method for determining the rate of synthesis and destruction of individual constituents of an organism. Organic substances synthesized in an aqueous medium containing heavy water will, in general, contain deuterium. The rate of appearance of deuterium in the organic compound will then be proportional to the rate of synthesis. In this communication we describe the application of this method to the fatty acids.

The experiments were conducted with mice placed on a bread diet, and given deuterium-containing drinking water to maintain a constant deuterium concentration in the body water. The deuterium content of the total fatty acids increased rapidly reaching a constant value in 6 or 7 days at which time all replaceable molecules were being replaced. Curve I of FIGURE 3 shows that the half-life of newly synthesized

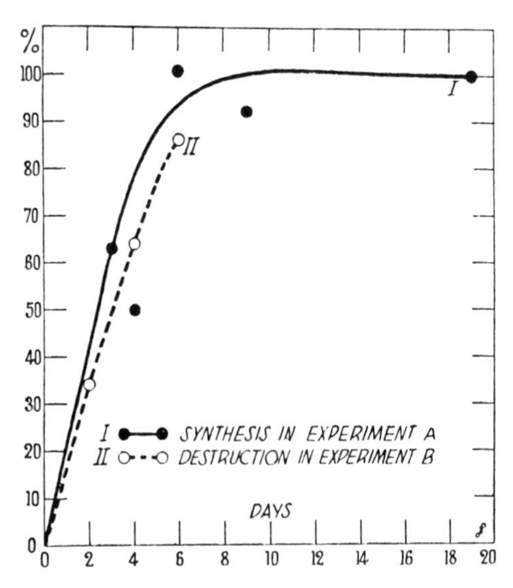

FIGURE 3. Synthesis and destruction of fatty acids in mice. (From Schoenheimer and Rittenberg.[26])

molecules was 2 to 3 days. To illustrate degradation of fatty acids, a group of mice were fed with deuterated fatty acids and received ordinary drinking water. After the deuterated fat feeding was discontinued, the deuterium content in the fatty acids (Curve II) fell at a rate almost identical with that found for the rate of synthesis. The rate of degradation given in Curve II is, in essence, a decay curve. This expression of the rate of a metabolic process in terms of the half-life represents the first kinetic treatment, I believe, of the turnover of a body constituent. "Turnover" and "half-life time" were indeed their terms, employed to describe the continual synthesis and destruction of fatty acids in well-nourished animals. Schoenheimer had been greatly excited by the metabolic implications of these extraordinary results since a year earlier he had shown with Rittenberg[27] that in partially starved mice, labeled dietary fat was deposited in fat depots before being utilized.

In 1933, Schoenheimer published an ingenious study, the first expression of his concept of a dynamic metabolism entitled "Synthesis and Destruction of Cholesterol in the Organism."[28] With the advent of deuterium, he saw the possibility of developing his ideas and turned his interests as soon as possible to probing fat metabolism in the work I have just mentioned. The plans for our amino acid studies must have been drawn up with similar ideas in mind. The obstacles encountered in the confusing scramble we came to recognize as the mobility of amino nitrogen was, however, an entirely unexpected finding; once perceived it proved an asset.

We turned next to the problems of protein specificity and protein turnover for we had seen that proteins of some tissues took up ^{15}N more rapidly than others. To gain specificity and the advantages of isolation, Schoenheimer turned to immunological procedures and to collaboration with Michael Heidelberger, in the Department of Medicine at P & S. Heidelberger was then engaged in his classical immunological studies on type-specific antigens of pneumococcus and had already developed, with Kendall and Kabat, the basis for the quantitative, specific precipitation of homogeneous antibody with the respective antigens. These were polysaccharides and free of nitrogen.

A rabbit was immunized with type III pneumococcus vaccine and then received a dietary supplement of ^{15}N-glycine for three days. As FIGURE 4 shows, ^{15}N concentration in antibody and plasma proteins increased during ^{15}N-glycine feeding and declined shortly after it ceased. Both the rise in ^{15}N content and the decline indicate protein turnover by continual synthesis and degradation. Thus synthesis was going on at the same time that the amount of antibody was decreasing. The rate of ^{15}N decline corresponds to a half-life of about 14 days for antibody protein and about the same for the remaining serum proteins.[18, 19]

As a control, antibody to type I pneumococcus was injected into a rabbit to produce passive immunization prior to ^{15}N-glycine feeding. We were asking whether antibody homologous to the species of animal used, but artifically introduced, would show the same behavior as antibody produced in active immunity. In this experiment, none of the samples of the passively introduced type I antibody contained ^{15}N, while ^{15}N was incorporated in other plasma proteins as before.

The striking contrast in the animal's capacity to synthesize antibody in active, as compared to passive, immunization was confirmed in a single animal actively immunized with type III pneumococcus and passively immunized with type I antiserum. The curves in FIGURE 5 show once more the uptake and decline in ^{15}N incorporation in type III antibody, while in the same animal the passively introduced type I antibody molecules showed no ^{15}N uptake. Since type I antigen was not given, synthesis could not occur. This contrast was confirmed in interesting experiments with liver and spleen slices by Ranney and London[29] in 1951, and again in 1954 by Humphrey and McFarlane[30] with intact animals. The half-life determined by the latter investigators proved to be 4 to 5 days, considerably lower than our value, as ^{14}C-labeling reduced the effects of amino acid reutilization.

The term "protein turnover" refers here to the general phenomenon

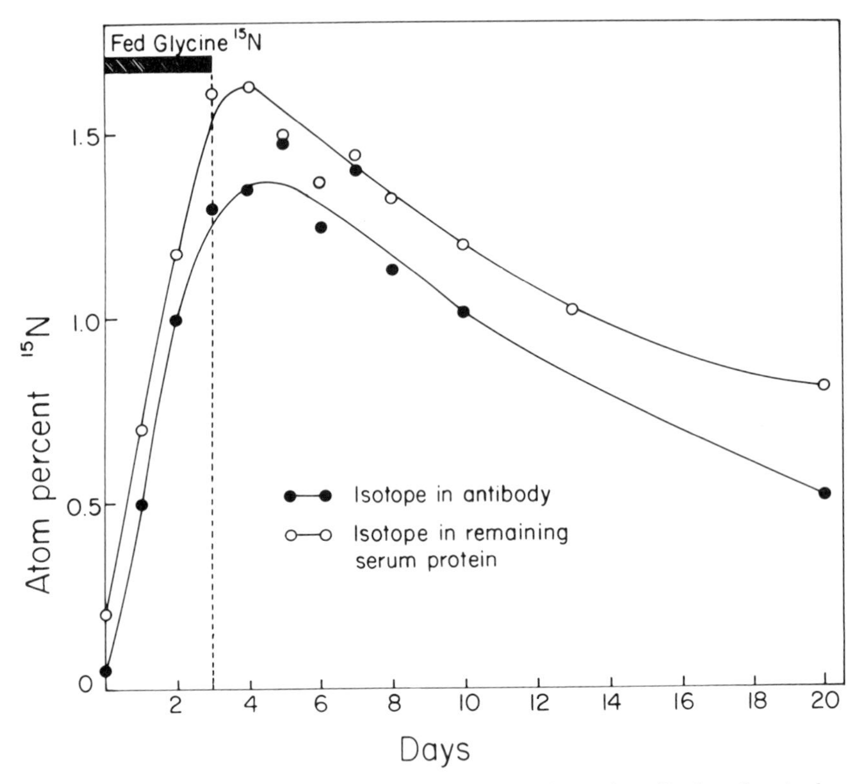

FIGURE 4. Concentration of ^{15}N in serum protein and antibody of actively immunized rabbit during and after feeding of ^{15}N-glycine (calculated for an isotope content of 100 atom percent in the glycine administered). Antibody concentration declined from 4 mg N per ml on day 0 to 1.25 mg on day 20. (Revised from Schoenheimer *et al.*[19])

disclosed by ^{15}N experiments whereby tissue proteins continually undergo synthesis and degradation. That Schoenheimer recognized these to be independent processes was made evident in the third Dunham Lecture[1] (Summary, p. 64):

> All regeneration reactions must be enzymatic in nature. The large molecules, such as the fats and the proteins, are under the influence of lytic enzymes, constantly being degraded to their constituent fragments. These changes are balanced by synthetic processes which must be coupled to other chemical reactions, such as oxidation or dephosphorylation. After death, when the oxidative systems disappear, the synthetic processes also cease, and the unbalanced degradative reactions lead to the collapse of the thermodynamically unstable structural elements. In general, every regeneration reaction involving an increase in free energy

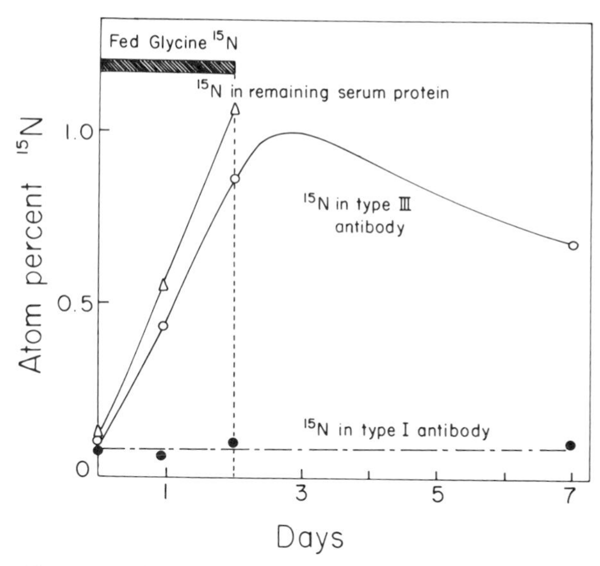

FIGURE 5. Concentration of ^{15}N in serum protein and in type III and type I antibodies of rabbit actively immunized against type III and passively immunized against type I antibody. Type III antibody concentration declined from 3.12 mg N per ml to 1.52, and type I from 1.09 mg N per ml to 0.18. (From Table III of Heidelberger *et al.*[20])

> must be coupled to another process. In order to maintain structure against its tendency to collapse, work has to be done. The replacement of a brick fallen from a wall requires energy, and in the living organism energy debts are paid by chemical reactions.

This was stated again in the first antibody paper[19] (p. 553):

> The introduction of isotopic nitrogen into antibody protein under these conditions must have involved both the opening and closing of peptide bonds in protein degradation and synthesis. Similarly the removal of isotopic nitrogen from the antibody after the 4th day could only have occurred by a continuation of the same reactions responsible for the introduction of isotope.

Despite these clear distinctions, changes in phrasing by others have introduced misinterpretations of the original meaning. The term "dynamic equilibria" crept into the literature as a substitute for "dynamic state," and the term "exchange reaction" has been substituted for protein synthesis.[31]

The application of ^{15}N to the study of protein metabolism slowly declined following Schoenheimer's death. Soon after the long-lived radioactive isotope of carbon was discovered by Ruben and Kamen[32] in 1941, ^{14}C became commercially available. Now that we have some understanding of the metabolic relationships between the nitrogen of amino acids and proteins, ^{14}C offers many advantages over ^{15}N for the study of protein turnover, but it would indeed make an entertaining exercise to speculate on what form our ideas would have taken if an isotope of carbon had become available before ^{15}N. The countless problems to which ^{14}C-labeled amino acids have been applied depended on the insight into metabolic behavior gained from ^{15}N. By stressing the significance of the continual operation of biosynthetic processes, the dynamic state concept focused a great deal of interest on biosynthetic pathways and related mechanisms. A number of studies involving amino acids were carried out with other collaborators during Schoenheimer's lifetime. Space does not allow more than this mention.

CONTINUED EXAMINATION OF PROTEIN TURNOVER

The concept of protein turnover has retained its original meaning for forty years, not without occasional refutation, and much more frequently with continuing reinforcement from new developments. A partial and brief accounting may be in order. In 1944, Shemin and Rittenberg[33] investigated the ^{15}N uptake of a number of body tissues at frequent intervals during prolonged ingestion of ^{15}N-glycine and described characteristic patterns of ^{15}N uptake in different tissues. Beginning in 1946, numerous laboratories, with ^{14}C- or ^{32}S-labeled amino acids, corroborated many times over that the amino acid incorporated was indeed bound in peptide linkage[34] and that all amino acids tested were thus incorporated.[35] Various improvements were introduced in the method and duration of labeled amino acid administration to attain higher specific activity.[34] Well-based kinetic treatments led to improvements in the determination of the half-life of a cellular protein,[36, 37] and attention was directed to sources of error due to amino acid reutilization.[37]

At an early stage in the search for mechanism of synthesis, attempts were made to determine whether synthesis is stepwise, involving peptide intermediates, or whether amino acids are arranged in sequence and then form the polypeptide chain all at once, in a single step. The papers of Simpson,[38] Muir *et al.*,[39] and Steinberg and Anfinsen[40] are fascinating as an approach to mechanism through a comparison of the specific radioactivity of an incorporated amino acid located in different positions of the same protein. To go into this any further would encroach on Paul Zamecnik's domain, if I have not already done so.

DO BODY PROTEINS TURNOVER?

Despite the weight of accumulated evidence, the existence of protein turnover as a general metabolic property of all biological systems was questioned in 1953 and 1955 by Hogness, Cohn, and Monod.[41] They had failed to obtain evidence of protein degradation in *E. coli* in the logarithmic phase of growth and went on to suggest that the Schoenheimer hypothesis of the dynamic state of body proteins should be seriously questioned on two grounds: (a) that only secreted proteins undergo amino acid incorporation, and (b) that the release of free amino acids represents the breakdown and dissolution of tissue cells while synthesis is confined to the formation of new cells. This was tantamount to limiting protein turnover to cell dissolution and replacement, which, in essence, revived the old replacement of wear and tear hypothesis.

In 1957, experiments by Mandelstam[42] with starved *E. coli* cells, or with cells in the stationary, nongrowing state, successfully demonstrated both protein synthesis (incorporation of amino acids) and cell degradation (release of amino acids). Mandelstam pointed out that the extrapolation of results obtained with *E. coli* cells in the logarithmic stage of growth to those obtained in adult animal tissues was unwarranted as the two represent physiological extremes. Thus a four year interval elapsed before the clarifying bacterial exidence by Mandelstam and others[43, 44] appeared. The influence of the paper by Hogness *et al.* was unduly prolonged since attention was given to their work in several detailed reviews[34, 45] of that period.

PROTEIN TURNOVER AND METABOLIC FUNCTION

It is now widely accepted that the cellular proteins of adult animals in a steady state are being continually degraded and replaced by synthesis at rates characteristic of specific proteins. The acceptance of the hypothesis came slowly, it seems to me, and not without hesitation, for what might be called teleological reasons: why does nature select a process so enormously wasteful of energy and without function? No one has ventured an explanation on theoretical grounds except perhaps to speculate that protein turnover persists as a survival of ancestral adaptive processes in lower species.

In 1964, Robert Schimke[46] described new developments which attributed metabolic functions to protein turnover. A history of the turnover concept would not be complete without some mention of his work. Schimke's contribution began with the finding that the level of liver arginase decreases in a change from a high to a low protein diet; however, in starvation, the level of arginase increases. He showed moreover that the capacity of an animal to adapt to dietary change through a change in enzyme level could be mediated through protein turnover, by effecting either the rate of synthesis or the rate of degradation. Experimentally he demonstrated both types of change.

The half-life of liver arginase is 4 to 5 days. Values for the half-life of other enzyme proteins vary greatly and may be as brief as 19 minutes. Berlin and Schimke[47] have pointed out that the shorter the half-life of an enzyme, the more quickly the enzyme concentration can change in response to hormonal or nutritional changes. FIGURE 6, which is taken from their paper, gives calculated rate curves for proteins of varying half-life. These show the fold-change brought about by an increased rate of synthesis without any change in the rate of degradation. By 1970 some 20 to 30 examples of the regulation of enzyme levels in animal tissues had been reported.[48] Berlin and Schimke[47] suggest:

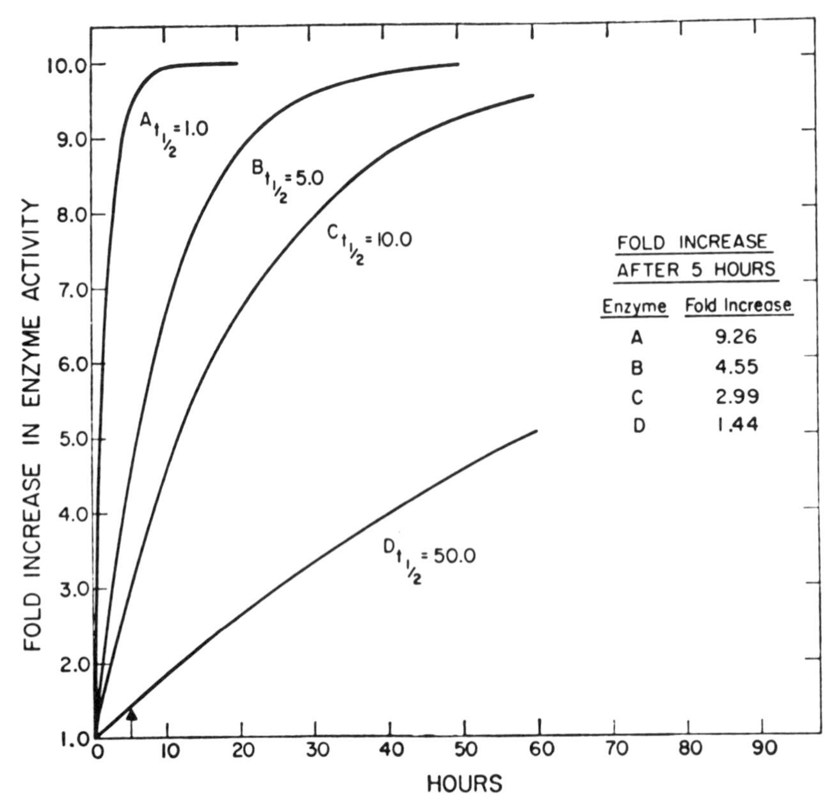

FIGURE 6. The effect of different half-lives (degradation rates) on the response of an enzyme to a 10-fold increase in the rate of synthesis. (From Berlin and Schimke.[47])

The marked heterogeneity of turnover rates of total liver protein and of various enzymes is striking. Although the biochemical basis for such heterogeneity is unknown, certain suggestions may be made regarding its physiological importance. Thus it is suggested that enzymes whose levels are rate limiting for a specific biochemical reaction *in vivo* (e.g., tryptophan pyrrolase) will have a rapid rate of turnover. Such enzymes would respond rapidly, both increasing and decreasing, in response to environmental alterations such as changes in diet, substrate levels, or hormonal levels. On the other hand, enzymes that are in constant usage, or whose physiological activity is controlled by feedback inhibition or by the availability of substrate (e.g., arginase), would not be required to have the ability to fluctutate rapidly, and hence would be those enzymes with slower rates of turnover.

Goldberg and Dice[49] have added the suggestion that protein catabolism must be especially important in an organism's adaptation to a poor environment, such as starvation, to provide essential amino acids for the synthesis of enzymes appropriate to such conditions or to provide an energy reservoir in muscle and liver for times of decreased caloric intake. In the future biochemists will undoubtedly expand even further the functions of the process of protein turnover.

ACKNOWLEDGMENT

This article was written during tenure of a John E. Fogarty International Center Scholarship-in-Residence, at the National Institutes of Health, from December 1, 1977, to March 10, 1978.

REFERENCES

1. SCHOENHEIMER, R. 1942. The Dynamic State of Body Constituents. Harvard University Press, Cambridge, Mass.
2. UREY, H. C., M. FOX, J. R. HUFFMAN, & H. G. THODE. 1937. A concentration of N^{15} by a chemical exchange reaction. J. Am. Chem. Soc. *59*: 1407–1408.
3. SCHOENHEIMER, R. & D. RITTENBERG. 1939. Studies in protein metabolism. I. General considerations in the application of isotopes to the study of protein metabolism. The normal abundance of nitrogen isotopes in amino acids. J. Biol. Chem. *127*: 285–290.
4. RITTENBERG, D., A. S. KESTON, F. ROSEBURY & R. SCHOENHEIMER. 1939. Studies in protein metabolism. II. The determination of nitrogen isotopes in organic compounds. J. Biol. Chem. *127*: 291–299.
5. SCHOENHEIMER, R. & S. RATNER. 1939. Studies in protein metabolism. III. Synthesis of amino acids containing isotopic nitrogen. J. Biol. Chem. *127*: 301–313.
6. KESTON, A. S., D. RITTENBERG & R. SCHOENHEIMER. 1939. Studies in protein metabolism. IV. The stability of nitrogen in organic compounds. J. Biol. Chem. *127*: 315–318.

7. Foster, G. L., R. Schoenheimer & D. Rittenberg. 1939. Studies in protein metabolism. V. The utilization of ammonia for amino acid and creatine formation in animals. J. Biol. Chem. *127*: 319–327.
8. Rittenberg, D. & R. Schoenheimer. 1939. Studies in protein metabolism. VI. Hippuric acid formation studied with the aid of the nitrogen isotope. J. Biol. Chem. *127*: 329–331.
9. Schoenheimer, R., S. Ratner & D. Rittenberg. 1939. Studies in protein metabolism VII. The metabolism of tyrosine. J. Biol. Chem. *127*: 333–344.
10. Schoenheimer, R., D. Ritenberg & A. S. Keston. 1939. Studies in protein metabolism. VIII. The activity of the α-amino group of histidine in animals. J. Biol. Chem. *127*: 385–389.
11. Rittenberg, D., R. Schoenheimer & A. S. Keston. 1939. Studies in protein metabolism. IX. The utilization of ammonia by normal rats on a stock diet. J. Biol. Chem. *128*: 603–607.
12. Schoenheimer, R., S. Ratner & D. Rittenberg. 1939. Studies in protein metabolism. X. The metabolic activity of body protein investigated with *l*(−)-leucine containing two isotopes. J. Biol. Chem. *130*: 703–732.
13. Bloch, K. & R. Schoenheimer. 1939. Studies in protein metabolism. XI. The metabolic relation of creatine and creatinine studied with isotopic nitrogen. J. Biol. Chem. *131*: 111–119.
14. Clutton, R. F., R. Schoenheimer & D. Rittenberg. 1940. Studies in protein metabolism. XII. The conversion of ornithine into arginine in the mouse. J. Biol. Chem. *132*: 227–231.
15. Ratner, S., R. Schoenheimer & D. Rittenberg. 1940. Studies in protein metabolism. XIII. The metabolism and inversion of *d*(+)-leucine studied with two isotopes. J. Biol. Chem. *134*: 653–663.
16. Ratner, S., D. Rittenberg, A. S. Keston & R. Schoenheimer. 1940. Studies in protein metabolism. XIV. The chemical interaction of dietary glycine and body proteins in rats. J. Biol. Chem. *134*: 665–676.
17. Weissman, N. & R. Schoenheimer. 1941. Studies in protein metabolism. XV. The relative stability of *l*(+)-lysine in rats studied with deuterium and heavy nitrogen. J. Biol. Chem. *140*: 779–795.
18. Schoenheimer, R., S. Ratner, D. Rittenberg & M. Heidelberger. 1942. The interaction of the blood proteins of rat with dietary nitrogen. J. Biol. Chem. *144*: 541–544.
19. Schoenheimer, R., S. Ratner, D. Rittenberg & M. Heidelberger. 1942. The interaction of antibody protein with dietary nitrogen in actively immunized animals. J. Biol. Chem. *144*: 545–554.
20. Heidelberger, M., H. P. Treffers, R. Schoenheimer, S. Ratner & D. Rittenberg. 1942. Behavior of antibody protein toward dietary nitrogen in active and passive immunity. J. Biol. Chem. *144*: 555–562.
21. Borsook, H. & G. L. Keighley. 1935. The "continuing" metabolism of nitrogen in animals. Proc. R. Soc. *118B*: 488–521.
22. Rose, W. C. 1938. The nutritive significance of the amino acids. Physiol. Rev. *18*: 109–136.
23. Braunstein, A. E. & M. G. Kritzmann. 1937. Uber den Ab- und Aufbau von Aminosauren durch Umaninierung. Enzymologia 2: 129–146.
24. Krebs, H. A. & K. Henseleit. 1932. Untersuchungen uber die Harnstoffbildung im Tierkorper. Z. Physiol. Chem. *210*: 33–66.
25. Bergmann, M. 1938. The structure of proteins in relation to biological problems. Chem. Rev. *22*: 423–435.

26. SCHOENHEIMER, R. & D. RITTENBERG. 1936. Deuterium as an indicator in the study of intermediary metabolism. VI. Synthesis and destruction of fatty acids in the organism. J. Biol. Chem. *114*: 381–396.
27. SCHOENHEIMER, R. & D. RITTENBERG. 1935. Deuterium as an indicator in the study of intermediary metabolism. III. The role of the fat tissues. J. Biol. Chem. *111*: 175–181.
28. SCHOENHEIMER, R. & F. BREUSCH. 1933. Synthesis and destruction of fatty acids in the organism. J. Biol. Chem. *103*: 439–448.
29. RANNEY, H. M. & I. M. LONDON. 1951. Antibody formation in surviving tissues. Fed. Proc. *10*: 562–563.
30. HUMPHREY, J. H. & A. S. MCFARLANE. 1954. Rate of elimination of homologous globulins (including antibody) from the circulation. Biochem. J. *57*: 186–191.
31. BORSOOK, H. 1953. Peptide bond formation. In Advances in Protein Chemistry. *8*: 127–174.
32. RUBEN, S. & M. D. KAMEN. 1941. Long-lived radioactive carbon: C^{14}. Phys. Rev. *59*: 349–354.
33. SHEMIN, D. & D. RITTENBERG. 1944. Some interrelationships in general nitrogen metabolism. J. Biol. Chem. *153*: 401–421.
34. TARVER, H. 1954. Peptide and protein synthesis. Protein turnover. In The Proteins. H. Neurath & K. Bailey, Eds. Vol. *11B*: 1199–1296. Academic Press, New York, N.Y.
35. BORSOOK, H., C. L. DEASY, A. J. HAAGEN-SMIT, G. KEIGHLEY & P. H. LOWY. 1950. Metabolism of C^{14}-labeled glycine, L-histidine, L-leucine, and L-lysine. J. Biol. Chem. *187*: 839–848.
36. REINER, J. M. 1953. The study of metabolic turnover rates by means of isotopic tracers. II. Turnover in a simple reaction system. Arch. Biochem. Biophys. *46*: 80–99.
37. KOCH, A. L. 1962. The evaluation of the rates of biological processes from tracer kinetic data. I. The influence of labile metabolic pools. J. Theoret. Biol. *3*: 283–303.
38. SIMPSON, M. V. 1955. Further studies of the biosynthesis of aldolase and glyceraldehyde-3-phosphate dehydrogenase. J. Biol. Chem. *216*: 179–183.
39. MUIR, H. M., A. NEUBERGER & J. C. PERRONE. 1952. Further isotopic studies on haemoglobin formation in the rat and rabbit. Biochem. J. *52*: 87–95.
40. STEINBERG, D. & C. B. ANFINSEN. 1952. Evidence for intermediates in ovalbumin synthesis. J. Biol. Chem. *199*: 25–42.
41. HOGNESS, D. S., M. COHN & J. MONOD. 1955. Studies on the induced synthesis of β-galactosidase in *Escherichia coli*: The kinetics and mechanism of sulphur incorporation. Biochim. Biophys. Acta *16*: 99–116.
42. MANDELSTAM, J. 1957. Turnover of protein in starved bacteria and its relationship to the induced synthesis of enzyme. Nature *179*: 1179–1181.
43. MOLDAVE, K. 1956. Intracellular protein metabolism in Ehrlich's ascites carcinoma cells. J. Biol. Chem. *221*: 543–553; *225*: 709–714.
44. BOREK, E., L. PONTICORVO & D. RITTENBERG. 1958. Protein turnover in microorganisms. Proc. Nat. Acad. Sci. U.S.A. *44*: 369–374.
45. KAMIN, H. & P. HANDLER. 1957. Amino acid and protein metabolism. Ann. Rev. Biochem. *26*: 419–490.
46. SCHIMKE, R. T. 1964. The importance of both synthesis and degradation in the control of arginase levels in rat liver. J. Biol. Chem. *239*: 3808–3817.
47. BERLIN, C. M. & R. T. SCHIMKE. 1965. Influence of turnover rates on the responses of enzymes to cortisone. Mol. Pharmacol. *1*: 149–156.

48. Schimke, R. T. & D. Doyle. 1970. Control of enzyme levels in animal tissues. Ann. Rev. Biochem. *39*: 929–976.
49. Goldberg, A. L. & J. F. Dice. 1974. Intracellular protein degradation in mammalian and bacterial cells. Ann. Rev. Biochem. *43*: 835–869.

DISCUSSION OF THE PAPER

Nachmansohn: In my view the elegant evidence offered for the dynamic state of cell constituents is a real landmark in the history of biochemistry. The idea goes far back in history, one could even quote the panta rhei [πάντα ῥεῖ] of Heraclitos. However, this was philosophy. But even in modern science the notion of a dynamic state has been repeatedly expressed. I remember when I joined Meyerhof's laboratory in 1926; I once asked him about the meaning of the fact that during muscular contraction the formation of lactic acid is greatly accelerated, up to several thousand times (according to the tension developed). He explained to me that in his view all cellular reactions are going on continuously. The cell is always in a dynamic, not in a static state. During activity certain reactions are greatly accelerated. Now this was a theoretical notion in which he believed. But it was the ingenious contribution by you, Schoenheimer, and the other associates that this essential feature of living cells was experimentally firmly established. My comment that the notion previously existed should by no means be interpreted as an inadequate appreciation. On the contrary, I have been a great admirer of Schoenheimer's scientific stature ever since we worked, in the twenties, in Rona's laboratory next to each other at the same bench. I just wanted to remind the audience, since we are in a session on the history of science, that the idea of the dynamic state existed before in the minds of several biologists. But in science it is decisive that an idea should be established by experimental facts.

I was also greatly interested when you mentioned that the concept existed in your minds since 1934. This fits well Einstein's claim that the concept *precedes* theory, although theory must be supported by experimental facts. After Werner Heisenberg gave a lecture on quantum mechanics in the spring of 1926 (he was then 24 years old), in a Nernst colloquium at the University of Berlin, Einstein, who was fascinated, had a long discussion with him. He told him: "Whether you find something or not depends on the theory you have. The theory decides your observations."

Ratner: I am sure many of you will want to remark on this point. I became very interested in the origin of the term "dynamic" and came across the book by Meyerhof entitled *The Chemical Dynamics of Life*

(1924). On going further back I found such views expressed among the Cambridge school of biochemists, led by Hopkins. In 1913 Hopkins gave an inspired lecture entitled "The Dynamic Side of Biochemistry" (reproduced in *Hopkins and Biochemistry* (Needham and Baldwin. 1949. Heffer), where with great originality, he presented the idea that reactions of low-molecular-weight compounds reveal the dynamic biochemical phenomena of cells. I assume that Meyerhof's interests account for a similar direction in his ideas. I began my search to see whether Schoenheimer, as a student, had been influenced by this older concept, but found little to support it. I tried to stress this morning that Schoenheimer's turnover concepts were concerned with the high-molecular-weight, complex constituents of the body tissues, in particular with the proteins of tissues and blood. Being a physiologist, he also avoided the term dynamic equilibrium.

F. Lipmann (*Rockefeller University, New York, N.Y.*): You only briefly mentioned fat synthesis, and I do not want to deviate from protein turnover but it was of enormous importance for me that in Schoenheimer's book, if I am not mistaken, he introduced the term active acetate; and it was Schoenheimer then who really told me to look for an activated acetate. Am I right?

Ratner: I have not recently reread all the papers on fatty acid turnover, and the whole question of active acetate came up again in connection with cholesterol biosynthesis, which I believe you were then also interested in. Schoenheimer may have used that expression. I remember that in 1941 in the first chapter in your article in *Advances in Enzymology* you referred to the concept of synthesis in Schoenheimer's work.

F. Portugal (*Carnegie Institution, Washington, D.C.*): Around 1939 Dr. Schoenheimer and Dr. Rittenberg, in studying amino acid metabolism using ^{15}N and deuterium did mention a point that you referred to in your talk, and that is, if the carbon isotope was available it might have been more useful for these studies. Around the same time, Rubin and Kamin were studying photosynthesis and they were using the first carbon isotope for metabolic studies, which was not ^{14}C actually but ^{11}C. I was curious if there might be an interesting story with regard to whether or not there might have been prior information about the use of that isotope or, if not, why there might not have been a transfer of information with regard to that as a possible isotope?

Also, at the time these studies were being done, relatively little was known about how the proteins are synthesized. That is, are they formed together like peptides or individual amino acids in a stepwise process? Was there much discussion in your group? Was there much discussion about the possibility that rather than an actual turnover, which would involve new synthesis and new degradation, that protein could conceivably open up and let an isotopic amino acid come into say the middle of the protein, so that your incorporation might represent that.

RATNER: In answer to your first question, I believe ^{11}C was then in short supply. We were using large animals and would require large amounts. I know the paper you referred to by Kamin and Rubin, a crucial experiment, but one requiring little isotope.

In answer to your other question, there was indeed speculation. In the leucine paper, which I have not had time to quote from, there was a passage that has attracted attention. Schoenheimer stated very clearly that there were the two simple alternatives: that is, either total synthesis de novo, or synthesis by transpeptidation, breakdown into polypeptide fragments and then recombination by transpeptidation. He gave the alternatives but abstained from making a choice. Since he referred to a review in which Bergmann suggested protein synthesis by transpeptidation (based on the work of Fruton in Bergmann's laboratory), the reference has sometimes been taken as an expression of preference. I point out that at that time there were no other theories on protein synthesis to refer to. I might mention that the leucine paper was published by the three of us in 1940, and Dr. Lipmann's very provocative review in *Advances in Enzymology* suggesting that protein synthesis could be driven by phosphate bond energy through amino acid carboxyphosphate formation did not come out until 1941. Dr. Lippmann was in Boston; Schoenheimer was in New York. I am not sure that they ever met each other, or, if they had, whether they would have discussed this problem. I think they very well might have.

EDSALL: I might remark in regard to the use of ^{11}C that that isotope has a half-life of 20 minutes, which makes it a rather difficult thing to work with. It was once used in studies on carbohydrate metabolism by Hastings and a group of his collaborators in 1940 and 41, but they had to work awfully fast to get their results. And ^{14}C, with its very long half-life had great advantages, when it became available.

I would like to ask one point about your state of mind at the time when this work began. You mentioned that the results on the amino acid incorporation came as a great surprise. On the other hand did you not already have serious doubts about the theory of separate exogenous and endogenous metabolism in view of Schoenheimer's findings on the dynamic state of the fatty acids? Were you not rather suspicious of the Folin theory already even before you began?

I think I should mention that in the biographical memoirs of the Royal Society André Lwoff has just published a biography of Jacque Monod and gives quite a fascinating account. In this he refers to Monod's work with David Hogness and Melvin Cohn on the lack of turnover in their studies of proteins of *E. coli* and their belief that this disproves the concept of the dynamic state. Lwoff apparently still holds to that view in his account from Monod's work, and he said that of course objections were raised to this work by Monod but, as he put it, the dead god could not be revived, and the idea of the dynamic state went down to the grave. A very dramatic statement, but I think it was in contradiction to many facts that have been found by other workers in recent years.

FRUTON: The Monod idea is a very good example I think of the way in which certain patterns of thought can on occasion have a national flavor, because in the writings of Claude Bernard there is a repeated reference to what he called eventually, *vitalité dual*, in which he emphasized the distinction between the degradation of substances like albumin and starch, which he considered to be processes independent of life, and the formation of such materials. Obviously he referred to glycogen in the first place but also included albumin. These were processes that were indissolubly linked to life and this *vitalité dual* reappeared again and has reappeared on many occasions, most recently in that particular case.

COMMENT: Novikoff and Holtzman have commented regarding the teleological reason for such a wasteful process as turnover. I think it is a little bit more general in your notion. They suggested that turnover may partly reflect the fact that enzymes and other molecules are subject to spontaneous changes that permanently inactivate them. Hence processes for their degradation and replacement seem advantageous for the cell's economy.

RATNER: To answer Dr. Edsall's question, it would have been entirely in accord with our expectations at that time if we had found the isotopic amino acid administered incorporated as such in tissue proteins. But to find instead that the ^{15}N of the administered amino acid was present in almost every amino acid isolated in addition to the one administered was altogether surprising and unexpected. A few further experiments showed us that what we had found was in accord with a new, more general expectation that most amino acids were replaced.

DAVID SHEMIN received his Ph.D. at Columbia in 1938. He remained at the Department of Biochemistry at Columbia until 1968 when he moved to Northwestern University. He is Chairman of the Department of Biochemistry and Molecular Biology at the Evanston Campus. His contributions include the discovery of the precursors of porphyrin, the discovery of δ-aminolevulinic acid and the relationship of porphyrin synthesis to the tricarboxylic acid cycle.

The Role of Isotopes in the Elucidation of Some Metabolic Processes*

DAVID SHEMIN

Department of Biochemistry and Molecular Biology
Northwestern University
Evanston, Illinois 62201

IN 1935, the year in which stable isotopes were introduced by Rudolf Schoenheimer as a tool to investigate cellular reactions, very little was known concerning the metabolism of amino acids. This was especially true in the area concerned with the biochemical pathways and reactions involved in the *de novo* synthesis of amino acids and their conversion to other nitrogenous cellular constituents. What was known about the metabolism of some amino acids was mainly deduced from the structural similarities of the amino acids to that of the possible product. The brilliant work of Garrod,[1] who related the metabolism of tyrosine and genetic mutations, was indeed helped by the structural relationships of tyrosine to the aromatic compounds that accumulated because of the lack of a functioning enzyme. In this context it is no surprise to find that arginine was considered to be the direct precursor of guanidinoacetic acid, that proline, pyrrolidine carboxylic acid, or the indole ring of tryptophane was suggested as a likely precursor of the pyrrole structure of porphyrins, and that histidine was a likely precursor of purines. The role of the simplest amino acid, glycine, in the elaboration of these somewhat more complicated molecules was not at all predicted, nor were the concepts of biosynthesis sufficiently developed for one to theoretically suggest the possible role of glycine or other apparently related amino acids in the synthesis of these compounds.

Part of the difficulty of conceiving the possibility that the synthesis of ring structures from aliphatic amino acids was the lack of appreciation that synthesis of so-called complicated structures from simple and readily available compounds is more readily understandable when one considered the formation of prebiotic compounds. Furthermore, the mecha-

* Over the years this research was supported by grants from the American Cancer Society on the recommendation of the Committee on Growth of the National Research Council, from the Rockefeller Foundation, from the National Institutes of Health, United States Public Health Service, and from the National Science Foundation.

0077-8923/79/0325-0211 $01.75/0 © 1979, NYAS

nistic difficulty of modifying ring structures by addition of groups also was not appreciated.

I mentioned above that glycine is utilized for the synthesis of creatine, purines, and porphyrins. Since this symposium has been organized to review past events, I thought it best that I deal with one area in which I was personally involved. Therefore, I plan to recall the events that led to the elucidation of porphyrin synthesis, rather than discuss the work of Konrad Bloch in creatine synthesis and John Buchanan's elaboration of purine synthesis.

Dr. Ratner has just given a warm account of the involvement of the laboratory at Columbia University on the concept of the dynamic state of body constituents and especially on protein turnover. In 1943 I undertook, together with David Rittenberg, an elaborate study on "Some Interrelationships in General Nitrogen Metabolism" in the rat.[2] This involved the labeling of the rat proteins with ^{15}N after the feeding of ^{15}N-labeled glycine. Having finished this study on a rat, we thought it might be some some interest to study the turnover of plasma proteins in man. To this end I synthesized 66 g of glycine labeled with 35 percent ^{15}N. On February 12, 1945, I started the intake of the glycine. Since one really did not know the effect of relatively large doses of ^{15}N and since we believed that the maximum incorporation of ^{15}N would be achieved by the ingestion of glycine in some continual manner, I ingested 1 g of the glycine at hourly intervals.

At stated intervals, blood was withdrawn and the ^{15}N concentrations of different fractions were determined. Since the utilization of glycine for heme synthesis was not anticipated, there appeared little urgency to prepare hemin samples and hemoglobin samples were merely stored. The ^{15}N concentrations of the plasma proteins were determined and the data obtained were consistent with the concept of the dynamic state of body constituents.[3] The ^{15}N concentration reached a maximum five days after the start of the experiment and then declined. The half-life of the plasma proteins were estimated to be less than five days. However, the ^{15}N concentration of the heme continued to rise and appeared to have reached a maximum value about 20 days after the start of the experiment. Since the data thus far were really not startling, the taking of blood samples became less frequent after the 18th day. Blood samples were taken on the 28th, 38th, 60th, and 77th day. The ^{15}N concentration of this latter sample was found to be similar to that obtained on the 18th day. Since Rittenberg and I were mentally programmed to the concept that body constituents are in a state of flux, we were astonished to find that the ^{15}N concentration on the 77th day was not appreciably lower than that on the 18th day. After a serious discussion in which each accused the other either for a faulty analysis or a contaminated sample, we analyzed another sample and found that the ^{15}N concentration was the same as the previous one. We then realized that hemoglobin, in contrast to all other proteins previously examined by these techniques, was not in the dynamic state and that its ^{15}N concentration reflected the life span of the red cell.

On continuation of the determination of the ^{15}N concentrations, the curve in FIGURE 1 was obtained. An analysis of this curve revealed that the average life span of the human red cell is about 127 days rather than the literature value of 30 days.[4] Furthermore, from the data one can readily conclude that glycine is the nitrogenous precursor of heme. For example, 10 days from the start of the experiment the ^{15}N concentration of the heme was 0.34 atom percent ^{15}N excess. Since the average lifetime of the red blood cell is about 127 days, approximately one-thirteenth of the cells were at this time newly formed and contained the isotopically labeled heme. These newly formed red cells must have contained heme with an average ^{15}N concentration 13 times as high as the total heme, or 4.4 atom percent ^{15}N excess (i.e., 13 × 0.34). The nitrogeneous source

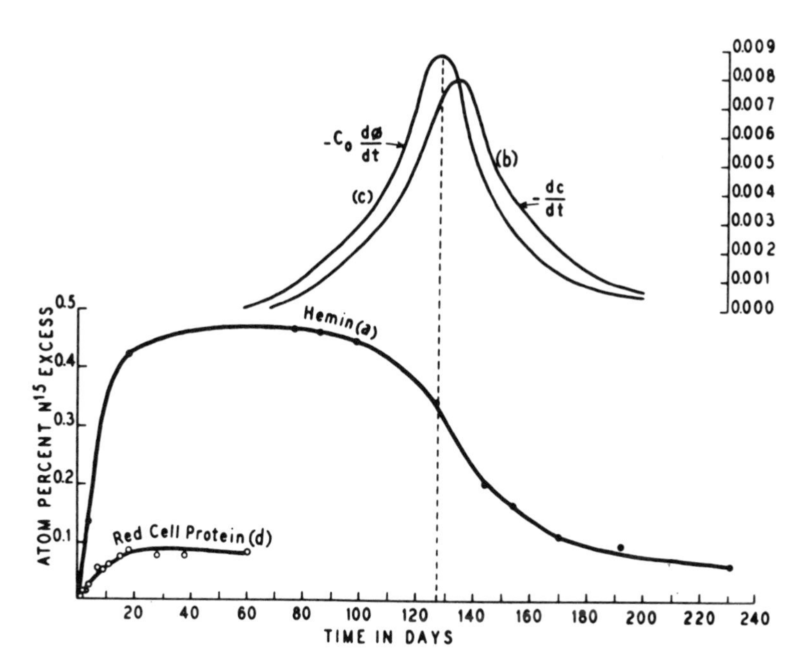

FIGURE 1. Concentration of ^{15}N in hemin after feeding ^{15}N-labeled glycine for 3 days.

of the heme in this 10-day period, therefore, must have had an average ^{15}N concentration of 4.4 atom percent ^{15}N excess. It was clear from these considerations that glycine, the isotopic amino acid fed, was the only compound that could have had as high an average ^{15}N concentration for the first 10 days. All other possible compounds (NH_3, glutamic acid, etc.) were eliminated by analysis of the ^{15}N concentration of plasma proteins of some of its amino acids and that of the urea nitrogen. Unless a special transamination occurred, we assumed that the nitrogen of glycine was utilized along with its carbon atom or atoms. Further support for the utilization of glycine for porphyrin synthesis was obtained by comparing the ^{15}N concentration of the heme of rat hemoglobin after the incorporation of a number of ^{15}N-labeled amino acids.[5] It is of interest to note that Hans Fischer in his last paper[6] found that formylacetone condenses with glycine to form a pyrrole. In this paper, Fischer speculates about the possibility of glycine as the nitrogenous precursor of porphyrins.

Later, in 1947, I worked in Einar Hammarsten's laboratory in the Karolinska Institute and, among other things, we used the technique I just described to determine the life span of the red blood cell of the chicken which was found to be about 28 days. On my return to Columbia, I sought a simpler system to continue our studies on heme synthesis. Probably my experience in Sweden led me to the suggestion that perhaps a nucleated red cell was capable of converting glycine to heme *in vitro*. On consultation with Z. Dische, we obtained a duck and found immediately that the duck red blood cell readily incorporated glycine into heme.[7] Furthermore, we found that globin synthesis occurred as well. Meanwhile, our new colleague, Irving London, who was measuring the life span of the red blood cell of patients with blood dyscrasias carried out a similar experiment with the blood of these patients and found that glycine was utilized readily for heme synthesis by the blood of patients with sickle-cell anemia. Although the reticulocytosis was similar in all these disorders, only the blood of the patients with sickle-cell anemia was capable of synthesizing heme from the glycine.[8] Apparently the sickle-cell blood contained more immature red cells than the blood of the patients with other anemias. This lack of correlation of heme synthesis with the reticulocyte count was borne out when London then introduced the use of immature reticulocytes of the rabbit as a biological system to study heme synthesis.[9]

We then turned our attention to the question of the utilization of the carbon atoms of glycine for porphyrin synthesis. Not only did we wish to determine which of the carbon atoms of glycine were utilized for heme synthesis, but also the particular position in the porphyrin molecule in which the carbon atoms of glycine are found. This became of extreme importance when Norman Radin in the laboratory incubated duck red cells with $^{15}NH_2$-$^{14}CH_2$-COOH and found that for each nitrogen atom, two α-carbon atoms of glycine were incorporated into the protoporphyrin.[10] Therefore, it appeared that eight carbon atoms of the porphyrin molecule arise from the α-carbon atom of glycine. Furthermore, earlier Grinstein, Kamen, and Moore found that the carboxyl carbon atom of glycine is not utilized for heme formation.[11] In order to get some clue as to the mechanism of glycine utilization, Jonathan Wittenberg set out to degrade the heme in such a manner that one can isolate unequivocally each carbon atom from a particular position in the porphyrin.

On degradation of protoporphyrin synthesized from [2-^{14}C]glycine, it was found that indeed eight carbon atoms are derived from the α-carbon atom of glycine and the positions were located: four bridge carbon atoms and one carbon atom in each of the four rings.[12] It was found that the carbon atom in the pyrrole rings, derived from the α-carbon atom of glycine, were in the α-position under the vinyl and propionic acid side chain. This suggested that a common precursor pyrrole for all four rings is first formed and that the vinyl groups arose from the propionic acid side chains by dehydrogenation and decarboxylation.

Since the α-carbon atom of glycine was found to be the source of eight carbon atoms of protoporphyrin, the source of the remaining 26 carbon atoms remained to be determined. Just prior to our finding that glycine was the nitrogeneous source of porphyrins, Konrad Bloch isolated hemin from a rat who had ingested deuterioacetic acid.[13] He found deuterium in the hemin and, although none of the carbon atoms of the pyrrole ring are bonded to hydrogen, it appeared that acetic acid may be involved. Jonathan Wittenberg then divided a batch of duck erthrocytes in half and added to one half ^{15}N-labeled glycine and [2-^{15}C]acetate and to the other half ^{15}N-labeled glycine and [1-^{14}C]acetate. The glycine was added in order to measure the degree of synthesis in each flask, and thus we could relate the data obtained with methyl-labeled acetate to that obtained with carboxyl-labeled acetate. It was found that all the remaining 26 carbon atoms were indeed derived from acetate.[14]

The distribution of the ^{14}C activities among the remaining twenty-six carbon atoms of the protoporphyrin allowed us to draw the following conclusions: (1) the acetic acid is utilized by its conversion to a four carbon atom unsymmetrical compound; (2) that each of the pyrrole is derived from the same four-carbon-atom compound; (3) that the common precursor pyrrole contained acetic and propionic acid side chains in the β-position; and (4) that the methyl groups in protoporphyrin arise by decarboxylation of the acetic acid side chains and that the vinyl groups arise by dehydrogination and decarboxylation of the propionic acid side chains.

The conclusions were arrived at mainly from the finding that comparable carbon atoms had the same ^{14}C activity in each of the experiments[14] (FIGURE 2).

Casting about for a mechanism to explain these data, it appeared that perhaps this distribution would occur if this four-carbon atom arose from an intermediate of the tricarboxylic acid cycle. Relative quantitative calculations concerning the distribution of the carbon atoms of acetate in intermediates of the tricarboxylic acid cycle had not previously been done. On the assumption that extraneous metabolic reactions would not obscure the theoretical distribution of the carbon atoms of acetate in members of the tricarboxylic acid cycle, we postulated the theoretical labeling pattern in the four-carbon-atom unsymmetrical compound that would result from the introduction of acetate into the cycle after each complete turn of the cycle. If one ignores any endogeneous dilution, one can calculate the relative distribution of ^{14}C activity in the four-carbon-atom compound that would theoretically be found each turn of the cycle after the entrance of the acetic acid. Let us first examine the result one could obtain from methyl-labeled acetate. Assuming a relative

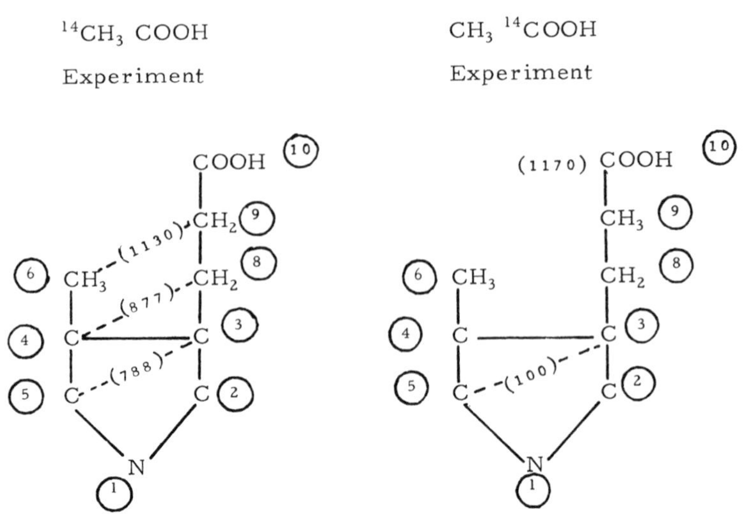

FIGURE 2. Labeling patterns found in each of the experiments. The ^{14}C activities are given in parentheses. The pyrrole unit represented contains a carboxyl group which is found only in two rings of protoporphyrin. Carbon atom numberings are encircled.

^{14}C activity of 10 in the methyl group of the acetic acid, the α-ketoglutaric acid formed the first turn of the cycle should have the relative ^{14}C distribution shown in TABLE 1. On formation of symmetrical succinate, the ^{14}C activity of the γ-carbon atom of the α-ketonglutaric acid, 10 cpm, would be distributed between the two methylene carbon atoms and thus their relative activities would be 5 cpm and 5 cpm. The newly formed oxaloacetate would therefore have this same distribution of radioactivity. The condensation of this newly formed oxaloacetate with more of the methyl-labeled acetate would give rise to α-ketoglutaric acid with the distribution of radioactivity shown in the second cycle. After each turn of the cycle the radioactivities of the γ and β carbon atom would increase by the same increment and approach that of the methyl group of the acetate or that of the γ-carbon atom of the α-ketoglutaric acid. Therefore, a four carbon atom unsymmetrical compound formed from α-ketoglutarate would contain three adjacent radioactive carbon atoms; the carbon atom adjacent to the carboxyl group being most radioactive, whereas the other two carbon atoms should have radioactivated somewhat less, but equal to each other. Let us compare the ^{14}C distribution theoretically predicted with that found after a finite number of cycles. The four-carbon-atom compound (FIGURE 2) utilized for pyrrole formation synthesized from methyl-labeled acetate does indeed have three adjacent carbon atoms arising from the methyl group of acetate; the one next to the carboxyl group has the highest activity and the other two carbon atoms although labeled are not quite equally labeled (877 cpm compared to 788 cpm). This slight inequality of activities

TABLE 1
RELATIVE DISTRIBUTION OF ^{14}C ACTIVITY IN CARBON ATOMS OF α-KETOGLUTARIC ACID RESULTING FROM UTILIZATION OF ^{14}C-LABELED ACETATE IN THE TRICARBOXYLIC ACID CYCLE*

α-Ketoglutaric Acid	From ^{14}C-Methyl-Labeled Acetate (Activity of Methyl Group = 10 cpm)				From ^{14}C-Carboxyl-Labeled Acetate (Activity of Carboxyl Group = 10 cpm)			
	No. of Cycles in Tricarboxylic Acid Cycle							
	1	2	3	∞	1	2	3	∞
COOH	0	0	0	0	10	10	10	10
CH_2	10	10	10	10	0	0	0	0
CH_2	0	5	7.5	10	0	0	0	0
C = O	0	5	7.5	10	0	0	0	0
COOH	0	0	2.5	5	0	5	5	5

*Results expressed in counts per minute.

would suggest that the tricarboxylic acid cycle is not functioning as postulated theoretically. However, a valid correction can be made so that these carbon atoms do indeed have the same radioactivity or arise equally from the methyl group of acetate: It can be seen from FIGURE 2 that carbon atoms 5 and 3 are in part also derived from the carboxyl group of acetate. In the experiment in which carboxyl-labeled acetate was the substrate, the carboxyl carbon atom labeled these same carbon atoms to the extent of 100 counts. Therefore, in the methyl-labeled acetate experiment carbon atoms 5 and 3 were diluted by the unlabeled carboxyl group by 100 counts. The correction by 100 counts to carbon atoms 5 and 3 raises the value to 888, a figure close to 877. It appeared, therefore, that acetate is utilized via the tricarboxylic acid cycle for the relative ^{14}C activities found were in excellent agreement with those theoretically predicted in the four-carbon-atom unsymmetrical compound derived from α-ketoglutarate.

A similar ^{14}C distribution pattern can be postulated with the rise of carboxyl-labeled acetate. It can be seen from TABLE 1 that the four-carbon-atom compound arising from carboxyl-labeled acetate would contain radioactivity only in its carboxyl group after one or an infinite

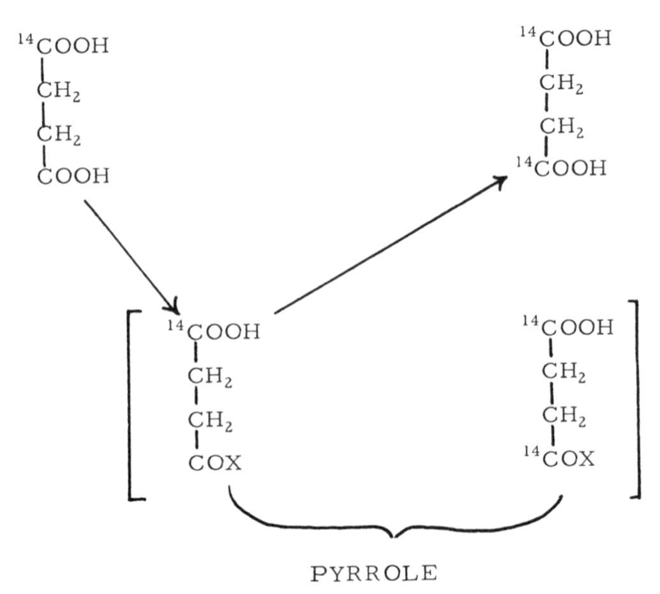

FIGURE 3. The formation of the 4-carbon-atom unsymmetrical compound from both α-ketoglutarate and from succinate. The ^{14}C distribution would suggest that in this biological system 90% of the succinyl derivative arises from α-ketoglutarate.

number of cycles. It can be seen from FIGURE 2 in which the labeling pattern arising from carboxyl-labeled acetate is shown that the four-carbon-atom compound is similar to that postulated in that most of the activity is in the carboxyl group. However, there was about 10 percent of the activity of the carboxyl group in the other terminal carbon atom. In order to explain this slight difference between the predicted ^{14}C distribution and that distribution of ^{14}C found, we postulated that perhaps the four-carbon compound can arise mostly from unsymmetrical α-ketoglutarate and in a lesser part from a symmetrical compound like succinic acid. In 1951 we suggested that the four carbon assymmetric compound was most likely a succinyl-coenzyme compound probably similar to acetyl-CoA. At that time succinyl-CoA was not known nor was it formation from succinate. However, if the four-carbon-atom compound did arise from both α-ketoglutaric and succinate, the ^{14}C distribution would be readily explainable. This postulated mechanism is illustrated in FIGURE 3.

The postulated relationship of the tricarboxylic acid cycle and porphyrin formation was then experimentally tested with labeled succinate. The relationship is shown in FIGURE 4. This scheme suggested the existence of a reaction that had not been described, namely, Reaction C, and furthermore suggested that this aspect of the tricarboxylic acid cycle could be studied by investigating porphyrin formation. This scheme postulated that the succinyl intermediate can be derived from both α-ketoglutarate and from succinate. This can be experimentally tested

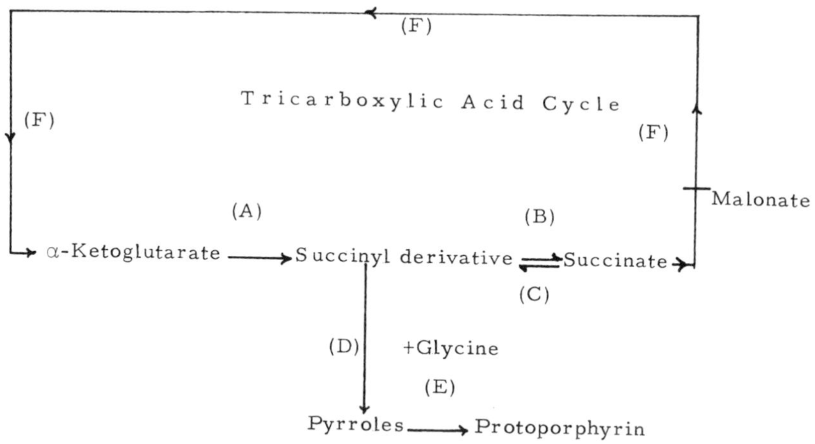

FIGURE 4. The relationship of the citric acid cycle and protoporphyrin formation.

by determining the labeling pattern in the protoporphyrin synthesized from either carboxyl-labeled succinate or methylene-labeled succinate and the utilization of the differently labeled succinates for porphyrin formation in the presence or absence of malonate. Theoretically ^{14}C-carboxyl-labeled succinate cannot give rise to ^{14}C-labeled protoporphyrin via Reaction F, but only via Reaction C. On entering the tricarboxylic acid cycle, i.e., in the oxidative direction of the cycle (Reaction F) carboxyl-labeled succinate should give rise to α-carboxyl-labeled α-ketoglutarate. The resulting succinyl intermediate arising by oxidative decarboxylation and utilized for porphyrin formation would therefore contain no ^{14}C. However, if Reaction C occurs, carboxyl-labeled succinate should produce labeled protoporphyrin. If this postulation is correct, then carboxyl-labeled succinate should produce labeled protoporphyrin only by Reaction C, and therefore malonate, which blocks Reaction F, should have little or no influence on the ^{14}C activity of protoporphyrin.

Methylene-labeled succinate, in contrast to carboxyl-labeled succinate, should produce labeled protoporphyrin via two pathways: (1) Reaction C, and (2) the oxidative direction of the tricarboxylic acid cycle (Reaction F). Methylene-labeled succinate should produce α-ketoglutaric acid labeled in its α- and β-carbon atoms by Reaction F. As a result the succinyl intermediate formed by this pathway, and utilized for porphyrin formation, should have 2 of its carbon atoms labeled with ^{14}C. Malonate should, in this case, have a marked inhibitory effect on the utilization of methylene-labeled succinate for porphyrin formation, and any ^{14}C activity in protoporphyrin made from methylene-labeled succinate in the presence of malonate will be the result of Reaction C.

The ^{14}C activity of the hemin samples obtained by incubating equal amounts of ^{14}C-carboxyl-labeled and ^{14}C-methylene-labeled succinate, with and without malonate, were determined. It was found that the ^{14}C activity of the hemin samples made from methylene-labeled succinate is much higher than that made from carboxyl-labeled succinate and, whereas malonate had a large inhibitory effect on the utilization of methylene-labeled succinate, it had essentially no inhibitory effect on the utilization of carboxyl-labeled succinate. These results were in complete agreement with the postulation that carboxyl-labeled succinate produces labeled protoporphyrin by Reaction C only and that methylene-labeled succinate forms labeled protoporphyrin by Reactions C and F.[15]

Further evidence for the existence of Reaction C and proof for the utilization of a four-carbon-atom compound for porphyrin formation were obtained by investigating the ^{14}C distribution in protoporphyrin synthesized from carboxyl-labeled succinate.. If Reaction C occurs and

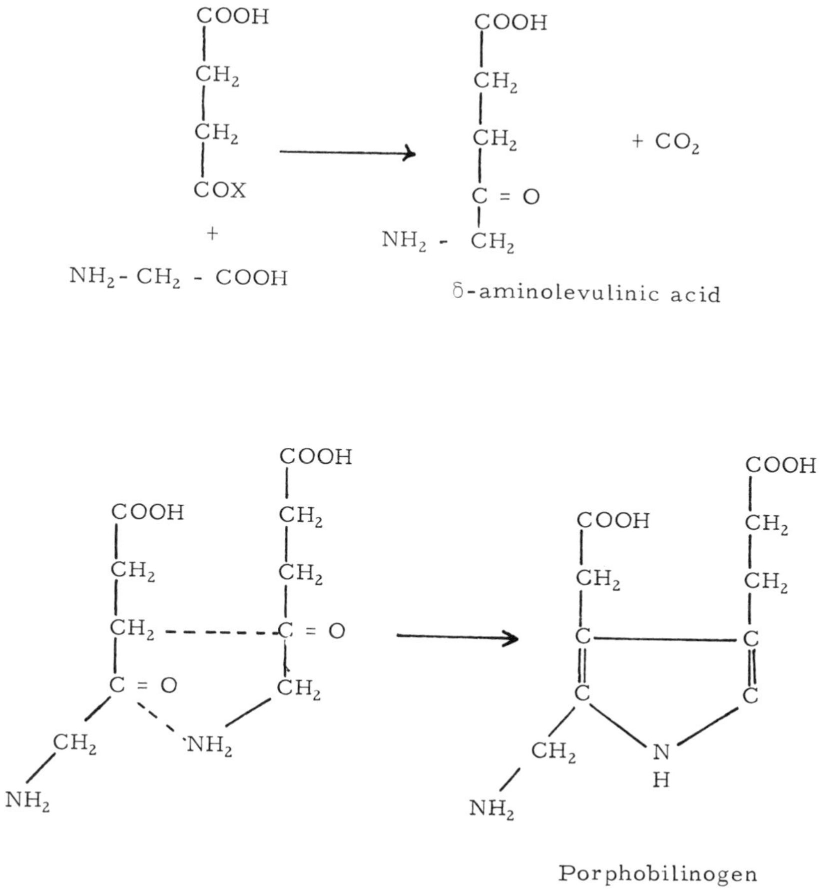

COOH
CH_2
CH_2
COX
+
NH_2 - CH_2 - COOH

→

COOH
CH_2
CH_2
C = O
NH_2 - CH_2
+ CO_2

δ-aminolevulinic acid

COOH
CH_2
CH_2
C = O
CH_2
NH_2

COOH
CH_2
CH_2
C = O
CH_2
NH_2

→

COOH
CH_2
C
C
CH_2
NH_2

COOH
CH_2
CH_2
C
C

N
H

Porphobilinogen

FIGURE 5. The formation of δ-aminolevulinic acid and its conversion to porphobilinogen.

if the succinate is utilized as a unit, carboxyl-lebeled succinate would not only give rise to labeled protoporphyrin, but the resulting porphyrin should have a particular ^{14}C activity pattern. The predicted ^{14}C activity pattern was found experimentally.

Having established that two molecules of a succinyl derivative and two α-carbon atoms of glycine and one nitrogen atom of glycine were involved in the formation of a pyrrole, we considered the possible intermediate. Since the carboxyl carbon atom of glycine is not utilized for porphyrin synthesis, a mechanism by which this carboxyl group is lost was sought. If the succinyl derivative condensed onto the α-carbon atom of glycine, a β-keto acid would be formed and would readily decarboxylate. Furthermore, the resulting aminoketone could readily undergo a Knoor type of condensation to yield a pyrrole. This pyrrole would be the precursor pyrrole for porphyrins and the porphyrin thus formed would contain the succinate and glycine carbon atoms in the positions previously established (FIGURE 5). To this end, we immediately developed methods to synthesize the δ-aminolevulinic acid with ^{15}N and ^{14}C and showed it to be the source of all the atoms of protoporphyrin.[16, 17]

While the experiments were in progress Westall[18] in England isolated the compound called porphobilinogen by Waldenstrom, from the urine of patients with acute porphyria. Our precursor pyrrole had the same empirical formula as that of Westall and very shortly Cookson and Rimington[19] published the structure of porphobilinogen. The δ-aminolevulinic acid also proved to be the source of all the atoms of the corrin ring of vitamin B_{12}.[20]

REFERENCES

1. GARROD, A. E. 1923. Inborn Errors of Metabolism. 2nd edit. Henry Frowde and Hodder and Stoughton, Publishers. London.
2. SHEMIN, D. & D. RITTENBERG. 1944. J. Biol. Chem. *153*: 401.
3. SCHOENHEIMER, R. 1941. The Dynamic State of Body Constituents. Harvard Press. Cambridge, Mass.
4. SHEMIN, D. & D. RITTENBERG. 1946. J. Biol. Chem. *166*: 627.
5. SHEMIN, D. & D. RITTENBERG. 1946. J. Biol. Chem. *166*: 621.
6. FISCHER, H. & E. FINK. 1944. Z. Physiol. Chem. *280*: 123.
7. SHEMIN, D., I. M. LONDON & D. RITTENBERG. 1948, 1950. J. Biol. Chem. *173*: 799, *183*: 757.
8. LONDON, I. M., D. SHEMIN & D. RITTENBERG. 1948. J. Biol. Chem. *173*: 797.
9. LONDON, I. M., D. SHEMIN & D. RITTENBERG. 1950. J. Biol. Chem. *183*: 749.
10. RADIN, N. S., D. RITTENBERG & D. SHEMIN. 1950. J. Biol. Chem. *184*: 745.
11. GRINSTEIN, M., M. D. KAMEN & C. U. MOORE. 1948. J. Biol. Chem. *174*: 767.
12. WITTENBERG, J. & D. SHEMIN. 1950. J. Biol. Chem. *185*: 103.
13. BLOCH, K. & D. RITTENBERG. 1945. J. Biol. Chem. *159*: 45.
14. SHEMIN, D. & J. WITTENBERG. 1951. J. Biol. Chem. *192*: 315.
15. SHEMIN, D. & S. KUMIN. 1952. J. Biol. Chem. *198*: 827.
16. SHEMIN, D. & C. S. RUSSELL. 1953. J. Am. Chem. Soc. *75*: 4873.
17. SHEMIN, D., C. S. RUSSELL & T. ABRONSKI. 1955. J. Biol. Chem. *215*: 613.
18. WESTALL, R. G. 1953. Nature *170*: 614.
19. COOKSON, G. H. & C. RIMINGTON. 1953. Nature *171*: 875.
20. BROWN, C. E., J. J. KATZ & D. SHEMIN. 1972. Proc. Nat. Acad. Sci. U.S.A. *69*: 2585.

DISCUSSION OF THE PAPER

KARLSON: You mentioned that very early during this work you had contacted medical people measuring the length of lifetime of blood cells.

I wonder if you ever came across the idea that porphyrias, or patients with porphyria, might be a good idea for studying the biosynthesis of porphyrins because these people excrete very large amounts and these could be isolated in a not too difficult way.

SHEMIN: Yes, except, in general, porphyrias are very rare. I did not have, as Hans Fisher had, a laboratory assistant from whom I could work up the porphyrins. But we chose subsequently an even simpler system when we went to bacteria. Photosynthetic bacteria have lots of good enzymes and one can study a lot of details, especially with *Rhodopseudomonas spheroides*. After taking a course with van Niel, I became interested in bacteria. Acute porphyria is also fairly rare but it is more prevadent in South Africa.

KIRSHENBAUM: Some 25 years or so ago I had a course with Dr. Shemin on advanced metabolism at Columbia. We all knew that Dr. Shemin would cover the lecture material because we had all heard the first lecture and it went quite rapidly. We knew he was brave because he was the subject of his experiments. But more than anything else, we knew he was a brilliant scientist because he could draw the whole structure of the porphyrin without stopping, putting in the methyl-vinyl, methyl-vinyl, methyl-propyl-propyl-methyl, all the double bonds, and the nitrogens in the right place all without one erasure. That convinced us after one lecture that we had the most brilliant man in the staff of Columbia.

PORTUGAL: This is in regard to some work that was going on about the same time at the Carnegie Institution of Washington, which I currently represent.

I just wanted to mention that the Department of Terrestrial Magnetism associated with the Carnegie Institution was one of the first to confirm the existence of tritium, which of course today is very important in amino acid metabolic studies. And their Plant Biology Department was one of the first to use carbon-11, which we talked about in the last discussion, in one of the earliest metabolic studies. The same department was also instrumental in some experiments which turned out, without their realizing it, to have some very important commercial value. The plant biology group was studying the growth of *Chlorella* and found that you could alter the environmental conditions and get *Chlorella* to reach a very high protein level.

They also found that you could use radioactive carbon dioxide to label the protein and by hydrolyzing it you would get a very good source of carbon-14-labeled amino acids at a time when carbon-labeleld amino acids were not available commercially. It is interesting that today many companies use that process commercially for furnishing isotopes to laboratories involved in investigation.

The Department of Terrestrial Magnetism had a biophysics section, which represented one of the earliest syntheses of people in physics and people in biology trying to work on current problems in metabolic studies. Using amino acid isotopes that group pioneered in the use of isotopic competition for tracing metabolic pathways. They showed in *E. coli*, for example, that the Krebs cycle was actually used for the production of amino acids rather than for oxidation and, in the same studies, gathered the initial evidence that in amino acid synthesis you had feedback inhibition or feedback control.

SHEMIN: I was going to bring that up, the work of R. B. Roberts, P. H. Abelson, D. B. Cowie, E. T. Bolton, and R. J. Britten.

FRUTON: I wonder if I may be permitted to add a note here which is implicit in what Dr. Shemin said and the way he told the story and that is that there was a large background of basic organic chemistry in the work of people like Nencki, Willstätter, and Küster and most prominently Hans Fischer, which was quite useful to him, especially in the development of the degradation scheme I suspect. I remember the conversations that I had with you about a paper by Hans Fischer, but I am never sure to what extent that possible mechanism was in your mind during the course of the unraveling of the labeling pattern.

SHEMIN: Well, first let me make a comment. Their chemistry did help, especially for the formation of the hematinic acid and the methylethyl maleimides. However, J. Wittenberg worked out the complete degradation of these compounds.

I did think of trying to make acetoacetic glycine. Well, I must admit you cannot readily make acetoacetyl chloride; it is very unstable. So I looked up a mechanism and found that if I made diketene it would give me acetoacetic glycine. I made a diketene, but I did not realize how explosive it could possibly be; we never had any accidents fortunately. I made diketene by polymerizing ketene, which was made by pyrolysis of acetone. I added diketene to glycine and in a few minutes had a beautiful crystalline material. I was very happy. I analyzed it, twisted it one way and the other, and thought we might have a pyrrol. It turned out to be a pyridine derivative.

D. SPRINSON (*Columbia University, New York, N.Y.*): Theories and concepts undoubtedly stimulate experiments. I have no doubt at all that this is true of many experiments in physics and occasionally in biology. However, deterium and ^{15}N were used in the study of intermediary metabolism by Schoenheimer and his colleagues mostly owing to the propitious availabillity of these isotopes. I do not think there was too much concern about proof for the dynamic state of body constituents. The concept of the dynamic state was supported by these experiments. Radioactive lead was used as a tracer by Hevesy and tritium was used by him to determine body water some years before. The unusual juxtaposition of Urey, Rittenberg, Schoenheimer, Foster, Ratner, and Shemin lead to the rapid recognition of the dynamic state and the use of isotopes in the study of intermediary metabolism.

SHEMIN: Well, the discovery of the pathway from glycine to prophyrin was just an accident. Similarly I think the same was true for the pathway from glycine to the purines. I talked to John Buchanan about it, but when he first did the experiment he had some concepts that may have been wrong. But nevertheless, what came out with a good degradation was then the beginning of purine synthesis.

Once you have the initial finding then you have to have some concepts, and sometimes when the concepts are there they hold you back. For example, we did not realize that hemoglobin was not in a dynamic state and did not keep to that line of research very long. But nevertheless, I think that the paper was sent in before that and I think we had to change a sentence over the phone.

RATNER: I think that the dynamic state of body constituents followed a somewhat different development. I have a feeling that some of the old hypotheses about nitrogen metabolism were being questioned and that the opportunity of using isotopes gave one a chance to test a hypothesis. One could identify the most common molecules, nitrogen, oxygen, hydrogen, the body is made up of. I would not take your point of view. Perhaps in some of the later stages, but I doubt that point of view started things off.

FRUTON: Dr. Ratner, may I expand just a little on your point in regard to Dr. Sprinson's remark? Is it fair to say that in the early days there were of course isotopes that became available, and then the choice of what kind of experiments to do was influenced by the kinds of hypotheses that were already around in regard to pathways of metabolism. So that it was natural to ask about creatin synthesis, for example, and other processes alike, because here was a tool for testing. As you said, one tested hypotheses that were around, and in the case of the Folin hypothesis, the hypothesis was disproved.

SHEMIN: May I just amplify this one thing? When I took the [^{15}N]-glycine, I also saved all my urine; this was before Dr. Buchanan's work. If I had the hypothesis of glycine going to uric acid I would have isolated the uric acid very quickly.

KARLSON: I doubt that the idea of the dynamic state of body constituents really came up only when isotopes were available. I draw your attention to Schoenheimer's old work in Freiburg where he investigated cholesterol biosynthesis by studies of the balance of what you have before and after some time. I think it was known at that time that cholesterol was converted to bile acids and therefore a certain amount of cholesterol must have been lost. And moreover, another point: some dynamic state of the proteins was already visible in the excretion of urea under fasting conditions, or under conditions of inadequate protein uptake where you lose a certain amount. You are out of nitrogen balance. And this observation must have lead to the idea that under these conditions, though the other nutrients are present, the state of the proteins, the cell constituents, is labile and is in a dynamic state.

Another point, many achievements were made through concepts that have been tested by experiments. But others are just discoveries. I would point out such a discovery was the finding just explained that you find in hemin a large amount of ^{15}N, which was unexpected, and this opened up the way for the whole series of studies not anticipated and not planned properly. But in the moment you have this discovery you see that you can enter that field, I think this was very great.

HOTCHKISS: In response to Sprinson's comment that the tools can bring forth the possibility, I would like to remind you that both the tools and the concepts have to interact. I have a very vivid memory of witnessing in graduate school at Yale University the time when deuterium became known, and three very particular conceptual reactions to its announcement within the hour. None of this is represented in the literature but I can vouch for the authenticity.

In the early 1930s, an expert from the Bureau of Standards, Dr. Washburn, lectured in the Department of Chemistry about deuterium and its relative abundance in all the hydrogen of the organic world. This new work stirred a lot of us, especially the organic chemists. Afterwards in a discussion, I went to my organic chemistry teacher, J. J. Donleavy. Aware of the early work of Hevesy labeling with bromine atoms, I said, "Isn't this wonderful? Now you can put hydrogen atoms into organic molecules and study their rearrangement?" (I had no concept of biochemistry at that point.) Donleavy said, the interesting thing to do would be to put hydrogen as one substituent on an asymmetric carbon atom and deuterium as another with two other groups and look for optical activity. Our nearest example of a biochemist was Prof. R. J. Anderson who was then Chief Editor of the *Journal of Biological Chemistry* and who had devoted much of his life to isolating lipids—just such lipids as Schoenheimer later worked with, purifying them from tubercle bacillus and other bacteria. He said, "Do you mean to say that all of those fatty acids and phosphatides that I've been preparing are still impure and randomly contaminated with deuterium?"

NACHMANSOHN: May I make a comment on Dr. Sprinson's remark? In stressing the paramount importance of concepts for the progress of science I do not imply that experimental facts are not essential. On the contrary, even if the concept comes first, it becomes reality only when supported by facts as I mentioned before. This is the basis on which modern science was built. I have frequently attended Planck's lectures on epistemology. He stressed time and again that you can have a vast number of facts, but there are always gaps. To connect the facts and to make a theory requires the imagination and intuition of the scieñtist. Theory can never be based on facts alone. The danger is that some scientists stick to theory until it becomes a dogma: they ignore the facts observed by other scientists. Then it becomes, as Planck called it, a pseudoscience and is bound to collapse. According to an expression of J. J. Thomson, a theory is a tool and not a creed. Einstein revolutionized epistemology when he postulated that all real progress in science is based on concepts and not, as Newton believed, on observation and experience. Newton did not realize that his revolutionary contributions were the result of his unique genius.

HAUROWITZ: In 1937, Hevesy, who was mentioned here a few times, paid a visit to us in Prague. He reported to us about his very interesting experiments with radioactive phosphorus. Deuterium was known already at that time. I asked Hevesy whether he as a physical chemist would not try to produce radioactive carbon. He explained to me in detail that it would never be possible to produce radioactive carbon.

WILLIAM C. ROSE received the Ph.D. degree from Yale University in 1911. Subsequently, he studied with Franz Knoop at the University of Freiburg, Germany. Following relatively brief appointments at the Universities of Pennsylvania and Texas, he joined the staff of the Department of Chemistry, University of Illinois, Urbana. There he remained as Head of the Division of Biochemistry until his retirement in 1955.

During most of his active career, Dr. Rose has been involved with the metabolism and growth significance of the amino acids. In 1934, he discovered, isolated, identified, and established the spatial configuration of the amino acid, threonine. Making use of mixtures of highly purified amino acids in place of dietary proteins, he determined the qualitative and quantitative amino acid needs of animals and adult man. Among other honors, he was a recipient of the 1966 National Medal of Science.

How Did It Happen?*

WILLIAM C. ROSE

Department of Biochemistry
University of Illinois
Urbana, Illinois 61801

ON SEVERAL OCCASIONS, I have been asked how I became interested in the biochemical role of the amino acids. The following is an answer to that question, even though I am fully aware that the source of one's inspiration or motivation may be of very little significance to anyone else. Indeed, as I write I sense an inner feeling of embarrassment lest the reader think that I attach too much importance to this bit of my life experience. In truth, the matter is discussed merely to set the record straight.

From time to time, some of my friends have stated or implied that my interest in amino acids stemmed from the contacts I had with Thomas B. Osborne and Lafayette B. Mendel during the years I was a graduate student at Yale University. This would be a simple explanation, but, perhaps unfortunately, it is totally incorrect. It is true that during the period in question (1907–11), Osborne and Mendel were engaged in their classical investigations of the growth of animals upon diets containing single, purified proteins; but all of their studies were carried out in the laboratories of the Connecticut Agricultural Experiment Station, and did not involve the graduate students in any way. As a matter of fact, we had little knowledge of the progrem until the findings appeared in print.

Perhaps a word about these two remarkable men may be in order for the benefit of those who are not acquainted with their early activities. Osborne, Director of the Station, had succeeded in isolating and purifying several vegetable proteins, particularly those that occur in grains and other seed crops. Some of the proteins, notably edestin, the globulin of hemp seed, were obtained in beautifully crystalline condition. There can be little doubt that Osborne's proteins were the purest that had ever been prepared up to that time. But he was not content merely to isolate and purify the proteins; his curiosity demanded that he establish their composition as well. Available methods for such work were not very

* Professor Rose was invited to the Symposium but unfortunately he was unable to attend; however, he was kind enough to prepare this article which we are happy to include in the *Annals*.

0077-8923/79/0325-0229 $01.75/0 © 1979, NYAS

satisfactory, but Osborne was a remarkable technician. He applied his skill in the use of all known procedures for the separation of amino acids, and improved several of the methods. The basic and dicarboxylic amino acids could be recovered with reasonable accuracy; it was the monoamino acids that presented the greatest difficulties. These were separated into groups by the fractional distillation of their ethyl esters as described by Emil Fischer (1901).[1] The free amino acids were then regenerated, and subjected to further purification by an elaborate process of fractional crystallization. Nothing was thrown away. The final supernatant fluids were combined and reworked. Every effort was made to recover every tiny bit of each amino acid. By running through the same methods with known mixtures of amino acids, Osborne established the loss at each step along the way, and thus was able to correct the findings in the protein analyses. The final results revealed a remarkably clear picture of the composition of each protein, and how it differed in make-up from other proteins in which he was interested. For many years, Osborne's analyses were accepted as reference standards for the composition of vegetable proteins.

The availability of purified proteins, together with reasonably accurate information concerning the distribution of the known amino acids in each, suggested the desirability of testing their comparative efficiencies as sources of nitrogen for growing animals. For this purpose, Osborne and his close friend, L. B. Mendel, pooled their intellectual resources. Mendel, as head of the Department of Physiological Chemistry, as biochemistry was denoted at the time, had already established an enviable reputation as a brilliant investigator and an inspiring teacher. His collaboration with Osborne marked the beginning of a prolonged association that resulted in the joint publication of more than one hundred papers.

It should be recalled that when Osborne and Mendel began their feeding experiments, very little was known concerning the mineral requirements of the growing rat, and even less about the existence and distribution of vitamins, and the needs of animals for these essential factors. It would take us too far to describe the frustrations and seemingly insurmountable difficulties that, for many months, plagued their efforts to formulate adequate rations. Suffice it here to state that by the ingenious use of liberal amounts of butter-fat and "protein-free milk," the dried residue obtained after the removal from milk of the fats and proteins, they succeeded in preparing a mixture of components that would sustain body weight, and permit moderate growth provided a protein of good quality (casein) was included. Subsequently, the authors observed that when zein of corn was the dietary protein, the animals rapidly lost weight, but that the quality of the food was enhanced and growth ensued when trytophan and lysine were incorporated in the ration. Thus, the indispensability of these two amino acids was clearly established. Earlier investigations elsewhere had suggested the possibility that single amino acids might be required by the animal organism, but the experiments of Osborn and Mendel provided the first clear-cut, irrefutable evidence for this fact.

The results of the zein experiments were published in 1914.[2] My first knowledge of them was when I saw the paper in print. By then, I had been away from Yale for three years, and had recently returned from the laboratory of Franz Knoop, University of Freiburg, where intermediary metabolism was the area of choice in biochemical research. Indeed, it was because of Knoop's productivity in metabolic research that I had gone there in the first place; consequently, I fully expected to devote my life to that type of endeavor.

Now that I was back home, and was beginning to become established in and adjusted to a new environment (University of Texas, 1913–22), my thoughts turned to the subject upon which I had worked as a graduate student, namely, creatine-creatinine metabolism. During succeeding years, several aspects of this problem were investigated and published. In the meantime, a widespread impression seemed to have developed to the effect that dietary arginine was the precursor of muscle creatine, and therefore of urinary creatinine. The evidence in support of this concept was contradictory and inconclusive. Some authors reported that an increased arginine consumption induced an exaggerated creatine production. Other investigators were unable to demonstrate any relationship between the arginine intake and the creatine-creatinine output in the urine (cf. Rose[3]).

The postulate upon which these experiments was based appeared to me to be unsound. Since creatine is an anabolic product, and exists in the muscles of a given species in quite constant amounts, one should not expect the administration of an excess of its precursor to increase the amount in the tissues. A more logical approach, I thought, would be to deprive an animal of arginine and see if this would inhibit creatine production and creatinine excretion. It should be recalled that at this period in our knowledge, no one knew that arginine can be synthesized by the animal organism.

In accord with the above reasoning, I decided to prepare a casein hydrolysate and remove the arginine as completely as available methods would permit. However, the plan had to be changed somewhat because of the appearance of a paper by Ackroyd and Hopkins (1916)[4] in which it was alleged that arginine and histidine are mutually interchangeable in metabolism, but that one or the other must be in the food in order to permit growth. Without questioning the validity of these assertions,

I decided to remove both arginine and histidine from the hydrolysate. Since, according to the British authors, the resulting material would not support growth, the effects of such a diet, if any, on creatine and creatinine would need to be compared with those induced by another type of deficiency in which a possible precursory relationship did not exist. Tentatively, a tryptophan deficiency was thought suitable for this purpose.

Obviously, the first requirement of the program was the preparation of the hydrolysate. In the light of current information, it is difficult to realize that sixty years ago many biochemists and physiologists did not believe that acid-hydrolyzed proteins could be rendered suitable for feeding purposes. No attempt has been made by the writer to discover the origin of this misconception. It may have arisen before the discovery of tryptophan in 1902. Of course, tryptophan is destroyed during the acid-hydrolysis of a protein. At any rate, to avoid any unforeseen complications, I decided to prepare the hydrolysate by enzyymatic action, followed by relatively mild treatment with acid. In subsequent investigations, acid was used as the hydrolytic agent.

Details of the procedures have been described elsewhere[5] and need not be repeated here. The hydrolysis was begun in Texas, and was continued much longer than was necessary to attain equilibrium, partly because I was occupied with other duties, and partly because toward the end of the period I was planning to move to Illinois. For its transportation, the solution was evaporated to dryness *in vacuo* at a low temperature, and the dry material was shipped along with my household goods. At the University of Illinois, the hydrolysis was completed by treatment with sulfuric acid. The arginine and histidine were removed from half of the material by the well-known method of Kossel and Kutscher.

To this point, my sole interest in these experiments was to determine, if possible, the relationship of arginine and histidine to creatine-creatinine formation. However, my attitude changed profoundly when preliminary growth tests were conducted to ascertain whether the two amino acids had been removed successfully, and whether the observations of Ackroyd and Hopkins could be duplicated. The animals lost weight rapidly when deprived of the two amino acids, and responded impressively when histidine was included in the food. In these respects, the findings agreed with those of Ackroyd and Hopkins. The data in both investigations demonstrated conclusively that histidine is an indispensable component of the food of the growing rat. On the contrary, the arginine tests failed completely to confirm the findings of the British scientists. This amino acid was totally incapable of replacing histidine, or of altering the rate of loss in weight. Clearly the two amino acids are not interchangeable in metabolism.,

The difference in the response of the animals to arginine and histidine —the first such experiments I had ever conducted—was, I thought, little short of sensational. The tests were repeated over and over again with different hydrolysates and different preparations of arginine. The results were always the same. The amazingly unequivocal findings stimulated my zeal for more information of a similar sort with respect to other amino acids. Indeed, why not undertake the classification of the remaining amino acids in regard to their dietary importance? Almost ten years had elapsed since the remarkable experiments of Osborne and Mendel, and yet progress had been made with only one other amino acid (histidine), and even that had been mixed with error respecting another amino acid (arginine).* Such studies would be important for another reason: they would afford an opportunity of determining the ability of animals to utilize the stereoisomers of the indispensable amino acids and of compounds more or less closely related in chemical structure. Could any imidazole derivative replace histidine in the diet? Could any indole derivative meet the growth needs in place of tryptophan? Here was an entirely new realm of possible intermediary metabolic investigations, in which growth would serve as the "indicator" of chemical change. And so I abandoned for the time being the creatine-creatinine investigations originally planned, though I came back to them with improved techniques on two occasions later. As for now, with a new dream and new eagerness I embarked on a new adventure. I cannot recall that it ever occurred to me that I might not be successful in establishing the growth significance of the other amino acids; in my enthusiasm, I was confident that methods could be found that would yield the desired answers.

The rest is history: how working with hydrolysates became unprofitable; how mixtures of pure amino acids were substituted as sources of nitrogen in the basal rations; how threonine was discovered and its chemical structure and spatial configuration were established; how ten amino acids were found to be necessary for the growing rat; how the stereoisomers and numerous derivatives of the essentials were tested as to their behavior in the animal organism; and how it all culminated in the establishment of the qualitative and *minimal* quantitative amino acid

* Based on experimental data that at the time appeared to be perfectly explicit, cystine was regarded as an indispensable amino acid (Osborne & Mendel, 1915).[6] Subsequent events revealed that the observed effects were really due to a sparing action exerted by cystine upon the methionine requirement (Womack, Kemmerer & Rose, 1937).[7]

needs of young men as measured by nitrogen equilibrium. For some thirty-three years the program has been a thrilling experience, in which many graduate students have participated, and to whom I am forever grateful.

How did it happen? It happened when I saw a lowly rat, on his way to oblivion because of histidine starvation, suddenly and dramatically change his course and begin to grow after consuming a tiny bit of the missing amino acid. Nothing in metabolism could be more spectacular, or arouse more reverence for the marvelous chemistry that takes place in living things.

REFERENCES

1. Fischer, E. 1901. Z. Physiol. Chem. *33*: 151.
2. Osborne, T. B. & L. B. Mendel. 1914. J. Biol. Chem. *17*: 325.
3. Rose, W. C. 1933. Ann. Rev. Biochem. (Stanford Univ.) *2*: 187.
4. Ackroyd, H. & F. G. Hopkins. 1916. Biochem. J. *10*: 551.
5. Rose, W. C. & G. J. Cox. 1924. J. Biol. Chem. *61*: 747.
6. Osborne, T. B. & L. B. Mendel. 1915. J. Biol. Chem. *20*: 351.
7. Womack, M., K. S. Kemmerer & W. C. Rose. 1937. J. Biol. Chem. *121*: 403.

HERBERT E. CARTER has worked in several areas, first on the chemistry and nutritional aspects of hydroxy amino acids, subsequently on the isolation and structure of antibiotics, and then on the isolation and structure of complex lipids in nerve tissue and plant seeds. Dr. Carter received his Ph.D. at the University of Illinois where he eventually became Professor and then head of the Department of Chemistry. During this period he received, among other honors, the Lilly Award. He subsequently assumed the position of Vice-Chancellor of Academic Affairs there. He also became chairman of the National Science Board. In 1971 he left Illinois and became Coordinator of Interdisciplinary Programs at the University of Arizona. Currently he is head of the University Department of Biochemistry.

Essential Amino Acids

HERBERT E. CARTER
Department of Biochemistry
College of Medicine
University of Arizona
Tucson, Arizona 85724

RECOGNITION OF THE dominant role played by amino acids in nitrogen metabolism began in the early 1900s. Prior to 1900 only twelve of the amino acids occurring generally in proteins were known and very little data existed as to their quantitative distribution. Proteins of different sources were for the most part regarded as essentially equivalent in nutritive value and considered as single entities in metabolism. There were some early indications that the proteins were not of equal nutritive value, and in 1860 Carl Voit developed the principle of N equilibrium and used this approach to show differences in nutritive value of proteins.[1] Even earlier in the 1850s at Rothamsted, England, farm animal feeding studies demonstrated that proteins are not of equal nutrient value.[2] However, these and other experiments were ahead of their time, and the concept of "quality" of proteins was not generally accepted until much later.

Shortly after the turn of the century a number of developments combined to shift attention dramatically to the role and importance of the individual amino acids. In 1901 Otto Cohnheim[3] clearly demonstrated the formation of amino acids by the action of intestinal enzymes on proteins. The following year Otto Loewi[4] showed that "biuret-free" autolysates of pancreatic protein would maintain an animal in nitrogen balance and could therefore substitute for all parts of the body protein that are degraded in protein turnover. These and other studies showed that proteins were almost completely cleaved to amino acids in the gastrointestinal tract and were not absorbed as such. These results were confirmed by Abderhalden using enzymic digests (acid hydrolysates would not support nitrogen balance).

A second important advance was the development of reasonably reliable procedures for the determination of individual amino acids. The Kossel and Kutscher[5] procedures for the basic amino acids and the Fischer ester distillation procedures for monoamino acids[6-8] not only demonstrated the tremendous variety in kind and amount of amino acids in various proteins, but the latter led to the discovery of several new

0077-8923/79/0325-0237 $01.75/0 © 1979, NYAS

amino acids (valine, proline, hydroxyproline). By 1904 seventeen amino acids were identified as protein constituents. The eighteenth (methionine) would not be added to the list until 1922.

The new analytical procedures revealed an astonishing diversity of protein composition. Gliadin was shown to be deficient in lysine[5]; zein to be practically devoid of lysine and tryptophan[9]; gelatin to be lacking in tryptophan, tyrosine, cystine, isoleucine and valine.[10] These observations directed the attention of investigators to the nutritive significance of the amino acids and it rapidly became apparent that nitrogen metabolism is concerned not with a single entity (protein) but with each of the individual amino acids.

In one of the first studies with incomplete proteins, Willcock and Hopkins[11] found that tryptophan improved the nutritive quality of zein (although not to the extent of supporting significant growth). However, it was the brilliant team of Thomas B. Osborne and Lafayette B. Mendel who provided the first irrefutable proof of the essential nature of an amino acid—lysine—and who formulated a concise definition of the concept of an essential or indispensable amino acid. Dr. William C. Rose has given an exciting account of these early developments in the paper "How Did It Happen," which is part of the record of this Conference. So I will give only a very brief account here of the major contributions of Osborne and Mendel.

In an epochal paper in 1914,[12] it was shown conclusively that lysine is an essential amino acid in feeding experiments in which gliadin was supplemented with lysine. This was, I believe, the first absolutely conclusive demonstration of the essential nature of an individual amino acid. This paper contains some fascinating and at times prophetic discussion of the role of individual amino acids in meeting "wear and tear" needs and growth needs. To quote, "wear and tear may be a need not for all 'bausteine' but for a particular amino acid needed to synthesize some NPN substance essential for normal metabolism"; and, "definite (as yet specifically undetermined) amino acids may serve special physiological functions." They pointed out further that whole body proteins ordinarily undergo degradation only because by this method some essential amino acid is liberated. They also noted that "in respect to nitrogen requirements of organisms growth sets a higher requirement than maintenance, for certain amino acids must serve both for maintenance and growth." The validity of this comment was amply substantiated by the erroneous conclusions reached by others on the basis of N-balance studies in adult animals.

Osborne and Mendel also confirmed and extended the work of Willcock and Hopkins with zein.[12] They showed that zein plus tryptophan maintained rats but only with the further addition of lysine was growth supported. This was the first successful attempt to raise animals on a diet in which zein was the sole protein. They also showed that tryptophan could be provided by another protein as well as in the free form. In the conclusion of this paper they state,

> The current trend of the investigation of the chemistry of nutrition is emphasizing the significance of the amino acids as the fundamental factors in all problems in which hitherto the role of the proteins has been involved. Obviously the relative values of the different proteins in nutrition are based on their content of these special amino acids which cannot be synthesized in the animal body and which are indispensable for certain distinct as yet unclearly defined processes which we express as maintenance or repair.

In 1915 Osborne and Mendel[13] showed that a diet containing only 9% casein would not support normal growth in young rats. The addition of cystine restored normal growth. Thus, cystine was considered to be an essential amino acid until some 15 years later.

In 1916 Ackroyd and Hopkins[14] reported that acid-hydrolyzed casein supplemented with tryptophan would support growth of rats but would not do so with arginine and histidine removed. Addition of either arginine or histidine checked the weight loss and "there may even be some growth." This unusual relationship could not be confirmed by Rose and Cox[15] some eight years later in experiments that aroused his intense interest in amino acid nutrition and started him on a road of brilliant accomplishment. Rose's interest in creatine metabolism led him to repeat the arginine-histidine experiments. To his surprise he found that histidine alone added to the deficient mixture supported good growth. Arginine was not required nor did arginine alone have any growth-promoting effects.

During this period there were many other studies reported, all too often of an inconclusive nature involving short-term N-balance studies. I will refer here only to the work of Abderhalden, who reported studies showing the indispensable nature of tryptophan,[16] phenylalanine or tyrosine,[17] histidine, and arginine.[17]

Thus, by 1924 only four amino acids had been firmly established as essential—cystine, histidine, lysine, and tryptophan—despite numerous attempts in many laboratories to determine the relationship of individual amino acids to maintenance and growth. It was clear that there were many difficulties in studies using deficient proteins or protein hydrolysates. Proteins completely devoid of a single amino acid were rare. There were few tests for specific amino acids delicate enough to insure that the removal of specific amino acids from protein hydrolysates was indeed quantitative. Ignorance of mineral salts and vitamin requirements also introduced uncertainties. Thus, it was possible in only a few cases to show that a specific amino acid was essential and almost impossible to prove that an amino acid was dispensable.

At this point William C. Rose began the brilliant studies that, over the next 30 years, established the complete classification of the amino acids as dispensable or essential for rat growth and for maintenance of nitrogen equilibrium in adult dogs and adult human males.

Many people believe Rose's interests stemmed from his contacts with Osborne and Mendel during his graduate days at Yale. Such is not the case and the true sequence is delightfully described in the paper Dr. Rose prepared for this Conference. I will mention here only that Dr. Rose's interest in the possible relationship of arginine to the biosynthesis of creatine led him to repeat the experiments of Ackroyd and Hopkins[14] on casein hydrolysates deprived of histidine and arginine. The dramatic response of the young rat to histidine supplementation (and the failure of any effect of arginine) so impressed Dr. Rose that he decided to undertake studies of the classification of all the remaining amino acids as to their dietary importance.

It was obvious at that time that the use of deficient proteins could never provide the capability of removing amino acids from the diet one at a time. Furthermore, the use of protein hydrolysates was also limited by inadequacy or lack of procedures for the complete removal of specific amino acids. Clearly the only effective approach to the problem was to replace completely protein in the diet with synthetic mixtures of purified amino acids.

Several previous investigators had attempted to maintain animals on diets containing amino acids as the sole or major source of protein. The literature on these unsuccessful attempts has been extensively reviewed by Rose.[18, 19] Usually the animals lost weight, refused food, sometimes vomited, and eventually died. Some of the failures were due to ignorance of vitamin requirements and lack of methionine. Whatever the cause, the idea was prevalent that animals could not subsist on amino acid diets and the failure of the animals to consume adequate food was attributed to rejection based on the taste (usually bitter) or texture rather than on nutritional inadequacy. On this point Rose, impressed by the rapid growth of his rats on the histidine-supplemented rations, never doubted that rats could indeed be supported on diets in which all protein was replaced by amino acids. His faith that loss of appetite followed inadequacy rather than causing it was richly, albeit in the beginning, slowly, rewarded. He did, however, remark at one time that "the chance still exists that this species (rat) might find the crude products present in hydrolyzed proteins more delectable than a mixture of purified amino acids."

Nonetheless, he decided against the protein-hydrolysate approach in favor of putting together a diet in which all protein was replaced by a mixture of purified amino acids. Formidable problems had to be solved in assembling such diets. In the early years the necessity of using crude vitamin preparations left room for uncertainty. Few amino acids were readily available and usually at high cost. Grants from the Rockefeller Foundation made the animal work possible and the human studies were supported by the Nutrition Foundation. From 1924 on, Rose and his students expended untold efforts in accumulating an adequate supply of the amino acids occurring generally in protein. The synthetic capabilities of the organic chemists at the University of Illinois were of tremendous help and provided many amino acids and related compounds. Many amino acids were isolated from proteins (a regular run was made to obtain hair from the campus barber shops—and never mind the cigarette butts) and a few were purchased. By 1930 the needed resources were available and the first experiments were undertaken. The first diet contained 19 amino acids (as the sole nitrogen source other than the vitamins) and simulated closely the composition of casein. In a second diet the proportion of tyrosine, glutamic acid, and aspartic acid was reduced (casein has a high content of these three amino acids). Although more nearly complete than any previous mixture, neither of these diets supported growth.[20] Food intake was low and the rats rapidly lost weight.

Comparison of these results with the adequate food consumption and good growth of rats on diets containing completely hydrolyzed casein (supplemented with tryptophan) convinced Rose that casein contained one or more as yet unknown components essential for growth. This view was proven correct when growth was obtained by the addition of 5 percent of various proteins to the basic amino acid diet. Gelatin gave poor growth, gliadin somewhat better, and casein gave good growth. These dramatic results culminating years of meticulous, painstaking work were published as the first two papers[20, 21] in the classic series on "Feeding Experiments with Mixtures of Highly Purified Amino Acids."

The next step was to show that casein hydrolysates were equally effective in supplementing the deficient amino acid mixture. This observation was highly significant since it established that rats would grow on protein-free diets whose amino acid composition was almost completely known and also provided a basic diet against which to compare fractions obtained from the "active" casein hydrolysate. It was soon determined that the insoluble amino acid fraction and the dicarboxylic and diamino acid fractions had no growth effects when added to the deficient diet. The unknown was present only in the monoamino fraction and this material gave good growth at a 2.5 percent level.

In the absence of the modern techniques for characterization and separation of amino acids, the final isolation of the new amino acid involved an enormous amount of work. To determine whether a given procedure produced further purification the products were fed to rats maintained on the basic deficient diet. It is really remarkable how such rats responded to the active material, gaining as much as 2–3 g overnight and giving reproducible results in 4-day feeding tests. However, it was always painful to have to feed your fraction to a rat in order to find out whether what you no longer had was what you hoped you had. Many fractionation procedures were employed (copper and zinc salt fractionation, ester distillation, and others).

Suspicion that more than one active substance was involved arose when activity decreased in procedures that were expected to concentrate the unknown. This possibility was finally confirmed when it was discovered that the monoamino fraction obtained by exhaustive butyl alcohol extraction on re-extraction from an aqueous solution could be separated into Unknown I (concentrated in first two extractions) and Unknown II (left in the aqueous solution). Neither fraction alone supported growth but together they gave good growth.[23] By testing amino acids known to be present in the monoamino fraction it was soon established that Unknown I was isoleucine. Two years of hard work could have been spared if better quantitative data had been available on the amino acid composition of casein.

It was now possible to make rapid progress on the further purification of Unknown II and a pure compound crystallizing as hexagonal plates from ethanol-water was obtained shortly.[24] This material was characterized by rigorous degradation procedures as one of the four optical isomers of α-amino-β-hydroxybutyric acid. Since its configuration was that of D-threose it was named threonine and designated as L-threonine since the amino group has the L-configuration. The fascinating account of this historical milestone was published in 1935.[25] It was now possible for the first time to support normal growth of young rats on completely defined diets in which protein was completely replaced with a synthetic mixture of purified amino acids.

The discovery of the last of the amino acids essential for rat growth made possible the simple direct classification of all the amino acids as essential or dispensable by observing the effect on growth of omitting the amino acids one at a time. By this procedure 10 amino acids—arginine, lysine, histidine, phenylanine, tryptophan, leucine, isoleucine, methionine, valine, and threonine—were shown to be indispensable. Of these arginine was unique in that rats would grow in its absence but not at an optimum rate. However, Rose defined an indispensable dietary component as one that cannot be synthesized by the animal organism out of the materials ordinarily available at a speed commensurate with the demands for normal growth and arginine qualifies as indispensable for the growing rat under this definition. These experiments have been described in detail by Rose in two excellent reviews.[19, 26]

Two other special relationships could now be conclusively established. For 20 years cystine had been considered an essential amino acid and indeed several laboratories had repeated the original experiments of Osborne and Mendel.[13] However, some doubt arose as a result of the observation of Jackson and Block[27] that methionine as well as cystine is capable of supplementing a low casein ration. Evidently part of the cystine may be replaced by methionine for growth. Further question was raised by the fact that methionine and cysteine (but not cystine) gave "extra" cystine in cystinurics.[28] This could be interpreted as indicating that methionine is converted to cysteine and on to cystine in cystinurics. To settle this matter in normal animals purified amino acid diets devoid of methionine and cystine were supplemented with the two amino acids separately and together. Methionine alone supported good growth in rats and cystine alone was completely ineffective.[29] Therefore methionine is converted to cystine in the animal body but the reverse step does not occur. However, cystine does spare the methionine requirement (about one-sixth).

In similar experiments it was shown that phenylalanine is essential and is converted into tyrosine.[30] However, tyrosine spares the phenylalanine requirement.

It was now possible to consider studies of human requirements, but before undertaking these Rose decided to determine the amino acid requirements of the adult dog as an intermediate step.[31] For this purpose nitrogen balance studies were employed. The limitations inherent in nitrogen balance experiments have been reviewed in detail by Rose.[32] However, no other method was available and the problems involved were adequately solved by meticulous attention to every detail. Adult female dogs were brought into nitrogen equilibrium on casein diets and were then transferred to similar rations in which casein was replaced by mixtures of highly purified amino acids.

Urine samples were collected at 24-hour intervals by catheterization and feces were divided into seven-day periods by carmine capsules. The first dog received the ten amino acids essential for the growing rat and manifested a slight positive nitrogen balance and continued to do so with slight weight gain for four weeks. Obviously the remaining 10–12 amino acids dispensable for the growing rat are also dispensable for the adult dog. At the fifth week, arginine was removed with no effect on nitrogen balance. Arginine, not unexpectedly, can be synthesized by the adult dog at a rate adequate to maintain nitrogen balance. In three other similarly maintained dogs removal of the amino acids essential for the rat in each case produced a pronounced negative nitrogen balance. Therefore (with the exception of arginine) the amino acid requirements of the dog are identical with those of the rat. The similarity of the two was reassuring and suggested that other species would behave similarly.

With these experiments completed in the fall of 1942 the human studies were undertaken and extended over the next ten years.[32, 33]

The subjects were healthy males, usually graduate students in biochemistry or related fields. The diet consisted of corn starch, sucrose, butter fat (protein-free) corn oil, inorganic salts, vitamins, and the amino acid mixture. Starch, salts, and part of sugar and butter were baked into wafers. The remainder of butter was spread on the wafers (only a partially successful attempt to render them palatable) and the remainder of the sugar plus lemon juice used to flavor the amino acid solutions. The carbohydrate to fat calorie ratio was 2.6, and the amino acid mixture provided 6.7–10 g of nitrogen daily. Celu flour was also given daily to provide bulk and counteract the laxative effect of the amino acids. Unidentified nitrogen amounted to 0.19–0.43 g daily. Urine was collected on a 24 hour basis and feces were marked by charcoal.

In preliminary studies an unexpected complication arose involving the relationship of the energy input to maintenance of nitrogen equilibrium. Extensive studies showed that in experiments comparing casein with hydrolyzed casein and with a mixture of purified amino acid the casein diet supported nitrogen balance at a 35 calorie per kilogram level whereas a substantially higher calorie input (at least 45 kcal/kg) was required with the amino acid diets. (This is still a puzzling problem.) In response to these studies it was decided to use diets providing 55 kcal/kg (and in one case 58 kcal/kg).

In the absence of information on which to compound amino acid mixtures suitable for man, the amino acid mixtures were composed of the ten amino acids necessary for the growing rat in approximately the ratios found to support optimum growth.[34] In other diets the basic amino acid mixture was supplemented with glycine and urea as sources of extra nitrogen for compounds synthesized by the organism.[34]

On diets containing the ten amino acids indispensable for the rat, the subjects readily attained nitrogen equilibrium in 3 to 4 days. These results, checked in many subjects, are of the highest significance since they establish conclusively that the amino acids that are dispensable for the rat[35] and dog[31] are dispensable also for adult human males.

With the basic procedures established and tested, the crucial experiments went forward. The ten amino acids were excluded from the diet one at a time with the total nitrogen intake held constant. The results showed unequivocally that eight of the ten amino acids were indispensable: valine, methionine, isoleucine, leucine, phenylalanine, threonine, lysine, tryptophan. Removal of any one of the eight gave a strongly negative N-balance. Return of the amino acid rapidly restored nitrogen equilibrium. The deficient diets caused profound failure of appetite (subjects had great difficulty in ingesting the full ration), extreme fatigue, and increased nervous irritability. Isoleucine deficiency gave particularly intense effects and a greater negative N balance than any other, amounting to as much as 3.9 g/day. In comparison, leucine gave a very small negative balance. In almost every case the symptoms disappeared in 24 hours after the missing component was returned to the diet. No specific symptoms were observed for any amino acid.

The removal of two amino acids—arginine and histidine—from the diet had no effect on N balance or on feeling of well-being. The arginine result was not unexpected in view of the results with adult dogs[31] and adult rats.[36] The histidine result was quite unexpected. The experiment was repeated many times with identical results. Careful tests revealed no impurity of histidine in the diet. The final conclusion, therefore, was that preformed histidine in the diet is not necessary for maintenance of nitrogen equilibrium in adult man. After these results, histidine was omitted from all later mixtures, and fifty subjects have been maintained in nitrogen balance on diets devoid of histidine. A number of interesting questions remain. Is histidine provided by bacterial synthesis in the intestine? Do growing children require histidine? What is the significance of the recent report of Kopple and Sevendseid[37] that longer continued experiments (25 days) demonstrated a histidine deficiency in adult males?

These experiments brilliantly fulfilled Rose's original commitment to determine the amino acid requirements of the human species. The results were published in a series of papers in the early 1950s and reviewed by Rose in masterful detail in *Nutrition Abstracts and Reviews*,[32] which stands as a classic contribution to the understanding of protein and amino acid requirement and metabolism.

The final classification of the amino acids as essential and nonessential for maintenance of nitrogen equilibrium in human males is given in Table 1.

There then remained the determination of the quantitative require-

TABLE 1

CLASSIFICATION OF AMINO ACIDS FOR ADULT HUMAN MALES*

Essential	Nonessential	
Valine	Alanine	Proline
Leucine	Glycine	Hydroxyproline
Isoleucine	Serine	Arginine
Threonine	Cystine	Histidine
Methionine	Aspartic Acid	Tyrosine
Phenylalanine	Glutamic Acid	Citrulline
Lysine		
Tryptophan		

*After Rose.[32]

ments for the essential amino acids in human males. The basic diet contained the eight essential amino acids supplemented with appropriate amounts of glycine to keep the nitrogen content constant. The amount of one amino acid was reduced progressively until a negative nitrogen balance was produced. This was considered to be the minimum quantitative requirement. This level was used in the next amino acid test and so on until the quantitative requirements for all eight had been established. Nitrogen balance procedures are far from ideal for this type of study. Daily fluctuations occurred and marked differences were also found in the minimum needs of certain subjects, making longer periods necessary. However, by careful control and by the use of several subjects, reasonably consistent figures were obtained. Some of the results were unexpected. Tryptophan was the first to be studied. Starting at a

TABLE 2

DAILY AMINO ACID REQUIREMENTS OF ADULT HUMAN MALES*

	Tentative Minimum Requirement and Range g/day
L-Isoleucine	0.70 (0.65–0.70)
L-Leucine	1.10 (0.50–1.10)
L-Lysine	0.80 (0.40–0.80)
L-Methionine	1.10 (0.80–1.10)†
L-Phenylalanine	1.10 (0.80–1.10)‡
L-Threonine	0.50 (0.30–0.50)
L-Tryptophan	0.25 (0.15–0.25)
L-Valine	0.80 (0.40–0.80)

*After Rose.[32]
†When cystine was added the methionine requirement was spared to 80–89%.
‡When tyrosine was added, the phenylanine requirement was spared to 70–75%.

level of 1.8 g/day the level for two subjects could be reduced to 0.15 g/day. Over the next three years fifteen subjects were maintained on 0.2 g/day, and finally one subject in phenylanine studies was found to require 0.25 g/day—the highest value for all the tests on tryptophan.

In all of these experiments there was no correlation between minimum requirement and body weight, body surface, metabolic rate, or creatinine output. Therefore, the data were presented simply on a g/day basis and the highest observed value was designated as the tentative minimum requirement. These values are shown in TABLE 2.

As Rose had pointed out previously,[19, 31] a mixture of highly purified amino acids compounded in accordance with the quantitative needs of the cells for each component may prove to be the most efficient type of nitrogen ever devised for the uses of the animal organism. The quantitative data obtained by Rose furnish a sound basis for determining human protein needs and have been an invaluable aid in evaluating world protein requirements.

REFERENCES

1. PIKE, R. R. & M. BROWN. 1975. Nutrition: An Integrated Approach. 2nd edit.: 16. John Wiley & Sons. New York, N.Y.
2. HALL, A. D. 1917. The Book of Rothamsted Experiments. 2nd edit. E. D. Dutton & Company. New York, N.Y.
3. COHNHEIM, O. 1901. Hoppe-Seyler's Z. Physiol. Chem. *33*: 451–465.
4. LOEWI, O. 1902. Arch. Exp. Path. Pharm. *48*: 303–330.
5. KOSSEL, A. & F. KUTSCHER. 1900–01. Hoppe-Seyler's Z. Physiol. Chem. 31:165–214.
6. FISCHER, E. 1901. Hoppe-Seyler's Z. Physiol. Chem. *33*: 151–176.
7. FISCHER, E. 1902. Chem. Ber. *35*: 2660–2665.
8. FISCHER, E. 1906. Chem. Ber. *39*: 2320–2328.
9. OSBORNE, T. B. & S. H. CLAPP. 1907–08. Amer. J. Physiol. *20*: 477–493.
10. DAKIN, H. D. 1920. J. Biol. Chem. *44*: 499–529.
11. WILLCOCK, E. G. & F. G. HOPKINS. 1906–07. J. Physiol. *35*: 88–103.
12. OSBORNE, T. B. & L. B. MENDEL. 1914. J. Biol. Chem. *17*: 325–349.
13. OSBORNE, T. B. & L. B. MENDEL. 1915. J. Biol. Chem. *20*: 351–378.
14. ACKROYD, H. & F. G. HOPKINS. 1916. Biochem. J. *10*: 551–576.
15. ROSE, W. C. & G. J. COX. 1924. J. Biol. Chem. *61*: 747–764.
16. ABDERHALDEN, E. 1912. Hoppe-Seyler's Z. Physiol. Chem. 77: 22–58.
17. ABDERHALDEN, E. 1922. Arch. Ges. Physiol. *195*: 199–215.
18. ROSE, W. C. 1931–32. Yale J. Biol. Med. *4*: 519–536.
19. ROSE, W. C. 1934–35. Harvey Lectures *30*: 49–65.
20. ROSE, W. C. 1931. J. Biol. Chem. *94*: 155–165.
21. ELLIS, R. H. & W. C. ROSE. 1931. J. Biol. Chem. *94*: 167–171.
22. WINDUS, W., F. L. CATHERWOOD & W. C. ROSE. 1931. J. Biol. Chem. *94*: 173–184.
23. WOMACK, M. & W. C. ROSE. 1935. J. Biol. Chem. *112*: 275–282.
24. MCCOY, R. H., C. E. MEYER & W. C. ROSE. 1935. J. Biol. Chem. *112*: 283–302.

25. Meyer, C. E. & W. C. Rose. 1936. J. Biol. Chem. *115*: 721–729.
26. Rose, W. C. 1938. Physiol. Rev. *18*: 109–136
27. Jackson, R. W. & R. J. Block. 1932. J. Biol. Chem. *98*: 465–477.
28. Brand, E., G. F. Cahill & M. M. Harris. 1935. J. Biol. Chem. *109*: 69–83.
29. Womack, M., K. S. Kemmerer & W. C. Rose. 1937. J. Biol. Chem. *121*: 403–410.
30. Womack, M. & W. C. Rose. 1946. J. Biol. Chem. *166*: 429–434.
31. Rose, W. C. & E. E. Rice. 1939. Science *90*: 186–187.
32. Rose, W. C. 1957. Nutr. Abs. Rev. *27*: 631–647.
33. Rose, W. C., W. J. Haines & J. E. Johnson. 1942. J. Biol. Chem. *146*: 683–684.
34. Rose, W. C., L. C. Smith, M. Womack & M. Shane. 1949. J. Biol. Chem. *181*: 307–316.
35. Rose, W. C., M. J. Oesterling & M. Womack. 1948. J. Biol. Chem. *176*: 753–762.
36. Wolf, P. A. & R. C. Corley. 1939. Amer. J. Physiol. *127*: 589–596.
37. Kopple, J. D. & M. E. Swendseid. 1973. Fed. Proc. *33*: 691.

DISCUSSION OF THE PAPER

S. Simmonds (*Yale University, New Haven, Ct.*): I want to tell a story that I am sure Dr. Carter knows as well as I do, and that is the story about whether homocysteine can replace methionine in rat diets. Both Dr. Rose and Dr. Du Vigneaud, who had been with Dr. Rose many years before, were interested in that problem, and they worked on it simultaneously; each knew what the other was doing and the two laboratories came up with different results: In one case homocysteine did replace methionine in the diet of growing rats, and in another case it did not. It was eventually traced back to the source of B vitamins that was being used in each diet. Very few of the B vitamins being known at the time, a crude preparation from rice polishings was used in each laboratory but they were different preparations. In the case where homocysteine did replace methionine in the diet it was found that the rice polishing preparation contained choline.

I would guess now by hindsight that that preparation also probably had a good bit of folic acid and B_{12} in it, which was helping matters. And indeed some years after the original homocysteine-methionine work Grace Mede in Philadelphia, with I think Warwick Sakami who was then a graduate student with her, had found that she could grow Philadelphia rats on homocysteine without any choline in the diet, and that was a puzzle for a number of years. Then Du Vigneaud's laboratory and Grace Mede's laboratory exchanged animals and rats were bred in New York from the Philadelphia ones and tested, and it was found that they would not grow on homocysteine unless choline was present. Whereas, the Cornell rats bred in Philadelphia and then tested *would* grow. This turned out to be the beginning really of the story of de novo methyl synthesis because it turned out that the mother rats in Philadelphia were fed a very complete diet with lots of fresh vegetables including lettuce and probably had very fine intestinal flora that was making folic acid, which the young acquired either by drinking mothers milk or because their intestinal bacteria were good too. This started the study of de novo methyl synthesis in mammals, a field to which Warwick Sakami has contributed so much.

CARTER: That is very interesting. I have a file about so thick of papers that it would have been a pleasure to present here, and that is indeed as you pointed out one of the most interesting. The relation of methionine and cysteine also gets involved in that.

Once threonine was discovered you could put together a diet from which you could remove one or more amino acids independently of all the others, and the relationship of methionine to cysteine and phenylalanine to tyrosine could be determined quantitatively. One could study the metabolism of alpha keto acids, alpha hydroxy acids, acetyl amino acids, and so on. You could study in a crude way metabolic transformations using growth in rats as a major measuring tool.

RATNER: I remember having a little correspondence with Dr. Rose because when his results came out that arginine was "semi-essential," it seemed to me that some experiments ought to be done with ornithine because we know that arginine could be made in the body, and I asked him if he had done any experiments with ornithine. We know something about the fact that ornithine can be made in the body, but it might have been rate-limiting for some reason, too slow for the needs of a growing animal. I several times looked for such a paper and did not find it. I wondered if you would have any comments?

CARTER: I do not recall whether it was determined what the rate-limiting step actually was.

SHEMIN: Sarah Ratner showed this morning that she isolated histidine after feeding a rat tyrosine with leucine. In a subsequent paper, when histidine from that rat was degraded, there was no ^{15}N in the imidazole group, agreeing with Dr. Rose's experiments, namely, that histidine is an essential amino acid for a rat.

I also mentioned this morning that I ingested glycine following up that type of experiment in which Dr. Ratner had isolated histidine. I isolated my histidine from my plasma proteins, degraded the histidine, and found ^{15}N in the alpha amino but none in the imidazole group. I have never published this, and I have been puzzled by this off and on for a number of years. But as biochemistry developed I found an explanation for it: Namely, when I was ingesting glycine I was also eating a normal diet and in essence one possible explanation would be that I was repressing histidine synthesis with histidine in my diet. And it is still unresolved whether, had I taken the glycine on a histidine-free diet, I would have ^{15}N in my imidazole nitrogen.

CARTER: Wixom's experiment says that you would.

SHEMIN: Well, I did not, and those experiments were all done carefully; that means that histidine may be repressed. In fact I would like to reinvestigate this on a histidine-free diet. It also means, in essence, that I did not have any bacteria.

CARTER: There is I think room for some really rigorous looks at the extent of the negative nitrogen balance that ensues when you remove a particular amino acid from the diet. With lysine it is very small; isoleucine deficiency gives as much as a 4 gram per day negative nitrogen balance in humans. I think one could perhaps begin to learn something about possible metabolic uses of the essential amino acids by looking more rigorously.

FRAENKEL-CONRAT: There is a paper in the literature, an amusing one, a single collaborative effort of Du Vigneaud and Stanley, namely, on giving rats TMV as their only dietary protein. TMV lacks both histidine and methionine. Therefore, I would surely assume the rats must have sadly decayed but I do not remember and I wonder whether you remember whether they complemented their TMV with those two amino acids or whether they were not rats.

CARTER: They could not have been albino rats at the end of that experiment, could they? They would have to be mosaic rats of some kind! I am sorry. I do not remember what the answer to that was.

HAUROWITZ: What is the symptomatology of a diet deficient in an essential amino acid as compared with a diet that is deficient in calories? Is there a fundamental difference in them because, if you give a rat for instance a diet that is deficient in calories, strongly deficient in calories, this does not bother them at all as far as life is concerned? They do not develop as well as other rats but they live as you know very well, much longer even than normal rats. With a diet that is deficient in one amino acid you would assume there would be a fundamental difference because you would finally come to an equilibrium where this one amino acid regulates the rate of the metabolism in all vital processes.

CARTER: I wonder whether you are recalling perhaps some statements that other people made, Holt and so on, about specific physiological effects of these deficiencies. Let me say that in every case a rapid decrease in appetite results, and it is then followed by loss of weight. In rats there is some irritability, and valine deficiency produces a kind of specific effect on agility and locomotion of the rat. But in the experiments with adult animals Dr. Rose makes a point of saying that there were no specific symptoms associated with any of those deficiencies. In every case appetite went down; there was muscular fatigue; there was loss of weight and nitrogen excretion; there was some irritability; but there were no significant special symptoms associated with one amino acid deficiency that he did not get with all the others.

NORMAN H. HOROWITZ received his Ph.D. from the California Institute of Technology for work in a laboratory established in 1928 by T. H. Morgan. In 1939, as a National Research Council Fellow, he began working at Stanford, where he met G. W. Beadle and E. L. Tatum, whom he joined in 1942 to work on *Neurospora* genetics. He later continued working with Beadle at Caltech and became one of the principal contributors to the "one gene–one enzyme" hypothesis.

Dr. Horowitz is now chairman of the Biology Division at Caltech. Recently he has been involved in the Viking-Mars project as an experimenter. He is currently interested in iron transport in *Neurospora*.

Genetics and the Synthesis of Proteins

N. H. HOROWITZ
Biology Division
California Institute of Technology
Pasadena, California 91125

GENES, PROTEINS, AND THE GENETIC SYSTEM

THE DISCOVERY AND elucidation of the connection between genes and proteins is one of the major accomplishments of 20th century science. Nothing in biology illuminates more clearly the fundamental organization of living systems than does the gene-protein relationship. We see that every organism has a genetic heritage that consists apparently entirely of specifications for the synthesis of an array of protein molecules, including their structure, time of appearance, and rate of production. From this initial input, all other aspects of the organism follow—its structure, development, metabolism, and behavior—insofar as these are genetically determined. The genetic specifications are an evolutionary product, generated by random mutations in DNA and screened by natural selection. They are a record of discovered solutions to the problems of survival encountered by the species during its long history. Without this historical record, life could not exist, because survival depends on the ability of the organism to synthesize a large variety of proteins, but proteins are highly improbable structures. If every generation had to discover for itself how to assemble amino acids in the correct sequences to produce useful proteins, survival would be impossible. Hence the need to preserve and transmit sequence information from generation to generation.

Until about 30 years ago, it was generally believed that proteins themselves perform this genetic function. Avery and those who followed him showed the error of this belief. Proteins cannot serve as templates for their own replication (at least they cannot do so very effectively) and therefore they cannot function as genes. Nucleic acids, however, can serve as their own templates, and hence they are suited to be carriers of hereditary information, but they cannot perform the many catalytic functions that are essential for the life of the cell and that proteins perform so readily. Hence the dual system, nucleic acids and proteins, interlocking and interdependent, each indispensable for the existence of the other. The properties of this remarkable system—the "genetic system"—coincide with the essential properties of living matter; that is to

0077-8923/79/0325-0253 $01.75/0

say, it is capable of duplicating itself, of mutating and duplicating its mutations, and of evolving adaptively.

GARROD'S DISCOVERY

The first intimation that biological inheritance is concerned with the synthesis of proteins came shortly after the rediscovery of the laws of Mendel. In 1902, just two years after the rediscovery, Archibald Garrod[1] published a paper in which he suggested that alcaptonuria in man is inherited as a simple Mendelian recessive. In this suggestion he had the support of William Bateson, the leading British proponent of Mendelism. By 1909, Garrod's investigations had brought him to the further conclusion that alcaptonuria results from the lack of an enzyme, present in normal individuals, that opens the ring of homogentisic acid. This he published in his classic treatise, "Inborn Errors of Metabolism," along with tentative evidence suggesting that several other inherited defects in man have a similar basis.[2]

Garrod's discovery, made practically at the outset of modern genetics, suffered a familiar fate: it was ignored and forgotten by geneticists for over 30 years, when the same principles were rediscovered in *Neurospora* by Beadle and Tatum. Although Garrod is now honored as the father of biochemical genetics, the fact is that his work had no influence on the development of genetics, which would have been the same had he never lived. The same thing, of course, is true of Mendel. Garrod's fate differed from Mendel's in one important respect, however: his work on alcaptonuria was highly regarded by biochemists and was accepted into the body of biochemical knowledge intact. It was discussed in a biochemistry course I took in 1935. What interested biochemists, however, were Garrod's findings in connection with the metabolism of phenylalanine and tyrosine; the possible implications of these findings for the nature of gene action completely escaped them.

Geneticists, too, had other, more urgent concerns in the early decades of the century. First was the question of the validity and generality of Mendelism as a description of heredity in plants and animals. Then came the chromosome theory, the proof of this theory, and the demonstration that genes can be mapped on the chromosomes. It was not until these matters had been settled that the question of how genes produce their effects became pressing.

In his book on the history of genetics, A. H. Sturtevant[3] gives two reasons for the failure of geneticists to appreciate Garrod's discovery. At first, they did not understand what Garrod was talking about because they were ignorant of biochemistry. Later, when they were more literate in biochemistry, they understood him but did not accept his findings as generally applicable; they were convinced that development was too complex to be explained by any simple theory of gene action. The view held by most geneticists until the 1950s was that genes are manifold, or pleiotropic, in their action. This view precluded serious consideration of Garrod's findings, and later it delayed acceptance of the extensive and convincing evidence obtained in *Neurospora* for a one-to-one relation between genes and enzymes. According to Sturtevant, the notion of pleiotropy arose from the early studies of De Vries on the so-called mutations in *Oenothera lamarckiana*, the evening primrose. These "mutations," which for a long time defied analysis, are now known not to be single-gene events, but multiple genetic changes resulting from recombination within the unusually complex chromosomes of *Oenothera*.

In a conversation I had with him a few years before he died, Sturtevant told me that a theoretical argument by E. B. Wilson had also been important in obscuring the significance of Garrod's findings. I believe that the argument Sturtevant referred to was the following from the 3rd edition of *The Cell*[4]:

> In what sense can the chromosomes be considered as agents of determination? By many writers they have been treated as the actual and even as the exclusive "bearers of heredity." . . . Many writers, while avoiding this particular usage, have referred to the chromosomes, or their components as "determiners" of corresponding characters; but this term, too, is becoming obsolete save as a convenient descriptive device. The whole tendency of modern investigation has been towards a different and more rational conception which recognizes the fact that the egg is a reaction-system and that (to cite an earlier statement) "the whole germinal complex is directly or indirectly involved in the production of every character." Genetic research is constantly bringing to light new cases of the cooperation of several or many factors in the production of single characters; and it is possible that all the chromosomes, or even all of the units which they contain, may be concerned in the production of every character.*

The view expressed in this quotation is one that was widespread among geneticists in the '30s and '40s. In a sense, this view is perfectly correct; but in another, and equally valid sense, it is totally wrong. It was a long time before these different ways of regarding gene action could be sorted out. In the meantime, Garrod's important discovery was forgotten.

* A. D. Hershey[34] quotes another, very similar, statement by E. B. Wilson in an interesting essay on the state of genetics written in 1970.

NEUROSPORA AND "ONE GENE–ONE ENZYME"

In the decades following the publication of *Inborn Errors* a number of starts were made on the biochemical analysis of mutant phenotypes in a variety of plants and animals, but the subject was not essentially advanced beyond the point where Garrod had left it until 1941, when Beadle and Tatum described the first nutritional mutants (also called "auxotrophs") in *Neurospora*.[5] (The initial experiments used both *N. sitophila* and *N. crassa*, but all subsequent work was with *N. crassa*.) In the interval, genetics had developed in almost total isolation from biochemistry and, in fact, from all physical sciences. This isolation was described in the following way by Sturtevant and Beadle in the preface to their *Introduction to Genetics*, published in 1939[6]:

> Physics, chemistry, astronomy, and physiology all deal with atoms, molecules, electrons, centimeters, seconds, grams—their measuring systems are all reducible to these common units. Genetics has none of these as a recognizable component in its fundamental units, yet it is a mathematically formulated subject that is logically complete and self-contained.

The 1941 paper of Beadle and Tatum marks the end of this isolation of genetics from the physical sciences. The recovery of single-gene mutants in which specific biosynthetic pathways were blocked opened a new dimension in the study of gene action. In place of the chemically undefined morphological mutations that up to that time had formed the main working material of genetics, there was now a wealth of inherited metabolic defects which were comprehensible in terms of known biochemistry. Unlike alcaptonuria in man, the *Neurospora* results could not be explained away as a singularity. Furthermore, the methods devised by Beadle and Tatum were applicable to other microorganisms, as Tatum soon showed. He was able to induce the same kinds of mutations in *Escherichia coli* (strain K-12, by lucky chance).[7] These mutants were later used by Lederberg and Tatum[8] to demonstrate sexual recombination in *E. coli*, itself a major event in the history of genetics.

Study of the *Neurospora* mutants soon made it clear that at the level of metabolic reactions genes are not pleiotropic at all, but are highly restricted in their range of action. Just as in alcaptonuria, the "biochemical" mutants of *Neurospora*, as they were then called, were blocked in single steps of metabolism.[9] The seeming exceptions turned out to be exceptions that proved the rule. For example, a single-gene mutant that required two amino acids—methionine and threonine—was found to be blocked not in two pathways, but in the synthesis of homoserine, a previously unrecognized common precursor. Again, a mutant that required both isoleucine and valine for growth was, after much investigation, shown actually to be blocked in two pathways, but these pathways —the last few steps in the synthesis of these two amino acids—are catalyzed by the same enzymes.[10]

The *Neurospora* results were summarized in the "one gene–one enzyme" hypothesis—i.e., a given gene is involved in the synthesis of a single enzyme or other protein. When the gene mutates, the enzyme is defective or is simply not made. As Beadle conceived it in 1945, the gene acted as a "master molecule or templet in directing the final configuration of the protein molecule as it is put together from its component parts."[11] This theory was advanced before any proof of enzyme involvement had been obtained and of course before anything was known about the chemical nature of the gene. It was an inference based on analysis of the growth requirements of the mutants, on their mode if inheritance, and on the identification of metabolic intermediates that accumulated in blocked pathways. The first direct demonstration of an enzymatic deficiency in a *Neurospora* mutant was of tryptophan synthetase in a tryptophan-requiring mutant, by Mitchell and Lein.[12] Many other examples followed.

Beadle's theory was greeted with hostility. It seemed that, despite the evidence, many geneticists preferred to believe Wilson's doctrine that every gene is concerned in the production of every character. The one gene–one enzyme hypothesis was denounced as unverifiable and also unfalsifiable. It was alleged to be based on a selection procedure that insured that only mutations supporting the theory would be detected. It was criticized for being too simple to explain all of the complexities of metabolism. Critiques published at the time are but pale shadows of the unpublished objections that were voiced in the '40s and '50s at the Cold Spring Harbor symposia and wherever else geneticists gathered.

The debate lasted on and off for years; it finally ended with the vindication of "one gene–one enzyme." We can date this milestone with the demonstration, simultaneously in Yanofsky's and Brenner's laboratories, of the colinearity of gene and protein in *E. coli* and bacteriophage, respectively.[13, 14] Yanofsky and his co-workers mapped a series of missense mutations in the structural gene for the A polypeptide of tryptophan synthetase of *E. coli*, and at the same time they mapped the corresponding amino acid replacements in the polypeptide. The two maps were superimposable within the limits of error of the measurements. The Brenner group mapped a series of nonsense mutations in the gene encoding the head protein of phage T4 and showed that the locations of the resulting interruptions in the elongation of the polypeptide were colinear with the genetic map. Colinearity is a sufficient condition for "one gene–one enzyme," although not a necessary one. In addition, since it was already becoming clear that the active configuration of proteins is determined by their amino acid sequence,[15] it followed that the gene carries all of the unique information needed to specify the enzyme. One gene–one enzyme was just a special case of the general relation: one gene–one polypeptide.

By 1964, when the colinearity papers were published, the terms of the discussion had changed considerably from what they had been in 1945 when Beadle advanced the notion of a simple relation between genes and enzymes. The chemical structure of the gene had by now been discovered, and great advances had been made in unravelling the mechanism of protein synthesis. It was no longer necessary to infer the nature of gene action from the results of genetic experiments, since gene action was rapidly becoming amenable to direct study at the molecular level. While the colinearity experiments of Yanofsky and of Brenner still contained recognizably classical features, they went far beyond classical genetics in their analysis of both the genotype and, especially, the phenotype. As will be seen later, we have reached the point today where the gene as well as its products are analyzable into their ultimate structural units, and new discoveries are being made at this level.

SICKLE-CELL HEMOGLOBIN

In the course of the one gene–one enzyme debate, a number of important results were obtained that should be mentioned here. Beadle's use of the word "templet" to describe the role of the gene suggests a mechanism of protein synthesis that is remote from the actual mechanism as we know it today, but his point, obviously, was not to propose a mechanism of protein synthesis, but to suggest that the gene determines the specific properties of the enzyme, not just its presence or absence. That this is so was first demonstrated not in *Neurospora* or *E. coli*, but in man. In 1949, Pauling and coworkers[16] showed that sickle-cell hemoglobin has a higher isoelectric point than normal hemoglobin, the difference amounting to 2–4 net charges per molecule. In 1957, Ingram[17] found that sickle-cell hemoglobin has valine in place of glutamic acid in position 6 of the β chains. This was the first demonstration that amino acid substitution can result from gene mutation.

TEMPERATURE-SENSITIVE MUTANTS

A class of mutants that was especially useful in establishing the relation between genes and proteins was the temperature-sensitive, or temperature-conditional, class. These mutants show their genetic defect at particular temperatures—usually above 30° C—while they are normal, or nearly so, at other—usually lower—temperatures. They are caused, as we know now, by mutations in the structural gene for an enzyme that result in the production of thermolabile forms of the enzyme. When they were first discovered, however, it was not known whether it was the enzyme or the enzyme-synthesizing mechanism that had become thermolabile. Temperature-conditional mutants were first found in the course of the mutant hunt that ran more or less continuously in Beadle's laboratory at Stanford in the '40s. The decision to search for such mutants had been made following publication of a paper by Stokes *et al.*[18] showing that one of the three original mutants of Beadle and Tatum, a pyridoxin-requiring strain, was pH-sensitive: its need for pyridoxin was displayed only when the pH of the medium was below 5.8. Above pH 5.8, it synthesized the vitamin. This finding suggested that a similar class of temperature-sensitive mutants might exist, and this was soon confirmed.[19, 20]

These mutants were especially valuable because they made it possible to detect, in the form of temperature-conditional alleles, genes whose ordinary mutations would be lethal and unrecoverable. This property made it possible to answer a fundamental criticism of the one gene–one enzyme hypothesis that had been advanced by Max Delbrück; namely, that the method of detecting nutritional mutants was such that it was inherently unlikely that any mutants with complex nutritional requirements would be recovered. If this criticism was valid, and if mutants with multiple metabolic defects were a significant fraction of the total, then this should be revealed by analysis of temperature-conditional mutants, which are selected only for their failure to grow at certain temperatures. Specifically, by placing the mutants at the temperature at which their mutant character is manifested, it could be determined what fraction of them failed to grow when supplied with the standard complete medium used in the standard selection procedure. When I applied this test to the known temperature-conditional mutants of *Neurospora* in 1950, I could find little evidence for selection of the kind Delbrück had postulated.[9] Leupold and I then examined a much larger number of temperature-conditional mutants in *E. coli* and found even less evidence for selection against multifunctional losses than in *Neurospora*.[21] This result was very reassuring for the one gene–one enzyme theory.

In 1952, Maas and Davis[22] described a temperature-conditional mutant of *E. coli* that required exogenous pantothenic acid at temperatures above 30°C, but not at 25°C. They were able to show that the mutant produces a thermolabile form of pantothenate synthetase, the enzyme that couples pantoic acid and β-alanine. This was the first evidence that temperature-sensitive mutations affect the enzyme, not the enzyme-synthesizing apparatus. A little later, Fling and I found a gene in *Neurospora* that determines both the thermostability[23] and the electrophoretic mobility[24] of the enzyme tyrosinase.

Occasionally, mutants were found with reversed temperature sensitivity—i.e., they were phenotypically mutant at low temperatures but normal at high temperatures. These were more difficult to account for, since no models were known to us of enzymes that were inactivated by lowering the temperature by a few degrees. In 1957, however, Fincham[25] found that a glutamic-acid-requiring mutant of *Neurospora* with reversed temperature sensitivity produced a glutamic dehydrogenase with precisely the properties needed to explain the phenotype; that is, the enzyme is active at temperatures above 25°C, but is inactivated reversibly at 20°C. All of this strengthened the idea that the structure of proteins is genetically determined.

RECENT DEVELOPMENTS

The foregoing is a brief summary of the investigations—with emphasis on those I had some personal involvement in—that led to the picture of gene-protein relationship that has been accepted for nearly two decades. According to this picture, a continuous segment of DNA, the gene, codes for a unique polypeptide which is a linear representation of the DNA. Until very recently, this picture was thought to be applicable throughout the living world. In the past year, however, findings made possible by powerful new methods for amplifying and sequencing nucleic acids have shown that the accepted model is far from universally applicable.

The new findings are of two kinds. First, several examples are now known of gene overlaps—i.e., stretches of DNA that are shared by two or even three genes.[26-28] A shared sequence may be translated in the same reading frame for two proteins, in which case the proteins have amino acid sequences in common, or it may be translated out of phase. In the latter case, one DNA sequence can have three different translations. These contradictions of the one gene–one polypeptide rule have been found so far only in viruses, where information compression presumably has a strong selective advantage.

The second kind of unexpected finding violates the colinearity rule. In a growing number of cases, eucaryotic structural genes are being found to contain interpolated DNA sequences that are not represented in the mRNA or tRNA that is read from these genes.[29-33] The inserts are known to be transcribed in at least some cases (and are presumed to be transcribed in all cases), but they are eliminated in the processing of the transcript. The significance of the interpolated DNA is unknown at this time, although there are some plausible suggestions.

It seems likely that these contradictions of the, until now, accepted model of the relationship between genes and proteins will turn out to be special evolutionary adaptations for life as a virus or as a eucaryote—that is, higher order refinements of the simple basic pattern. In any case, these fascinating results are doubtless just the forerunners of discoveries that the new techniques now available will make possible. It is clear that the story of genes and proteins is not yet over.

REFERENCES

1. Garrod, A. E. 1902. Lancet *1902*(ii): 1616–1620.
2. Garrod, A. E. 1909. Inborn Errors of Metabolism. Frowde, Hodder & Stoughton. London.
3. Sturtevant, A. H. 1965. A History of Genetics. Harper & Row. New York, N.Y.
4. Wilson, E. B. 1925. The Cell in Development and Heredity. 3rd edit. :975–976. Macmillan. New York, N.Y.
5. Beadle, G. W. & E. L. Tatum. 1941. Proc. Nat. Acad. Sci. U.S.A. *27*: 499–506.
6. Sturtevant, A. H. & G. W. Beadle. 1939. An Introduction to Genetics. W. B. Saunders Co. Philadelphia.
7. Gray, C. H. & E. L. Tatum. 1944. Proc Nat. Acad. Sci. U.S.A. *30*: 404–410.
8. Lederberg, J. & E. L. Tatum. 1946. Cold Spring Harbor Symp. Quant. Biol. *11*: 113–114.
9. Horowitz, N. H. 1950. Advances in Genetics *3*: 33–71.
10. Myers, J. W. & E. A. Adelberg. 1954. Proc. Nat. Acad. Sci. U.S.A. *40*: 493–499.
11. Beadle, G. W. 1945. Chem. Rev. *37*: 15–96.
12. Mitchell, H. K. & J. Lein. 1948. J. Biol. Chem. *175*: 481–482.
13. Yanofsky, C., B. C. Carlton, J. R. Guest, D. R. Helinski & U. Henning. 1964. Proc. Nat. Acad. Sci. U.S.A. *51*: 266–272.
14. Sarabhai, A., A. O. W. Stretton, S. Brenner, & A. Bolle. 1964. Nature *201*: 13–17.
15. Anfinsen, C. B. 1973. Science *181*: 223–230.
16. Pauling, L., H. A. Itano, S. J. Singer & I. C. Wells. 1949. Science *110*: 543–548.
17. Ingram, V. 1957. Nature *180*: 326–328.
18. Stokes, J. L., J. W. Foster & C. R. Woodward, Jr. 1943. Arch. Biochem. *2*: 235–245.
19. Mitchell, H. K. & M. B. Houlahan. 1946. Fed. Proc. *3*: 370–375.
20. Mitchell, H. K. & M. B. Houlahan. 1946. Amer. J. Bot. *33*: 31–35.
21. Horowitz, N. H. & U. Leupold. 1951. Cold Spring Harbor Symp. Quant. Biol. *16*: 56–65.
22. Maas, W. K. & B. D. Davis. 1952. Proc. Nat. Acad. Sci. U.S.A. *38*: 785–797.
23. Horowitz, N. H & M. Fling. 1953. Genetics *38*: 360–374.
24. Horowitz, N. H., M. Fling, H. Macleod, & N. Sueoka. 1961. Genetics *46*: 1015–1024.
25. Fincham, J. R. S. 1957. Biochem. J. *65*: 721–728.

26. SANGER, F., G. M. AIR, B. G. BARRELL, N. L. BROWN, A. R. COULSON, J. C. FIDDES, C. A. HUTCHISON III, P. M. SLOCOMBE & M. SMITH. 1977. Nature *265*: 687–695.
27. SHAW, D. C., J. E. WALKER, F. D. NORTHROP, B. G. BARELL, G. N. GODSON & J. C. FIDDES. 1978. Nature *272*: 510–515.
28. CRAWFORD, L. V., C. N. COLE, A. E. SMITH, E. PAUCHA, P. TEGTMEYER, K. RUNDELL & P. BERG. 1978. Proc. Nat. Acad. Sci. U.S.A. *75*: 117–121.
29. GOODMAN, H. M., M. V. OLSON & B. D. HALL. 1977. Proc. Nat. Acad. Sci. U.S.A. *74*: 5453–5457.
30. VALENZUELA, P., A. VENEGAS, F. WEINBERG, R. BISHOP & W. J. RUTTER. 1978. Proc. Nat. Acad. Sci. U.S.A. *75*: 190–194.
31. TILGHMAN, S. M., D. C. TEIMEIER, J. G. SEIDMAN, B. M. PETERLIN, M. SULLIVAN, J. V. MAIZEL & P. LEDER. 1978. Proc. Nat. Acad. Sci. U.S.A. *75*: 725–729.
32. TILGHMAN, S. M., P. J. CURTIS, D. C. TEIMEIER, P. LEDER & C. WEISSMAN. 1978. Proc. Nat. Acad. Sci. U.S.A. *75*: 1309–1313.
33. WEINSTOCK, R., R. SWEET, M. WEISS, H. CEDAR & R. AXEL. 1978. Proc. Nat. Acad. Sci. U.S.A. *75*: 1299–1303.
34. HERSHEY, A. D. 1970. Nature *226*: 697–700.

DISCUSSION OF THE PAPER

KARLSON: I would like to say a few words in favor of the insects in the role of the development of molecular genetics. This started with the discovery by Caspari in the laboratory of Kühn, in 1933, a couple of years before *Neurospora* came up, that a transplantation of a piece of the wild strain into the mutant strain of *Ephestia* and *Ptychopoda* had the effect that the color of the wild type arose in the mutant strain, the A strain of *Ephestia.* This was followed up later by Kühn and collaborators and a similar mutant was detected and investigated in the laboratory of Ephrussi and later by Ephrussi and Beadle, or Ephrussi and Tatum.

And Butenandt and his collaborators joined Alfred Kühn in investigating these color mutants. It is remarkable that it was realized that a diffusible substance could be produced by the wild type transplant to make the pigment of the wild type in the mutant strain.

However, under the influence of the preoccupation with hormones they were called gene hormones instead of substrates, which is what they really are. In further experiments searching for the chemical nature of these gene products, it was found that kynurenine was one of the key substances for *Ephestia* as well as for the vermilion strain of *Drosophila*, by Weidel and Butenandt in 1939 or 1940. Soon afterwards Kühn himself detected that the kynurenine was incorporated into the pigments so that it was really a precursor for the pigment and not of the mutant.

Similar experiments to identify the vermilion substance—it was called the gene product—were made by Tatum, Ephrussi, and Hammersmith, and they came up with an additional compound of sucrose with kynurenine, which they had isolated due to the fact that the medium contained sucrose. This later led to a very elegant separation of synthetic kynurenine into its D & L modifications by Butenandt. So in this whole concept, I would say the insect was the very beginning of the story and that Tatum and Beadle later shifted to *Neurospora* in searching for the proper organism in which to study the relationship of genes to metabolism in a more rigorous and elegant way.

HOROWITZ: Yes, when I was at Stanford in 1939 and 40, Beadle and Tatum were isolating what they called the vermilion-plus substance, that is, the substance made by the wild-type fly that was not made by the vermilion mutant. Tatum obtained these crystals that eventually turned out to be a sucrose ester of kynurenine but in the meantime Butenandt had taken the bull by the horns and he had systematically tested every known metabolite of kynurenine. It was Tatum who originally found that it was a derivative of tryptophan because he had to grow a bacterium on tryptophan-containing medium to produce this activity. Butenandt tested all the known metabolites of tryptophan and discovered that kynurenine was it, and then later that 3-hydroxy-kynurenine was the next step. As you say, this was the first indication that two different steps in a sequential pathway are controlled by different genes. But, Beadle then found that he was blocked in doing any more along these lines because there were no other mutants that they could use in *Drosophila.* He and Ephrussi systematically tested large numbers of *Drosophila* mutants. These were the only ones that behaved in this nonautonomous way; all other mutants were autonomous in the sense that if you transplanted the mutant part to the wild type, the mutant organ, the eye or whatever it was, just grew up as a mutant. It was self-contained; it was uninfluenced by the environment it was growing in; and that is why he turned to *Neurospora*. Your point is well taken.

HOTCHKISS: I just wanted to say a little in defense of the original presentation, that when we come to the pigments of course we are again talking about products of enzymes, and therefore we are not really talking about the basic developmens that Horowitz was trying to get at.

There were, of course, in the intervening period between Garrod and Beadle and Tatum, very elegant propositions that plant pigments, as visualized by Scott-Moncrieff and Robinson, were all traced to gene products, because these are, after all, the products of the primary gene product. Therefore I think you are going a little bit away from the gene-to-enzyme relation.

HOROWITZ: As I said in my paper, and I still think it is true, no one advanced essentially beyond what Garrod had done until they got on to *Neurospora*.

COMMENT: I wonder whether you could explain certain difficulties with this concept that the gene is the determinant of the enzyme. It is difficult to reconcile this idea that the genes also control development, in that in higher organisms the final form of the organism and its behavior during its life are also supposed to be dependent on the genetic endowment.

In this case genes must be much more complex systems than simply substances of nucleotides. And in fact I think there is a whole school of embryology now, like Kaufman and Wolpert, who assume that genes are a kind of information mechanisms which produce amounts of information like a computer for instance. If it is instructed in a specific way, it produces information.

How can you reconcile the fact that the development of an organism is also genetically conditioned in this very simplistic idea of the production of certain substances?

HOROWITZ: Well, that of course is the essential problem of modern developmental biology. How do the genes express themselves, and what is the program? But the way it works, we think, is that the genes control their own expression and there is a program in the genome that tells a given gene when it is to be turned on and when it is to be turned off. The essential difference between, let's say, a man and a horse is not that they make different proteins—they make very much the same proteins—but when one gene turns on and turns off and how long it is. It is a matter of timing. That is the current idea, but how this happens is very much now under investigation in many different laboratories.

ROBERT OLBY (*University of Leeds, Leeds, England*): I was fascinated by Dr. Horowitz's paper and I would like to take up just one or two points if I may. It was not quite clear to me whether in looking at the reception of Garrod's work you took the view that it was wrongly neglected, or whether your view was that in the context of the time it was very justifiably and understandably put on one side?

HOROWITZ: I think both statements are true. I think one can understand what was happening but on the other hand there were people who made a specialty of what was then called physiological genetics, who were writing about it and influencing students, and who really wrongly ignored Garrod's work. I might mention Goldschmidt for one. Goldschmidt was the leading physiological geneticist in the 1920s and 30s. He wrote a book published in the 30s on physiological genetics which had an enormous bibliography. Every paper of any relevance to physiological genetics is mentioned, except Garrod. Garrod's name does not appear in the bibliography.

Goldschmidt said later that it was obvious that Garrod's name should have been in that bibliography. Garrod's work was the most important up to that time in physiological genetics.

OLBY: I gather that the work on anthocyanin pigments was much better known, and, despite what was said a little earlier, I had the impression that the recognition of the clear differences between plant forms with slightly different pigments was attributed to different factors lacking in the metabolic machinery at some specific stage, so that surely that work should have been regarded as equivalent in impact to Garrod's work.

HOROWITZ: No, it was not. First of all, it was well known because Haldane was behind it. I think Haldane was the intellectual father of that. And of course they had the resources of the John Innes Institution in Britain, and the work is very elegant on the pigments of ornamental plants. But they were dealing with changes at the far end of the metabolic machinery, far from the enzyme itself. In fact as far as I know, nobody knows now how one anthocyanin is changed to another one in the course of development. The enzymatic mechanisms are not worked out. Thus, Haldane later felt that the *Neurospora* was a big advance over what they were trying to do. He had the right idea but it was just the wrong material. It was just not as favorable as a microorganism.

OLBY: Please do not have the impression that I was trying to compare the anthocyanin work with the *Neurospora* work. I was comparing it with the alcaptonuria.

HOROWITZ: Well, in the alcaptonuria work the big thing was that homogentisic acid had been isolated and there was an enormous amount of experimentation on the feeding of homogentisic acid to man. It was done in Germany and in Britain, and it was quite clear that homogentisic acid was an intermediate in the catabolism of phenylalanine and tyrosine. And these compounds never got past that stage in this mutant. The case I think was much clearer and much simpler to understand. There was no equivalent that I am aware of in the plant pigment work.

OLBY: You have mentioned a number of reasons for the rejection or the neglect of this early work, You did not, as far as I recall, mention any philosophical reasons, and I draw this to our attention because Nils Hansen in Norway has done a rather nice little paper on several people in biology who had what he calls holistic views about the organism.

In his analysis he includes Morgan and Delbrück, and he draws particularly on the quite late paper of Delbrück at the Connecticut Academy in 1947, showing some very clear examples of the view that it is not possible by scientific techniques to arrive at an analysis that will satisfactorily account for processes that are under the control of the whole of the organism. You seem to be suggesting that this was true of Wilson. Did you take this view?

HOROWITZ: I wanted to add an epilog to this paper but I realized I would run way over time. The history of genetics is nothing if not theatrical. I think that people are afraid of genetics in a sense. Genetics displays the fundamental determinism of living matter and many brilliant geneticists who have made important contributions pull back from that final recognition. Bateson, for example, who was the leading mandarin of genetics in the early part of the 20th century, refused to accept the chromosome theory. He would never believe that the genes were material substances that could be identified with chromosomes. For him it was always too materialistic.

The Lysenko controversy, the present argument we have about genetics and I.Q., about sociobiology, about recombinant DNA, these are all part I think of this fabric, and for some reason human beings resist the implications of genetics. About Wilson, I am not quite sure. I do sense a tremendous resistance on the part of people who should know better about the meaning of genetics.

PAUL ZAMECNIK graduated from Harvard Medical School in 1936. After completing his internship and residency, he began work in the basic sciences with Linderstrøm-Lang, and then Max Bergmann at the Rockefeller Institute. He then returned to Harvard, eventually becoming a Professor of Medicine. Dr. Zamecnik also worked with Linus Pauling and A. R. Todd. He has been involved with protein synthesis from its beginnings.

Historical Aspects of Protein Synthesis*

PAUL C. ZAMECNIK†
National Institutes of Health
Bethesda, Maryland 20038

Huntington Memorial Laboratories
of Harvard University
at the Massachusetts General Hospital
Boston, Massachusetts 02114

EARLY CONCEPTS

DETERMINATION OF THE alpha peptide bond as the linkage mechanism joining amino acids into the long linear chain comprising the primary structure of the protein molecule provided the cornerstone for all subsequent work on the structure and synthesis of proteins.[1] In the early days of awareness of enzymes as biological catalysts there was a firmly fixed concept of the reversibility of their activity. It was considered that the same enzymes were involved in the synthesis as well as the degradation of proteins.[2] In later pursuits of this possibility, Wastenys and Borsook[3] found that concentrated partial hydrolysates of egg albumin would form "plasteins," or large, protein-like masses, in the presence of pepsin or trypsin. Voegtlin and his colleagues[4] also reported extensively that in the presence of sulfhydryl compounds and oxygen, soluble peptide-like components of ruptured cells precipitated out of solution readily, in what they considered to be conditions favoring resynthesis of proteins by intracellular proteolytic enzymes. In repeating these experiments, however, Linderstrøm-Lang and Johansen[5] stated firmly that during this aeration of protein-proteinase digests no protein synthesis had taken place. In retrospect it would appear that in the case of Voegtlin's experiments, disulfide cross linking of existing long peptide chains and aggregation of partial digests into insoluble products were taking place.

The development of new methods for synthesizing synthetic peptide substrates by the Bergmann-Fruton School[6] made it possible to demonstrate that intracellular proteolytic enzymes had sharp specificities, and

* This is publication No. 1549 of the Harvard Cancer Commission. This work was supported by Contract No. 1-CP-71007 of the Virus Cancer Program, Contract Ey-76-5-02-2404 of the Energy Research and Development Administration, and a Grant-in-Aid from the American Cancer Society.

† Fogarty Scholar, N.I.H.

0077-8923/79/0325-0269 $01.75/0 © 1979, NYAS

to examine in greater detail the question of whether proteolytic (i.e., intracellular catheptic) enzymes might synthesize, as well as cleave, peptide bonds under favorable circumstances. Thus it was in fact shown that peptide chain elongation took place in the presence of mixtures of peptides and catheptic enzymes.[6]

In the field of carbohydrate metabolism at this time, however, mechanisms were being discovered whereby energy derived from degradation of glycogen could be stored and transferred by means of phosphate anhydrides and esters.[7] The enzymes involved in the hydrolysis of phosphate esters were found to be distinct from those capable of synthesis of phosphate esters. Thus Lipmann, who had been a student of Meyerhof's, and an expert in the intricacies of the Pasteur effect, was mentally prepared to consider that in the less understood field of protein synthesis an undiscovered phosphorylating enzyme might be present and responsible for this endergonic activity within the living cell,[8] rather than the well-known proteolytic enzymes. In the same year, Kalckar, who had recently coined the term "oxidative phosphorylation" had also suggested the possibility of a phosphorylated intermediate.[9] Borsook and Huffman[10] had also at this time pointed out from calorimetric data that the equilibrium point in the reaction dipeptide $\rightleftarrows$ amino acid lay far on the side of the amino acid.

THE CARBON-14 ERA

World War II interrupted thought and experimentation on this problem. In 1946 came information that a new isotope of carbon was available, one with a long half-life in contrast to the 20-minute half-life of ^{11}C, the latter too short for convenient use. In a brief time, ^{14}C-amino acids were synthesized by several laboratories, including by our colleague, Robert Loftfield.[11] Loftfield also resolved the D- and L-isomers of the chemically synthesized labeled amino acids, and supplied us with the highly purified L-form. ^{35}S-labeled cysteine and methionine were also synthesized and used at this time,[12, 13] but the propensity of these amino acids to engage in reactions other than those involved in alpha peptide bond synthesis made results obtained with them difficult to interpret. Within the next year it was shown that ^{14}C-labeled amino acids were built into proteins and were degraded in whole animals and in surviving tissue slices.[14-18] It was further demonstrated in tissue slice experiments that incorporation of labeled amino acids into protein required oxygen,[17] and was inhibited by azide.[15] Particularly compelling was the finding in 1948[19] that the addition of dinitrophenol to incubating liver slices abolished the uptake of labeled amino acids into protein. At this time it had just been observed by Loomis and Lipmann[20] that dinitrophenol abolished oxidative phosphorylation. Thus a connection between the energy of phosphorylation and isotopic labeling of protein was highly suggestive.

THE TERM "INCORPORATION"

In the above statement was embodied the equivocation that incorporation of a labeled amino acid into protein might not be the same as protein synthesis. There were several reasons for this circumspection in terminology, as follows. In the first place, the pioneering work of Schoenheimer, Rittenberg, Ratner, Shemin, Bloch, and their colleagues[21] left undecided the question of whether amino acids labeled with heavy isotopes made their way into the proteins of animals by way of *de novo* synthesis of protein molecules or by exchange of individual amino acids in existing peptide chains, with reason favoring the former mechanism. Some contemporary experiments had, however, suggested that both mechanisms might apply, depending on the availability of the supply of amino acids.[22, 23] Secondly, there was the nagging consideration that the bonding of the labeled amino acid might be in other than alpha peptide linkage. There might be covalent attachment to the side chains of the amino acid members of a long peptide chain, a thought later found under certain conditions to be a reality by Schweet[24] in the case of lysine incorporation. Finally, there was the uncertainty that the labeled atom in an amino acid remained in that same molecule during the incorporation experiment.

THE USEFULNESS OF STARCH COLUMN CHROMATOGRAPHY

In answering the last mentioned imponderable, the new and elegant amino acid separation procedure of Stein and Moore,[25, 26] introducing starch column chromatography, provided a useful tool. Before their method was even published, Moore and Stein tutored us in its use. My colleague Ivan Frantz and I found, as had been demonstrated earlier by the Schoenheimer group, that certain amino acids were extraordinarily labile metabolically (alanine, glycine, and aspartic and glutamic acids in particular), and the ^{14}C label turned up in carbohydrate, lipid, and in other amino acids at the termination of a slice (or later a homogenate) experiment.[27] Other amino acids such as leucine, valine, and isoleucine were much more stable metabolically, and therefore came to be preferred. Starch column chromatography separated nearly every amino acid unambiguously, and represented a great advance over the chemical derivative isolation procedures elaborated by Bergmann, and slaved over in earlier days by Stein and Moore.[28] The point of emergence of each amino acid from the starch column under specified elution conditions had been established by means of single amino acids and varied mixes. Moore and Stein allowed us to provide the first protein hydrolysate (nonlabeled, they specified), which behaved in a completely predictable way in its elution pattern. Thus, when we returned home, experiments on labeled protein hydrolysates showed that the same labeled amino acid was present in the protein at the end of the incubation as the one we had added.[27] As a result of partial hydrolysis of labeled proteins, it was later shown that labeled amino acids were located in the interior of long peptide chains.[29, 30]

THE COMPLEXITY OF THE SEQUENCE OF INSULIN POSES A PROBLEM

At a Cold Spring Harbor Symposium in 1949 a remarkable paper was presented. Fred Sanger related that the sequence of amino acids that he had thus far deduced in insulin was far more complex than had been anticipated.[31] Previous hypotheses of cyclol cages[32] and repeating blocks of peptide subunits[33] were blown away. In their place emerged the concept that each protein might have a unique structure, consisting of a complex and unpredictable sequence of amino acids. Resolution of the complete primary structure of insulin,[34, 35] instead of clarifying its mechanism of action, unexpectedly presented the deeper mystery of why the amino acid sequence was necessarily so intricate. It was at this point that biochemists began to wonder at the possible relationship between the transforming DNA of Avery, MacLeod, and McCarty[36] and the amino acid sequence of proteins. There was also the early finding of Caspersson[37] and Brachet[38] to be taken into account—namely, that tissues such as the pancreas, which synthesized protein in great quantity for export, had a high concentration of nucleic acid in their cytoplasm, the nucleic acid being different from that in the nucleus. In the interests of greater understanding of the mystery of protein synthesis (cf. the uncertainty as depicted in FIGURE 1), it appeared imperative at this time to search for conditions under which protein might be synthesized in a cell-free system.

This quest occupied the attention of several laboratories[39-41] in the late '40s and early '50s. A more modest and logical objective was simultaneously being pursued by other investigators—Lipmann and Bloch in particular. In the case of the former, the discovery of coenzyme A was in a way a by-product of the desire to study the mechanism of formation of a simple peptidic bond.[42] Johnston and Bloch[43] investigated the syn-

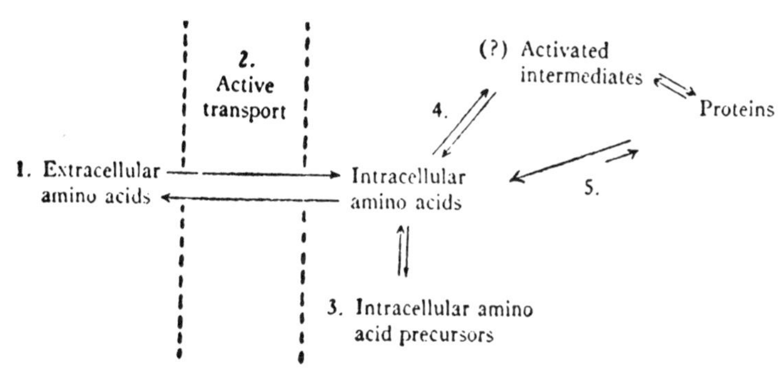

FIGURE 1. 1949 conception of protein synthesis. (From Zamecnik & Frantz.[27])

thesis of the tripeptide glutathione. In both instances, the necessity for phosphate bond energy participation was found in the formation of the peptide bond.

THE DEVELOPMENT OF A RELIABLE CELL-FREE SYSTEM

But energizing the formation of a peptide bond was one thing, and arranging the sequence of amino acids in a protein was another. Somehow those interested in finding a cell homogenate still capable of synthesizing protein hoped to find both answers, possibly intertwined. For the period between 1948 and 1952 several laboratories including ours added labeled amino acids to broken cell preparations, ranging from *E. coli* to rat liver, with indecisive results. At this time our colleague Robert Loftfield asked the question whether the proteolytic enzymes inside the cell had the requisite specificity to achieve a structure as unique as Sanger had found insulin to be. He and his associates[44] found that although an extract of rat liver could degrade α-aminobutyrylglycine as well as it did valylglycine, a rat liver slice excluded [^{14}C]-α-aminobutyrylglycine, and yet readily incorporated [^{14}C]-valine into protein. Thus the specificity of the intracellular enzymes was insufficient to qualify them for the job of synthesizing biologically active proteins.

Our colleague Siekevitz eventually perfected conditions in a rat liver homogenate such that incorporation of amino acids into trichloracetic-acid-precipitable protein was dependent on oxidative phosphorylation.[41, 45] At this same time, our colleague Bucher[46] had also discovered that cholesterol could be synthesized in a cell-free rat liver homogenate, provided that the glass mortar and motor driven pestle fitted loosely, thus breaking up a fraction of the minced liver fragments gently rather than disrupting all of the cells completely. In retrospect, this "gentle homogenization" probably avoided rupturing most of the lysosomes, thus keeping the concentration of degradative enzymes in the preparation within tolerable bounds for study of a synthetic reaction. We therefore employed this gentle homogenization technique, but initially no evidence of protein synthesis was found, even in the presence of added ATP. When, however, an ATP-regenerating system (phosphocreatine plus creatine kinase, or phosphopyruvate plus pyruvate kinase) was added to the broken cell preparation and its various additives, a reproducible incorporation of labeled amino acid into protein was observed, one strictly dependent on the presence of ATP.[47, 48] In time this incorporation was found to pass the requisite criteria, including irreversibility,[49] and it became safe to call the process protein synthesis.[50]

DISSECTION OF THE CELL-FREE SYSTEM

This cell-free system now made it possible to investigate the interaction of its essential components. It had already been determined in intact cellular experiments that ribosomes were the site at which new protein was being synthesized, beginning with suggestive studies by Borsook and collaborators,[18] Hultin,[51] and Keller.[52] This was later demonstrated with certainty by Keller *et al.*[53] It became clear from the cell-free studies mentioned above[48] that at least four components of the cell-free homogenate were required for protein synthesis: amino acid(s), ATP (and an ATP-generating system), the microsome fraction, and enzyme(s) from the soluble, cytosol fraction of the cell. The question was thus posed as to how the amino acids were activated. Was it at the ribosome polypeptide polymerization site, or by the soluble protein-containing cytosol fraction? At this time, Hoagland returned to our laboratories after a period of study in Linderstrøm-Lang's and Lipmann's laboratories. With Novelli in Lipmann's laboratory he had found that the peptidic bond between β-alanine and pantoic acid was formed by a reaction in which a pyrophosphate cleavage of ATP occurred.[54] Contemporaneously, Berg[55] was finding that acetate was activated by an enzyme that interacted with ATP to release pyrophosphate to form acetyladenylate. Thus acetate activation occurred via acetyladenylate rather than by way of the previously favored acetylphosphate. Berg then found,[56] in confirmation of Hoagland,[65] that amino acid activation occurred similarly. Novelli and colleagues synthesized aminoacyl adenylates[57] and found them to function as activated intermediates. Cantoni,[58] studying methionine activation, also observed that an S-adenosylmethionine residue was formed by interaction with ATP, via a pyrophosphorolytic reaction. Other synthetic reactions involving pyrophosphate cleavage of ATP were also becoming known at this time.[59-63] There was also an appealing earlier suggestion by Chantrenne[64] that it was more likely that an organic phosphate compound might be an intermediate in protein synthesis as an aminoacyl anhydride rather than an unsubstituted phosphoryl group itself, as previously postulated,[8] since the former would be a more stable compound. At this time in 1953, our view of protein synthesis was as depicted in FIGURE 2. A more mechanistic model (FIGURE 3) was offered by Lipmann.[71] Hoagland looked for an activated amino acid by two independent methods—the dependence of pyrophosphate exchange into ATP on the presence of added amino acids in preparations containing in addition either the cytosol or the microsome fraction of the cell, in either case the protein-containing fraction being freed of small molecules by repeated precipitation or by dialysis. The results were dramatic,[65] and to some extent puzzling. In the first place, the soluble enzyme-containing fraction, and not the microsome fraction, contained the activation principle. In the second place, separation of the proteins in the soluble protein fraction of the cell revealed the presence of more than one activating enzyme, suggesting the existence of one for each amino acid,[66] a situation which at first blush appeared to be cumbersome.

In order to inquire more directly into the question of whether aminoacyladenylate was an intermediate in the activation step, tryptophan was labeled with ^{18}O in the carboxyl group, and was interacted with purified tryptophan synthetase and ATP. Transfer of an ^{18}O atom from the

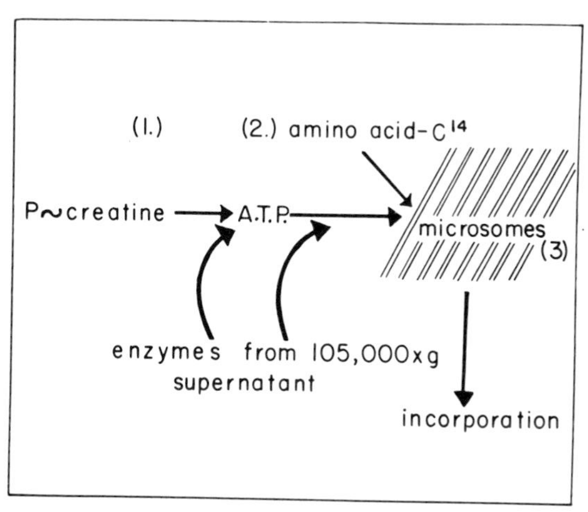

FIGURE 2. A 1953 model of protein synthesis. (From Zamecnik.[89])

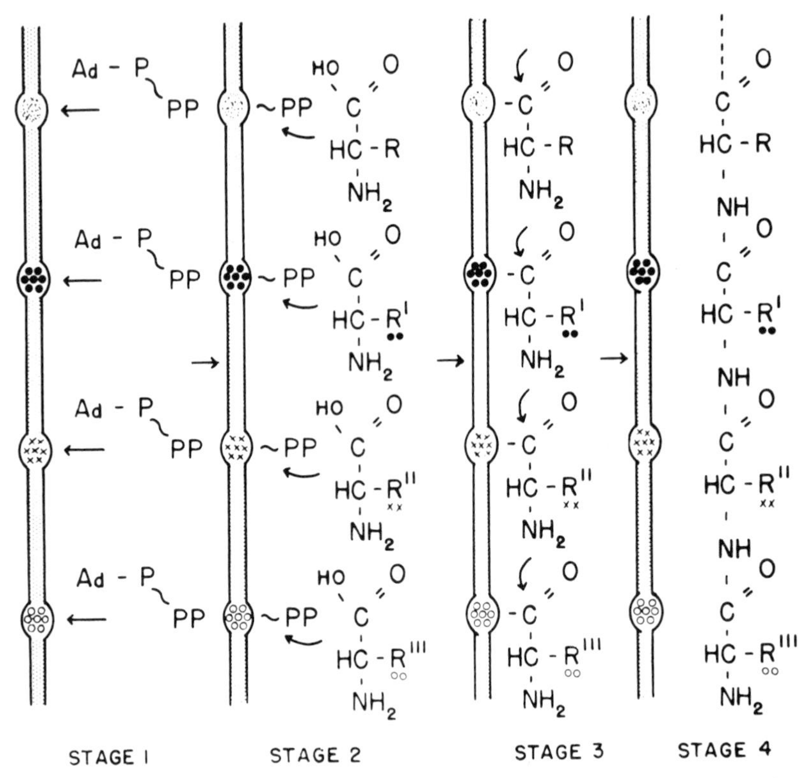

FIGURE 3. 1954 model of amino acid activation and sequence determination. (After Lipmann.[71])

amino acid to AMP was found at the termination of the experiment, indicating that these two molecules had shaken hands in a direct molecular interaction.[67] The possibility of a concerted reaction mechanism, in which aminoacyladenylate is not an obligatory intermediate, was later raised by Loftfield,[68] and remains undecided.

THE ROLE OF THE RIBOSOME IN THE SEQUENCE ARRANGEMENT OF THE NASCENT PEPTIDE CHAIN

It was known that the microsome was composed of both protein and of nucleic acid. Electron microscopic studies of Palade[69] and of Sjöstrand and Hanzon[70] had identified the microsome as an important feature of eukaryotic cells engaged in rapid protein synthesis. The work of Keller *et al.*,[53] referred to previously, pointed out that in the whole animal, injected intravenously with a ^{14}C-labeled amino acid and allowed to

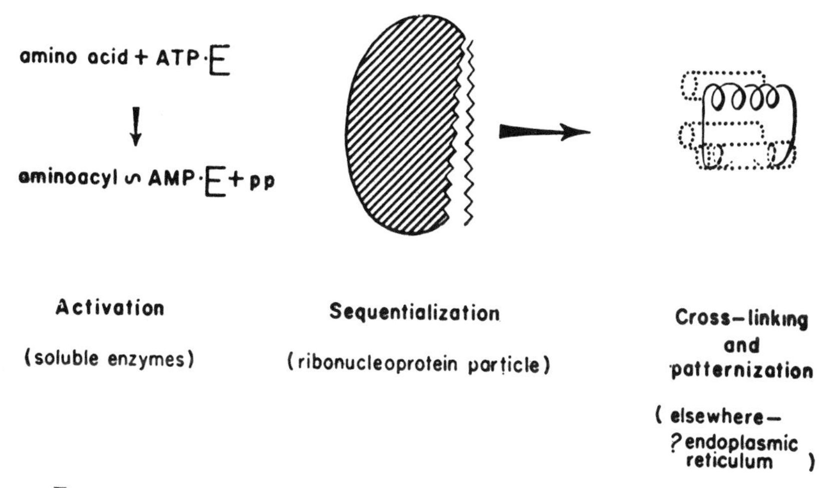

FIGURE 4. 1955 model of sequence determination. (From Zamecnik.[89])

survive for a few minutes, protein associated with the microsomes of the liver initially had the highest specific radioactivity, and then could be released or chased into other cell fractions. This situation is expressed in FIGURE 4, from that time.[89] But what was the mechanism of the sequence arrangement? A second model (FIGURE 5) was published by Koningsberger and Overbeek[72] and a modification of that model as depicted by Loftfield[73] is shown here. Its essence is that the side chains of the amino acids are proposed to fit pre-existing cavities in the ribonucleoprotein particle, and to determine the sequence of the growing peptide chain. The function of the nucleotide part of the aminoacyl-tRNA was not appreciated.

AMINO ACIDS AND NOT PEPTIDES AS PRECURSORS

Another open question concerning the mechanism of protein synthesis at this time was whether the polymerization occurred from single activated amino acids, or whether smaller definitive peptides might first be formed[74, 75] and then linked up to form the larger chain. Loftfield[73] induced the synthesis of ferritin *in vivo* in the whole rat, and then maintained a constant intracellular specific activity of ^{14}C-labeled leucine, isoleucine, and valine. After one to three days, ferritin isolated from the rat livers had these three amino acids with the same specific radioactivity as the intracellular free amino acid pool, and had much higher radioactivity than other liver proteins. Thus the ferritin could not have been derived significantly from proteins that had existed previously, and

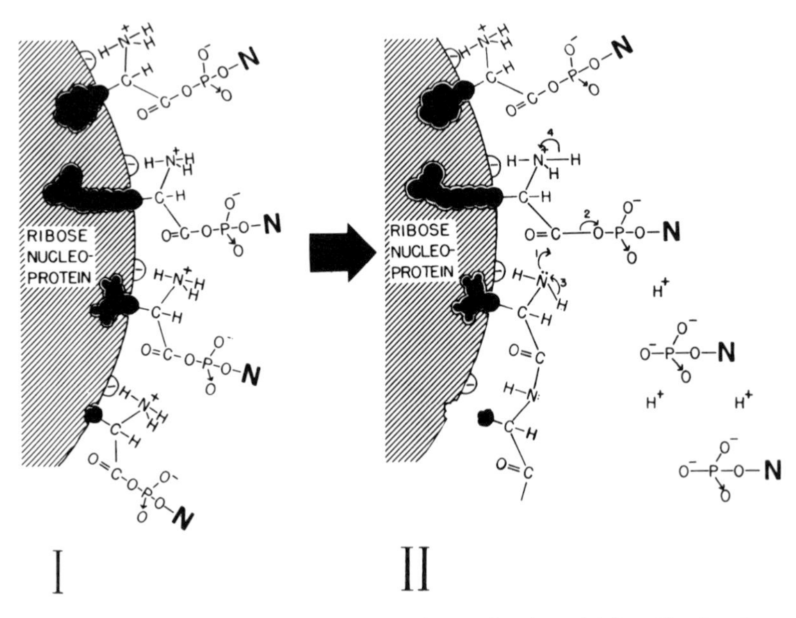

FIGURE 5. A 1956 model of sequence determination. (After Koningsberger & Overbeek.[72])

protein synthesis appeared to proceed from polymerization of individual aminoacyl units without appreciable participation of intermediate peptide blocks.

THE DISCOVERY OF TRANSFER RNA

There were in the years 1954–56 several hints in the air that something might be missing between the activation of amino acids in the cytosol of the cell and the determination of sequence on the ribosomes. In the first place, the suggestions offered in FIGURES 3 and 5 seemed wanting in a sound theoretical basis for sequence specificity. In an experimental sense, Hultin[76, 77] then offered evidence for the presence of an unknown, ribonuclease-sensitive, intermediate step. It seemed also that a labeled amino acid on the way to a completed protein could not be diluted out by unlabeled amino acid. We had furthermore been disappointed in 1955 (in unpublished experiments) that a combination of purified tryptophan-activating enzyme (kindly furnished by Davie *et al.*[78]), ATP, [^{14}C]-tryptophan, and washed ribosomes failed to incorporate tryptophan into protein.

At this time, in November 1955, we were aware of the discovery of the formation of polyadenylic acid by Ochoa and his colleagues.[79] We wondered whether our same cell-free conditions that were sufficient to achieve protein synthesis might not also work for RNA synthesis. We were also suspicious that the synthesis of a ribonucleic acid of complex sequence might follow a different path from that of poly(A) synthesis. In addition, since protein synthesis and DNA synthesis proceeded with the energy being furnished by a pyrophosphate splitting of the energizing ATP, we speculated on why this might not also happen in the case of heterocomplex RNA synthesis. We therefore added ^{14}C-ATP to our cell-free liver system, and following a short incubation, isolated the total RNA. The RNA was labeled with the ^{14}C that had originated in the precursor ATP. We repeated the experiment, and in addition added [^{14}C]-leucine to one flask as a control on the post-incubation washing procedure. When the RNA was isolated from this flask, it was also found to be labeled. In spite of vigorous washing procedures, the RNA remained labeled with the [^{14}C]-leucine, and we concluded that there must be a covalent attachment. But to which cellular RNA fraction? We omitted the ribosome fraction in the next experiment, and found that the residual RNA that was isolated was labeled with amino acid. At this point, our colleague Jesse Scott pointed out to us that a small RNA fraction in the cell, amounting to perhaps 10–15 percent of the total, did not centrifuge down with the ribosomes, was of smaller molecular weight and of obscure function. Currently, a number of investigators were becoming aware of its presence.[80-84] We immediately considered that this amino acid moiety attached exclusively to the low-molecular-weight RNA might be an intermediate between amino acid activation in the cytosol and the sequence determining step on the ribosomes. It was, however, also possible that following activation, energized aminoacyl entities were stored in covalent attachment to this soluble RNA, or "sRNA" as we dubbed it, on a side path, which fed into the direct path between free amino acid and completed protein as needed. When Mahlon Hoagland finished working on the amino acid activation step, he undertook to investigate this amino acid-sRNA combination, and to define whether or not it played an important role in protein synthesis.[85] FIGURE 6 depicts the precursor-product relationship[86] between aminoacyl-tRNA and newly synthesized protein on the ribosome, which clearly emerged from this study. We therefore considered the sRNA, or transfer RNA (tRNA) as Schweet subsequently named it, to serve as a translation piece between language of the gene and that of the protein.

In the summer of 1956, when these thoughts were going through our heads, we heard that Francis Crick had suggested on theoretical grounds

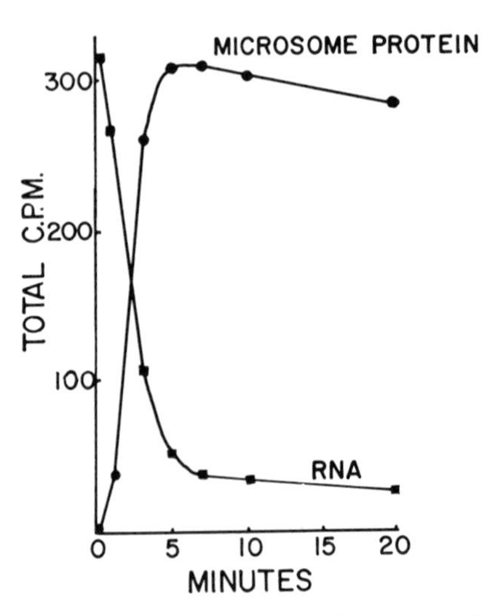

FIGURE 6. The precursor-product relationship of tRNA and nascent protein. (After Hoagland *et al.*[86])

that a trinucleotide might serve as an adaptor piece attached to an amino acid to direct it to the proper site on an RNA template to establish the specific sequence of a growing peptide chain.[87] We were curious as to why the tRNA we had found was so large—why 75 nucleotides rather than 3? It was possible that the tRNA was chopped to a smaller size before associating with the ribosomal RNA template. This was found not to be the case,[88] however, and we then looked for an explanation in the second type of coding operation that was present in this puzzle—namely, how the aminoacyl synthetase recognized its cognate tRNA—since there was an aminoacyl synthetase for each amino-acid-specific tRNA.[66] The problem of how these two different types of macromolecule recognize each other has remained an elusive mystery of molecular biology, in spite of two decades of speculation and probing by numerous workers (see, for example, References 50, 89–94). In the summer of 1956 we also became aware that Holley had found a ribonuclease-sensitive intermediate step between the activation of alanine (but not of the other amino acids he had tested) and the ribosomal polypeptide forming step in protein synthesis.[95] At this same time, unbeknown to us, Ogata and Nohara[96] in Japan were also finding evidence for an aminoacyl-RNA intermediate.

If, as predicted by the adaptor hypothesis[97] the placing of an amino acid in the proper sequence in a growing peptide chain rested with the transfer RNA to which it had become attached, then if one were to attach the wrong amino acid to a particular transfer RNA, it should become positioned incorrectly in a growing peptide chain. Such an experiment was devised by Lipmann and colleagues,[98] who formed cysteinyl-tRNA, then converted it to alanyl-RNA, and added the latter to a system which synthesized a specific protein containing a cysteinyl residue. An alanyl residue turned up in this particular location, thus proving that the specificity for positioning in the nascent peptide chain resided with the tRNA.

THE COMMON-CCA END OF ALL TRANSFER RNAS AND THE SITE OF AMINO ACID ATTACHMENT

As mentioned above, we found that [^{14}C]-ATP, when incubated with the cell-free protein-synthesizing system, transferred its label to the RNA in the system. This turned out also to be the sRNA (i.e., tRNA) fraction, and the reaction proceeded by a pyrophosphate cleavage with AMP addition, and catalyzed by a special enzyme.[50] Berg and Ofengand[99] demonstrated that the aminoacyl synthetases were also responsible for the aminoacylation of the tRNAs. Initially we wondered whether the mononucleotide additions to tRNA might not be part of a total RNA synthesis, different from the Grunberg-Manago and Ochoa mechanism, one with greater specificity. Curiously enough, however, only labeled CTP and ATP were able to be incorporated into the tRNA, and the stoichiometry indicated that two C residues and one A residue were added to the 3′-end of each tRNA molecule.[100, 101] Another interesting finding was that the A residue was terminal, and its presence was required in order for the amino acid to become covalently attached to the tRNA.[102] It seemed likely therefore that the site of aminoacylation was on the terminal adenosine moiety. But where on the adenosine? On the ribosyl group or on the adenine base? Zachau *et al.*[103] and our group[102] looked for the answer contemporaneously by different routes. The former investigators attached a labeled amino acid to tRNA, hydrolyzed the RNA with pancreatic ribonuclease, isolated labeled aminoacyl-adenosine, and showed that the aminoacyl group was attached to the 2′ or 3′ position of the ribose. We found that periodate oxidation, which converts to a dialdehyde the vicinal 2′,3′-ribosyl hydroxyl groups of the terminal adenosine residue, abolished aminoacylation.[102] We also observed that borate, which complexes with vicinal hydroxyl groups, inhibited aminoacylation. Thus the same conclusion was reached by both groups. The question of whether the 2′- or the 3′-ribosyl position was the site of initial aminoacylation remained unsettled for many years because of the rapid equilibration of the aminoacyl groups between these positions.[104-106] It appears likely at present, however, that certain amino acids are initially aminoacylated preferentially at the 2′-position, and others at the 3′-position.[107-110]

THE DISCOVERY OF THE ROLE OF GTP IN POLYPEPTIDE CHAIN ELONGATION

When microsomes were separated from the $100,000 \times g$ supernatant fraction of disrupted liver cells, then redissolved in fresh suspension medium, and repelleted, they were freed from the preponderance of low-molecular-weight substances trapped in the microsomal pellet. The $100,000 \times g$ supernatant fraction of the cell was adjusted to pH 5, the precipitated protein was collected as a pellet, then redissolved at pH 8, and once more adjusted to pH 5, and the protein precipitate redissolved in fresh medium. This soluble-enzyme-containing fraction was in this way also freed from low-molecular-weight constituents. When now the "washed" microsomes and supernatant proteins were used along with other components of the protein synthesizing mix including ATP, there was little or no evidence of synthetic activity. If, however, we employed a crude preparation of ATP from a commercial source, rather than a purified one, the ability of the system to synthesize protein was restored. We suspected that another nucleoside triphosphate in addition to ATP was necessary. Dr. Rao Sanadi had at this time just found that GDP played a role in intermediate carbohydrate metabolism, and he was kind enough to send us a little of this precious compound, not available commercially. Dr. Keller and I observed that GTP (or GDP, we couldn't quite distinguish which at that time) restored the ability of our preparations to synthesize protein.[111] It was clear that the point at which GTP acted was in the polypeptide polymerization step, not earlier.[86] We also thought that there were probably unidentified enzymes involved in the peptide polymerization step(s), and later adduced the first evidence for their presence (FIGURE 7).[112]

DIRECTION OF GROWTH OF THE NASCENT CHAIN

Although it could be shown that a labeled amino acid was built into the interior of a growing peptide chain in a cell-free system, there were in the early days two troublesome uncertainties about the observation. The first was whether new chains were being initiated, or whether existing ones were merely being extended. The second imponderable was whether the growth of the chain occurred from the amino terminal end or from the carboxyl terminal end. Loftfield puzzled about this question,[68] and pointed the way for others to follow. Bishop, Leahy, and

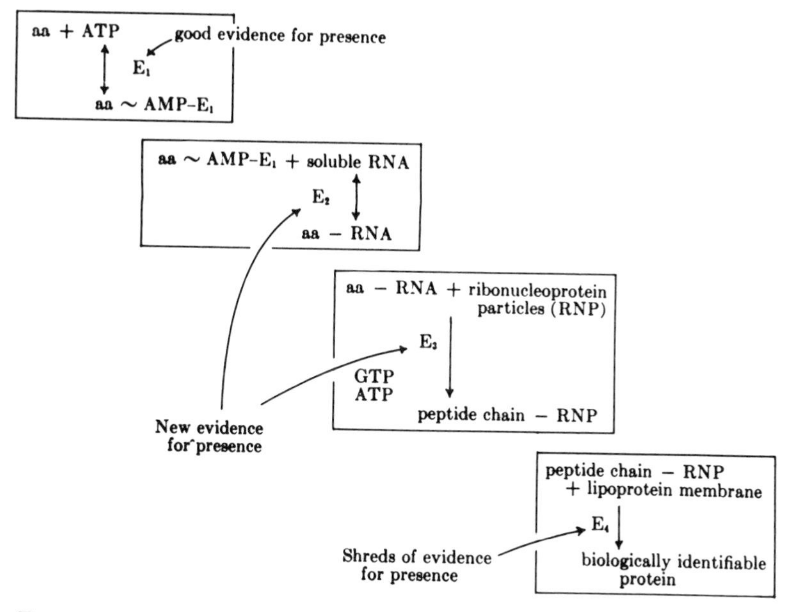

FIGURE 7. A 1958 conception of steps in protein synthesis. (From Zamecnik *et al.*[112])

Schweet[113] and Dintzis[114] then showed that the chain grew from the amino terminus to the carboxyl end. Schweet and colleagues[113] employed the reticulocyte cell-free system of the Borsook group, which synthesized hemoglobin principally, and concluded that the cell-free systems suffered from two lesions—difficulty in initiating new chains, and premature terminations. The first-mentioned difficulty remained for some years before being resolved by the discovery and use of messenger RNA to initiate new chains in the presence of special added factors, and the second was solved in part by the addition of special factors to promote chain elongation.

THE MORE SPECIFIC TERM "RIBOSOME" REPLACES "MICROSOME" IN PROTEIN SYNTHESIS

The term microsome was an operational one, designating that fraction of disrupted cell contents which pelleted upon centrifugation (at 100,000 × *g* for an hour) of the 7500 *g* supernatant portion in the cell disruption medium. This microsome fraction contained lipids, glycogen, cytochrome, ribonucleoprotein particles, and other unknown constituents.[49] Sodium deoxycholate had been found to solubilize much but not all of the microsome fraction.[115] Littlefield *et al.*[49] found the highest specific radioactivity in the ribonucleoprotein, or ribosome, fraction shortly after a brief pulse of radioactive amino acid either to a whole rat (i.e., in the rat liver) or to a cell-free protein-synthesizing system prepared from rat liver. The ribonucleoprotein particles appeared identical to the cytoplasmic particles seen by the electron microscopists[69, 70] in cells such as those of the pancreas engaged in synthesis of proteins for export. It was concluded that the ribonucleoprotein particles were the site of initial incorporation of activated amino acids into protein, and that only a very small fraction of the total amino acids present in the proteins of these particles was involved in the formation of new peptide chains.[49, 116]

CELL-FREE PROTEIN SYNTHESIS IN HIGHER PLANTS AND IN BACTERIA

After the cell-free system had been elaborated for rat liver, we wondered whether one might not establish a similar system for plants. As a graduate student, Mary Stephenson undertook to work on this problem. She chose tobacco leaves as the object of this study, since they provided the added virtue of susceptibility to tobacco mosaic virus (TMV), and one might inquire whether TMV could be synthesized under cell-free conditions. In the light of 25 years hindsight, this latter thought seems optimistic, but that long ago the complexity of plant viruses was less appreciated. Two useful things came out of this arbeit. A cell-free system for higher plants was established, whose essentials were very similar to those of the animal system.[117] Secondly, it became clear that chloroplasts were, in terms of the requirements for protein synthesis, a semiautonomous part of the higher plant.[117]

It was evident in the late '50s that there would be a great advantage if a cell-free system could be worked out for bacteria, particularly for *E. coli.* The central difficulty, we thought, as a result of earlier unsuccessful attempts of ours (Zamecnik, Stephenson, and Novelli, unpublished data, 1951–52), was the problem of getting rid of all intact bacterial cells, while still being sufficiently gentle in cell disruption to preserve parts of the cell essential for the synthetic activity. Our colleague Lamborg worked for three years on this problem. He finally evolved conditions that were successful, and the critical features were the following: (1) use of early log phase cells, which had high synthetic activities and minimal degradative enzymatic propensities; (2) gentle cell rupture technique, consisting then of hand grinding of the *E. coli* pellet with sand and alumina, to a thick, tacky paste; and (3) constant checking of the proteosynthetic incubation medium for its intact bacterial concentration. The latter had to be below 1×10^5 live bacteria per milliliter in order to avoid the possibility that the observed protein synthesis in the incubation medium was not due in large measure to the live cells present, in which the rate of protein synthesis per mg protein far exceeded that in the cell-free part of the same incubation medium. In respect to this last criterion and safeguard, previous bacterial broken cell preparations had not been trustworthy. This bacterial cell-free system[118] was accepted and adopted by our neighbors Tissieres and Watson (who were also studying protein synthesis in bacterial systems[119]) while still in manuscript form, and they immediately made significant advances with it.[120] As they had predicted, this system for the study of protein synthesis would far outstrip in usefulness the earlier animal and plant systems, and the subsequent cracking of the genetic code by Nirenberg and Matthaei[121] grew out of it.

DEVELOPMENT OF THE CONCEPT OF MESSENGER RNA

There was so much ribonucleic acid in the ribosome particle that initially it was taken for granted that this RNA served as the template for the synthesis of the nascent peptide chain. Beginning in 1953 with the experiments of Hershey and colleagues,[122] followed by the work of Volkin and Astrachan,[123] of Riley *et al.*,[124] of Jacob and Monod,[125] of Brenner, Jacob, and Meselson,[126] and of Watson and colleagues,[127] the concept grew that a segment of RNA might be transcribed from the DNA genome, associate with ribosomes for a brief time while being translated, and then be degraded. The ribosomes on the other hand had a longer life span, and provided a framework to which the messenger RNA became bound. The rather abundant RNA of the ribosome served simply in a structural role.

In answer to a direct query by the writer as to whether the ribosomal RNA might not serve as mRNA, at a meeting in the Vatican in Rome in 1961 Jacob replied,[128] "studies of bacterial enzyme systems in the late '50s began to show that the synthesis of more and more proteins was under the control of some inducible or repressible system. Previously, the synthesis of enzymes belonging to biosynthetic pathways was considered to be constitutive, that is, to be independent of environmental conditions until their repressible character was discovered. In systems under study," Jacob continued, "i.e., catabolic systems such as lactose or galactose utilization, and biosynthetic pathways such as tryptophan or arginine synthesis, the breakdown of one regulation system by mutation results in an enormous production of the corresponding proteins.

This production amounts to nearly 10 percent of the total protein in the investigated cases. It is clear therefore that a bacterial cell could not survive the breakdown of many regulation systems. I would like to think that all structural genes work in a similar way and in particular that they all eventually may be expressed at a similar maximal rate." The second argument for a special messenger RNA, Jacob stated, "comes from studies of protein synthesis in phage-infected bacteria. When *E. coli* are infected with T even phages, protein synthesis proceeds linearly, but no detectable amount of new ribosomes is made. The messenger RNA of the phage becomes associated with the ribosomes which were formed before infection, and which, at this time, were manufacturing bacterial proteins. After 10 or 12 minutes, some 70–80 percent of the total proteins produced correspond to one molecular species, the phage head building block. It seems that after infection, the amount of ribosomes present is the factor limiting the rate of protein synthesis and that the bulk of the ribosomes is engaged in the production of phage head proteins." It would be hard to state the case for the existence of messenger RNA more succinctly.

Brenner *et al.*[126] had also carried out heavy isotope labeling experiments with ribosomes to see how long a ribosome once synthesized was conserved as such. They found that most of the 50S and 30S ribosomes of *E. coli* remained intact for many generations after being synthesized.

DECIPHERING THE GENETIC CODE

Thus by the summer of 1961 the concept of messenger RNA was accepted by all but nucleotide biochemists, who were reluctant to embrace the concept completely in the absence of messenger polynucleotide chains which could be isolated and subjected to chemical examination (see the discussion in Reference 128). It was at this time that Nirenberg presented to a Biochemical Congress in Stockholm the remarkable finding of the translation properties of polyuridylic acid, which he and his colleague Matthaei discovered to code specifically for phenylalanine, producing a polyphenylalanine chain.[121] Thus the messenger RNA concept was placed on firm ground, and the same experiment introduced the first break in the genetic code. Ochoa and his colleagues[129] then found that polyadenylic acid codes for polylysine, and in a brief time these two laboratories identified the nucleotide composition of the code words of virtually all the amino acids.[120, 131] Khorana and his associates[132] then synthesized complex, repeating triplet units of the four ribonucleotides, and removed the uncertainties from a complete specification of the genetic code. This signal achievement in molecular biology took place in just four years.

At this time the messenger RNA was also visualized by electron microscopy, running like a clothes line between ribosomes, thus providing a clear picture of the interaction, and supporting the concept that a number of polypeptide chains might be in the process of formation at different points on a single messenger RNA chain,[133-135] thereby introducing the new term "polysome." The translational movement of the mRNA relative ot the two tRNA binding sites on a single ribosome was also appreciated at this time.[134, 136]

THE ERROR LEVEL IN PROTEIN SYNTHESIS

The high degree of accuracy of protein synthesis was the subject of admiration of Loftfield, who purified L-amino acids, which he synthesized and resolved to a degree not obtainable from commercial sources. He then determined the extent to which one amino acid could replace its higher or lower homolog (i.e., L-valine replace L-leucine or L-isoleucine, for example, in a well-characterized protein being synthesized). The error level was of the order of one part in 3000 or even less. The lower limit could not be determined with accuracy. It appeared to Loftfield[137] that the triplet coding mechanism could not in itself account for this degree of accuracy, and that there must be other unknown reading steps, a point of view unpopular at the time. This consideration has however lately been taken up and developed by other investigators.[138] The hypothesis that ageing represents in part a progressive loss in accuracy in specificity of amino acid placement in protein synthesis[139] also grew out of these quantitative studies of accuracy of protein synthesis.

DETERMINATION OF THE PRIMARY STRUCTURE OF ALANINE TRANSFER RNA

After the discovery of transfer RNA, it became important to find a method of preparing it conveniently on a large scale. The most accessible source was yeast, which we disrupted initially by hand grinding in a large mortar and pestle with sand and alumina. This was however a tedious process, and one lazy day we added 50 percent aqueous phenol to intact yeast rather than to broken yeast. This treatment appeared to permeabilize the yeast cell wall, and cell contents containing tRNA leaked out, while ribosomes and ribosomal RNA were retained within the yeast cell. In this way an easy purification of tRNA was obtained, as elaborated by our colleague Monier.[140] This procedure was employed with a little modification by Holley and his colleagues[141] to prepare the large quantities of starting material necessary for his countercurrent distribution separation of various subspecies of tRNA specific for esterification of individual amino acids. In the period of time 1960–65 other talented investigators engaged in the important pursuit of sequencing a single pure tRNA molecule.[142-144] The countercurrent distribution separation of sufficient quantity of a pure tRNA put the Holley team in a leading position for the difficult and largely uncharted task of sequencing such a large nucleotide, which we shall describe in a little detail, as follows. By means of specific nucleases, smaller fragments of tRNA were produced. These were separated by means of the new 7-molar-urea-containing columns introduced by Tener *et al.*[145] The smaller fragments were degraded stepwise by exonucleases, and their sequence determined. A similar technique was followed on a second sample of the tRNA, but with varied initial enzymatic cleavages producing smaller fragments different from those mentioned above. When these latter were sequenced, the points of overlap of the fragments in the two sets could be seen, and the linear relationship of the sequenced fragments determined. Thus the complete primary structure of yeast alanine tRNA became known, including the location of the numerous minor base constituents present.[146] The complete structure of phenylalanyl tRNA soon followed.[142] Once again, as in the previous instance of the elucidation of the primary amino acid structure of insulin, knowledge of the primary structure of a 77 nucleotide containing tRNA did not immediately clarify its varied complex functions. It is also interesting to note, 13 years later, that a double-stranded DNA, the SV40 genome, consisting of over 10,000 nucleotides, has now been sequenced, using new techniques.[147, 148]

A LIGHT BRUSH OVER A FEW MORE RECENT DEVELOPMENTS

We shall leave for others[149] discussion of many developments which have occurred since 1965, which becomes with a few exceptions the arbitrary cut-off point of this chronicle. A more complete, less personalized, description of the scientific events and the contributors to them in this earlier time period may be found in Medvedev's detailed book on protein biosynthesis.[150] By the mid-1960s, the main participants in the machinery of protein synthesis had been identified, at least by our standards. As for the subsequent years, the crystallization of tRNAs by several groups, almost simultaneously,[151-155] opened the door for x-ray crystallographic studies. It should be mentioned that such studies were greatly helped by the large-scale purification and provision of samples of individual aminoacylating tRNAs by G. D. Novelli of the Oak Ridge National Laboratories and by F. Bergmann of the National Institute of General Medical Sciences.

Determination of the three-dimensional structure of tRNAs has been a major event.[156,157] It is interesting to note that the three-dimensional conformation of a tRNA molecule is determined both by stacking interactions of adjacent bases, and by hydrogen bonding attractions to base pairing areas in nonadjacent parts of the same polynucleotide chain. This situation differs from that determined by Anfinsen[158] for protein structure, in which the primary amino acid sequence largely determines the final tertiary structure.

The identification of a great number of factors, and the complexity of their interplay, in the initiation and propagation of a peptide chain has become an area of enormous interest.[149] The isolation of the many proteins of the 30*S*[159] and 50*S*[160] subunits of the ribosome, and study of their interaction, with reconstitution of the intact subunits, has been a difficult but largely successful major task. Investigations of kinetic proofreading steps have become a major focus.[149]

Among the unsolved questions that remain for the future is the knotty one of how each aminoacyl synthetase recognizes its cognate tRNA. Another is the nature of the biological ratchet mechanism which moves the messenger RNA along with respect to a ribosome, renewing the A or amino acid and P or peptide sites to permit continued growth of the nascent peptide chain.[161] Study of the regulation of protein synthesis has now moved to the center of the stage. The effects of gene splicing, posttranscriptional modification of messenger RNA precursors, and posttranslational unmasking of proteins by proteolytic enzymes all play roles in controlling the integrated behavior of the living cell. The mechanisms of orchestration of these multileveled modifiers of the machinery of growth still await exploration.

REFERENCES

1. Fruton, J. S. & S. Simmonds. 1958. General Biochemistry. John Wiley and Sons, Inc., New York, N.Y.
2. Thierfelder, H. 1903. Hoppe-Seyler's Handbuch der Physiologisch-und-Pathologisch Chemischen Analyse. :321. Berlin.
3. Wastenys, H. & H. Borsook. 1930. Physiol. Rev. *10*: 110–145.
4. Voegtlin, C., M. E. Maver & J. M. Johnson. 1933. J. Pharmacol. *48*: 241–265.
5. Linderstrøm-Lang, K. & G. Johansen. 1939. Enzymol. 7: 239–240.
6. Bergmann, M. J. S. Fruton. 1941. *In* Advances in Enzymology and Related Subjects. F. F. Nord & J. S. Fruton, Eds. Vol. 1: 63–98.
7. Cori, C. F., G. T. Cori & A. H. Hegnauer. 1937. J. Biol. Chem. *121*: 193–202

8. Lipmann, F. 1941. *In* Advances in Enzymology and Related Subjects. F. F. Nord & J. S. Fruton, Eds. Vol. *1*: 99–162.
9. Kalckar, H. M. 1941. Chem. Rev. *28*: 71–178.
10. Borsook, H. & H. M. Huffman. 1938. *In* Chemistry of Amino Acids and Proteins. C. L. A. Schmidt, Ed. Thomas. Baltimore.
11. Loftfield, R. B. 1947. Nucleonics *1*(3): 54–57.
12. Melchior, J. B. & H. Tarver. 1947. Arch. Biochem. *12*: 301–308.
13. Melchoir, J. B. & H. Tarver. 1947. Arch. Biochem. *12*: 309–315.
14. Frantz, I. D., Jr., R. B. Loftfield & W. W. Miller. 1947. Science *106*: 544–545.
15. Winnick, T., F. Friedberg & D. M. Greenberg. 1947. Arch. Biochem. *15*: 160–161.
16. Anfinsen, C. B., A. Beloff, A. B. Hastings & A. K. Solomon. 1947. J. Biol. Chem. *168*: 771–772.
17. Zamecnik, P. C., I. D. Frantz, Jr., R. B. Loftfield & M. L. Stephenson. 1948. J. Biol. Chem. *175*: 229–314.
18. Borsook, H., C. L. Deasy, A. J. Haagen-Smit, G. Keighley & P. Lowy. 1948. J. Biol. Chem. *174*: 1041–1042.
19. Frantz, I. D., Jr., P. C. Zamecnik, J. W. Reese & M. L. Stephenson. 1948. J. Biol. Chem. *174*: 773–774.
20. Loomis, W. F. & F. Lipmann. 1948. J. Biol. Chem. *173*: 807–808.
21. Schoenheimer, R. 1942. The Dynamic State of Body Constituents. Harvard University Press. Cambridge, Mass.
22. Gale, E. F. & J. P. Folkes. 1953. Biochem. J. *55*.
23. Gale, E. F. 1955. *In* Symposium on Amino Acid Metabolism. W. D. McElroy & H. B. Glass, Eds. :171–192. The Johns Hopkins Press. Baltimore.
24. Schweet, R. 1955. Fed. Proc. *14*: 277.
25. Stein, W. H. & S. Moore. 1948. J. Biol. Chem. *176*: 337–365.
26. Moore, S. & W. H. Stein. 1948. J. Biol. Chem. *176*: 367–388.
27. Zamecnik, P. C. & I. D. Frantz, Jr. 1950. Cold Spring Harbor Symp. Quant. Biol. *14*: 199–208.
28. Stein, W. H. & S. Moore. 1946. Ann. N.Y. Acad. Sci. *47*: 95–118.
29. Zamecnik, P. C. & E. B. Keller. 1954. J. Biol. Chem. *209*: 337–354.
30. Zamecnik, P. C., E. B. Keller, J. W. Littlefield, M. B. Hoagland & R. B. Loftfield. 1956. J. Cell. Comp. Phys. *47*(Suppl. 1): 81–102.
31. Sanger, F. 1949. Cold Spring Harbor Symp. Quant. Biol. *14*: 153–160.
32. Wrinch, D. 1948. Science *107*: 445–447.
33. Bergmann, M. & Niemann, C. 1937. Science *86*: 187–190.
34. Sanger, F. & E. O. P. Thompson. 1963. Biochem. J. *53*: 353–374.
35. Sanger, F. & H. Tuppy. 1961. Biochem. J. *49*: 483–490.
36. Avery, O.T ., C. M. Macleod & M. McCarty. 1944. J. Exp. Med. *79*: 137–158.
37. Caspersson, T. O. 1950. Cell Growth and Cell Function. W. W. Norton and Co. New York, N.Y.
38. Brachet, J. 1950. Chemical Embryology. Interscience Publishers. New York, N.Y.
39. Winnick, T., F. Friedberg & D. M. Greenberg. 1948. J. Biol. Chem. *175*: 117–126.
40. Borsook, H., C. L. Deasy, A. J. Haagen-Smit, G. Keighley & P. H. Lowy. 1949. Fed. Proc. *8*: 589.
41. Siekevitz, P. & P. C. Zamecnik. 1951. Fed. Proc. *10*: 246.
42. Lipmann, F. 1945. J. Biol. Chem. *160*: 173–190.
43. Johnson, R. B. & K. Bloch. 1949. J. Biol. Chem. *179*: 493–494.

44. LOFTFIELD, R. B., J. GROVER & M. L. STEPHENSON. 1953. Nature *171*: 1024–1025.
45. SIEKEVITZ, P. 1952. J. Biol. Chem. *195*: 549–565.
46. BUCHER, N. L. R. 1953. J. Am. Chem. Soc. 75: 498.
47. ZAMECNIK, P. C. 1953. Fed. Proc. *12*: 976.
48. ZAMECNIK, P. C. & E. B. KELLER. 1954. J. Biol. Chem. *209*: 337–353.
49. LITTLEFIELD, J. W., E. B. KELLER, J. GROSS & P. C. ZAMECNIK. 1955. J. Biol. Chem. *217*: 111–124.
50. ZAMECNIK, P. C. 1960. Harvey Lect. *54*: 256–281.
51. HULTIN, T. 1950. Exp. Cell Res. *1*: 376–381.
52. KELLER, E. B.1 951. Fed. Proc. *10*: 206.
53. KELLER, E. B., P. C. ZAMECNIK & R. B. LOFTFIELD. 1954. J. Histochem. & Cytochem. *2*: 378–386.
54. HOAGLAND, M. B. & G. D. NOVELLI. 1954. J. Biol. Chem. *207*: 767–773.
55. BERG, P. 1955. J. Am. Chem. Soc. 77: 3163–3164.
56. BERG, P. 1956. J. Biol. Chem. *222*: 1025–1034.
57. DEMOSS, J. A., S. M. GENUTH & G. D. NOVELLI. 1956. Proc. Nat. Acad. Sci. U.S.A. *42*: 325–332.
58. CANTONI, G. & J. DURRELL. 1957. J. Biol. Chem. *225*: 1033–1048.
59. MAAS, W. 1955. 3rd Congress International de Biochime, Brussels.
60. HILZ, H. 1955. Fed. Proc. *14*: 227.
61. ROBINSON, W. G., B. K. BACHHAWAT & M. J. COON. 1955. Fed. Proc. *14*: 270.
62. SCHACTER, D. & J. V. TAGGART. 1954. J. Biol. Chem. *208*: 263–275.
63. DEMOSS, J. A. & G. D. NOVELLI. 1955. Bact. Proc. (of 55th General Meeting) :125.
64. CHANTRENNE, H. 1948. Biochim. Biophys. Acta *2*: 286–293.
65. HOAGLAND, M. B. 1955. Biochim. Biophys. Acta *16*: 288–289.
66. HOAGLAND, M. B., E. B. KELLER & P. C. ZAMECNIK. 1956. J. Biol. Chem. *218*: 345–358.
67. HOAGLAND, M. B., P. C. ZAMECNIK, N. SHARON, F. LIPMANN, M. P. STULBERG & P. D. BOYER. 1957. Biochim. Biophys. Acta *26*: 215–217.
68. LOFTFIELD, R. B. 1972. *In* Progress in Nucleic Acid Research and Molecular Biology. Vol. 12: 87–128. Academic Press. New York, N.Y.
69. PALADE, G. E. 1955. J. Biophys. & Biochem. Cytol. *1*: 59–68.
70. SJÖSTRAND, F. S. & V. HANZON. 1954. Exp. Cell Res. 7: 393–414.
71. LIPMANN, F. 1954. *In* Mechanisms of Enzyme Action. W. D. McElroy & B. Glass, Eds. Johns Hopkins Press. Baltimore.
72. KONINGSBERGER, V. & J. T. G. OVERBEEK. 1955. Proc. Koninkl. Ned. Acad. Wetenschap. B57 :248.
73. LOFTFIELD, R. B. 1957. Prog. Biophys. Biophys. Chem. *8*: 347–386.
74. BORSOOK, H., C. I. DEASY, A. J. HAAGEN-SMIT, G. KEIGHLEY & P. H. LOWY 1948. J. Biol. Chem. *174*: 1041–1042.
75. STEINBERG, D. M. VAUGHAN & C. B. ANFINSEN. 1956. Science *124*: 389–395.
76. HULTIN, T. 1956. Exp. Cell Res. *11*: 222–224.
77. HULTIN, T., A. VON DER DECKEN & G. BESKOW. 1957. Exp. Cell Res. *12*: 675–677.
78. DAVIE, E. W., V. V. KONINGSBERGER & F. LIPMANN. 1956. Arch. Biochem. Biophys. *65*: 21–38.
79. GRUNBERG-MANAGO, M., P. J. ORTIZ & S. OCHOA. 1955. Science *122*: 907–910.
80. HEIDELBERGER, C., E. HARBERS, K. C. LEIBMAN, Y. TAKAGI & V. R. POTTER. 1956. Biochim. Biophys. Acta *20*: 445–446.
81. PATERSON, A. R. P. & G. A. LE PAGE. 1957. Cancer Res. *17*: 409–417.

82. Canellakis, E. S. 1957. Biochim. Biophys. Acta *25*: 217–218.
83. Edmonds, M. & R. Abrams. 1957. Biochim. Biophys. Acta *26*: 226–227.
84. Zamecnik, P. C., M. L. Stephenson, J. F. Scott & M. B. Hoagland. 1957. Fed. Proc. *16*: 275.
85. Hoagland, M. B., P. C. Zamecnik & M. L. Stephenson. 1957. Biochim. Biophys. Acta *24*: 215–216.
86. Hoagland, M. B., M. L. Stephenson, J. F. Scott, L. I. Hecht & P. C. Zamecnik. 1958. J. Biol. Chem. *231*: 241–257.
87. Crick, F. H. C. 1958. Symp. Soc. Exp. Biol. *12*: 138–163.
88. Hoagland, M. B. & L. T. Comly. 1960. Proc. Nat. Acad. Sci. U.S.A. *46*: 1554–1563.
89. Zamecnik, P. C. 1969. Cold Spring Harbor Symp. Quant. Biol. *34*: 1–16.
90. Haines, J. A. & P. C. Zamecnik. 1967. Biochim. Acta *146*: 227–238.
91. Chambers, R. W. 1971. *In* Progress in Nucleic Acid Research and Molecular Biology. J. N. Davidson & W. E. Cohn, Eds. Vol. 11: 489–525. Academic Press. New York, N.Y.
92. Lagerquist, U. & L. Rymo. 1970. J. Biol. Chem. *245*: 435–438.
93. Schulman, L. H. & H. Pelka. 1977. J. Biol. Chem. *252*: 814–819.
94. Schoemaker, H. J. P. & Schimmel, P. 1977. Biochemistry *16*: 5461–5464.
95. Holley, R. 1957. J. Am. Chem. Soc. *79*: 658–662.
96. Ogata, K. & H. Nohara. 1957. Biochim. Biophys. Acta *25*: 659–660.
97. Hoagland, M. B., P. C. Zamecnik & M. L. Stephenson. 1959. *In* A Symposium on Molecular Biology. R. E. Zirkle, Ed. :105–114. University of Chicago Press. Chicago, Ill.
98. Chapeville, F., F. Lipmann, G. von Ehrenstein, B. Weisblum, W. J. Ray, Jr. & S. Benzer. 1962. Proc. Nat. Acad. Sci. U.S.A. *48*: 1086–1092.
99. Berg, P. & E. J. Ofengand. 1958. Nat. Acad. Sci. U.S.A. *44*: 78–86.
100. Hecht, L. I., M. L. Stephenson & P. C. Zamecnik. 1958. Biochim. Biophys. Acta *29*: 460–461.
101. Hecht, L. I., P. C. Zamecnik, M. L. Stephenson & J. F. Scott. 1958. J. Biol. Chem. *233*: 954–963.
102. Hecht, L. I., M. L. Stephenson & P. C. Zamecnik. 1959. Proc. Nat. Acad. Sci. U.S.A. *45*: 505–518.
103. Zachau, H. G., G. ACS & F. Lipmann. 1958. Proc. Nat. Acad. Sci. U.S.A. *44*: 885–889.
104. Zamecnik, P. C. 1962. Biochem. J. *85*: 257–264.
105. Griffin, B. E. & C. B. Reese. 1965. Proc. Nat. Acad. Sci. U.S.A. *51*: 440–444.
106. Wolfenden, R., D. H. Rammler & F. Lipmann. 1964. Biochemistry *3*: 329–338.
107. Hussain, Z. & J. Ofengand. 1973. Biochem. Biophys. Res. Commun. *50*: 1143–1151.
108. Hecht, S., J. W. Kozarich & F. J. Schmidt. 1974. Proc. Nat. Acad. Sci. U.S.A. *71*: 4317–4321.
109. Sprinzl, M. & F. Cramer. 1975. Proc. Nat. Acad. Sci. U.S.A. *72*: 3049–3053.
110. Fraser, T. H. & A. Rich. 1975. Proc. Nat. Acad. Sci. U.S.A. *72*: 3044–3048.
111. Keller, E. B. & P. C. Zamecnik. 1956. J. Biol. Chem. *221*: 45–59.
112. Zamecnik, P. C., M. L. Stephenson & L. I. Hecht. 1958. Proc. Nat. Acad. Sci. U.S.A. *44*: 73–78.
113. Bishop, J., J. Leahy & R. Schweet. 1960. Proc. Nat. Acad. Sci. U.S.A. *46*: 1030–1038.
114. Dintzis, H. M. 1961. Proc. Nat. Acad. Sci. U.S.A. *47*: 247–261.
115. Strittmatter, C. F. & E. G. Ball. 1952. Proc. Nat. Acad. Sci. U.S.A. *38*: 19–25.

116. LITTLEFIELD, J. W. & E. B. KELLER. 1957. J. Biol. Chem. *244*: 13–30.
117. STEPHENSON, M. L., K. V. THIMANN & P. C. ZAMECNIK. 1956. Arch. Biochem. Biophys. *65*: 194–209.
118. LAMBORG, M. R. & P. C. ZAMECNIK. 1960. Biochim. Biophys. Acta *42*: 206–211.
119. TISSIERES, A. & J. D. WATSON. 1958. Nature *182*: 778–780.
120. TISSIERES, A., D. SCHLESSINGER & F. GROS. 1960. Proc. Nat. Acad. Sci. U.S.A. *46*: 1450–1463.
121. NIRENBERG, M. W. & J. H. MATTHAEI. 1961. Proc. Nat. Acad. Sci. U.S.A. *47*: 1588–1602.
122. HERSHEY, A. D., J. DIXON & M. CHASE. 1953. J. Gen. Physiol. *36*: 777–789.
123. VOLKIN, E. & L. ASTRACHAN. 1956. Virology *2*: 149–161.
124. RILEY, M., A. B. PARDEE, F. JACOB & J. MONOD. 1960. J. Mol. Biol. *2*: 216–225.
125. JACOB, F. & J. MONOD. 1961. J. Mol. Biol. *3*: 318–356.
126. BRENNER, S., F. JACOB & M. MESELSON. 1961. Nature *190*: 576–581.
127. GROS, F., H. HIATT, W. GILBERT, C. G. KURLAND, R. W. RISEBROUGH & J. D. WATSON. 1961. Nature *190*: 581–585.
128. JACOB, F. 1962. *In* Pontificiae Academiae Scientiarum Scripta Varia. Vol. 22: 85–95.
129. LENGYEL, P., J. F. SPEYER & S. OCHOA. 1961. Proc. Nat. Acad. Sci. U.S.A. *47*: 1936–1942.
130. NIRENBERG, M., T. CASKEY, R. MARSHALL, R. BRIMACOMBE, D. KELLOGG, B. DOCTOR, D. HATFIELD, J. LEVIN, F. ROTTMAN, S. PESTKA, M. WILCOX & F. ANDERSON. 1966. Cold Spring Harbor Symp. Quant. Biol. *31*: 11–24.
131. SPEYER, J. F., P. LENGYEL, C. BASILIO, A. J. WAHBA, R. S. GARDNER & S. OCHOA. 1963. Cold Spring Harbor Symp. Quant. Biol. *28*: 559–567.
132. KHORANA, H. G., H. BUCHI, H. GHOSH, N. GUPTA, T. M. JACOB, H. KOSSEL, R. MORGAN, S. A. NARANG, E. OHTSUKA & R. D. WELLS. 1966. Cold Spring Harbor Symp. Quant. Biol. *31*: 39–49.
133. WARNER, J. R., A. RICH & C. E. HALL. 1962. Science *138*: 1399–1403.
134. NOLL, H., T. STAEHELIN & F. O. WETTSTEIN. 1963. Nature *198*: 632–638.
135. WATSON, J. D. 1963. Science *140*: 17–26.
136. ALLEN, D. W. & P. C. ZAMECNIK. 1962. Biochim. Biophys. Acta *55*: 865–874.
137. LOFTFIELD, R. B. 1963. Biochem. J. *89*: 82–92.
138. HOPFIELD, J. J. 1974. Proc. Nat. Acad. Sci. U.S.A. *71*: 4135–4139.
139. ORGEL, L. 1963. Proc. Nat. Acad Sci. US.A. *49*: 517–521.
140. MONIER, R., M. L. STEPHENSON & P. C. ZAMECNIK. 1960. Biochim. Biophys. Acta *43*: 1–8.
141. HOLLEY, R. W., J. APGAR, B. P. DOCTOR, J. FARROW, M. A. MARINI & S. H. MERRILL. 1961. J. Biol. Chem. *236*: 200–202.
142. ZACHAU, H. G., D. DUTTING, H. FELDMANN, F. MELCHERS & W. KARAU. 1966. Cold Spring Harbor Symp. Quant. Biol. *31*: 417–424.
143. INGRAM, V. M. & J. A. SJOQUIST. 1963. Cold Spring Harbor Symp. Quant. Biol. *28*: 133–138.
144. RAJ BHANDARY, U. L., A. STUART, R. D. FAULKNER, S. H. CHANG & H. G. KHORANA. 1966. Cold Spring Harbor Symp. Quant. Biol. *31*: 425–434.
145. TOMLINSON, R. V. & TENER, G. M. 1962. J. Am. Chem. Soc. *84*: 2644–2645.
146. HOLLEY, R. W., J. APGAR, G. A. EVERETT, J. T. MADISON, M. MARQUISEE, S. H. MERRILL, J. R. PENSWICK & A. ZAMIR. 1965. Science *147*: 1462–1465.
147. REDDY, V. B., B. THIMMAPPAYA, R. DHAR, K. N. SUBRAMANIAN, B. S. ZAIN, J. PAN, P. K. GHOSH, M. L. SELMA & S. WEISSMAN. 1978. Science *200*: 494–502.
148. FIERS, W., K. CONTRERAS, G. HAEGMAN, R. ROGIERS, A. VAN VOORDE, H. VAN

Heuverswyn, J. van Herriweghe, G. Volckaert & M. Ysebaert. 1978. Nature, in press.
149. Weissbach, H. & Pestka, S., Eds. 1977. Molecular Mechanisms of Protein Biosynthesis. Academic Press, New York, N.Y.
150. Medvedev, Z. A. 1966. *In* Protein Biosynthesis and Problems of Heredity, Development, and Aging. Translated by A. Synge. Oliver and Boyd. London.
151. Hampel, A. M., M. Labanauskas, P. G. Connors, L. Kirkegard, U. L. Rajbhandary, P. B. Sigler & R. Bock. 1968. Science *162*: 1384–1387.
152. Kim, S. H. & A. Rich. 1968. Science *162*: 1381–1384.
153. Cramer, R., F. V. D. Haar, W. Saenger & E. Schlimme. 1968. Angew. Chem. Int. Ed. Engl. 7: 895.
154. Fresco, J. R., R. D. Blake & R. Langridge. 1968. Nature *220*: 1285–1287.
155. Clarke, B. F. C., B. P. Doctor, K. C. Holmes, A. Klug, K. A. Marcker, S. J. Morris & H. H. Paradies. 1968. Nature *219*: 1222–1224.
156. Suddath, F. L., G. J. Quigley, A. McPherson, D. Sneden, J. J. Kim, S. H. Kim & A. Rich. 1974. Nature *248*: 20–24.
157. Robertus, J. D., J. E. Ladner, J. T. Finch, D. Rhodes, R. S. Brown, B. F. C. Clark & A. Klug. 1974. Nature *250*: 546–551.
158. Anfinsen, C. B. 1972. Biochem. J. *128*: 737–749.
159. Nomura, M. 1976. *In* Reflections on Biochemistry, in Honour of Severo Ochoa. A. Kornberg *et al.*, Eds. :317–324. Pergamon Press. New York, N.Y.
160. Wittman, H. 1976. *In* Reflection on Biochemistry, in Honour of Severo Ochoa. A. Kornberg *et al.*, Eds. :325–336. Pergamon Press. New York, N.Y.
161. Spirin, A. S. 1978. *In* Progress in Nucleic Acid Research and Molecular Biology. W. Cohn, Ed. Vol. 21: 39–62. Academic Press. New York, N.Y.

DISCUSSION OF THE PAPER

Bernstein: Pasteur is well known for his statement that chance favors only the prepared mind. I was wondering if you could possibly tell us something more about why you think that maybe you missed seeing the polysomes while Rich did not a couple of years later? What was he looking for that maybe you were not looking for.

Zamecnik: Well, he already knew about messenger RNA, and the Nirenberg and Matthaei experiment had been done. In earlier years, I was dazzled by how much RNA was present in the ribonucleoprotein particle, and I wondered what it was doing if it was not serving as a messenger. It is still a puzzle as to why all that so-called structural RNA is in the ribosome. I was not thinking of a separate species of messenger RNA at that time.

Haurowitz: You mentioned at the beginning of your talk the plastein synthesis. At the same time of your later work, 1955 or so, I was interested in this problem and tried to investigate it with radioactive amino acid esters. I repeated the classical plastein experiment, which means digestion of the protein with pepsin, evaporation, and I think pepsin again, and got a precipitate like Dilevsky in Russia.

If we repeated the same experiment with radioactive amino acid, none of the radioactive amino acid was incorporated, but if we took amino acid ethyl esters in the presence of pepsin or of chymotrypsin, we found incorporation of these radioactive amino acid residues; it was a transpeptidation of ester to protein, a very simple explanation.

Well, we published it in a short paper, 1955 or 56, the same time you mentioned in your work. I forgot about all this until about March 1978 when I got a visit from a Japanese scientist who wanted to see me about this. I was very surprised. They now use the incorporation of amino acid by means of ethyl esters to improve the biological quality of proteins which are deficient in, for instance, methionine or of phenylalanine. They find that this works better than the proteins which are devoid of these amino acids and they plan to feed many more people now with these biologically more valuable proteins. It is a very unexpected result. This is a publication of the American Chemical Society, a monograph. I do not remember precisely the title but it was published in 1977.

ZAMECNIK: Very interesting. There was a conference about two months ago at the NIH, under the auspices of Fogarty Center, on post-translational modifications of proteins, which is becoming an enormously interesting field. It would not surprise me if transpeptidations might play a role in that. After all, if you can splice genes why not proteins?

FRAENKEL-CONRAT: You mentioned my brief stay in Edman's and in Bergmann's laboratory. At that time actually we synthesized the first peptide bonds. Bergmann was very unhappy about me because I found that it was a function only of the insolubility of the resulting bonds. We used papain and what we got when both polar groups at the end were blocked was benzoyl, glycyl, phenylalanyl, anilide. It was a solubility problem and Bergmann fussed with me because it did not support the specificity concepts of enzymes that were being pushed at the Rockefeller Institute at the time, and that is why I stayed at Rockefeller only one year. But I think in regard to plasteins and all these things that this means that when the equilibrium can be shifted sufficiently by having an insoluble product you can go in the opposite direction with all of these enzymes and get the small amount of synthesis that may always be present in a equilibrium situation. You push the equilibrium towards the insoluble product and thereby get synthesis.

So I wonder whether with the ethyl esters, since they also form more insoluble products, synthetic reactions would be favored.

ZAMECNIK: There is a paper by Max Brenner, who had also been at the

Rockefeller Institute. He used methionyl ethyl esters and he eluted them on a thin-layer chromatographic system. As they became more concentrated they condensed, so that he synthesized methionylmethionine, methionyl tripeptides, and so on.

PHILIP SIEKEVITZ (*Rockefeller University, New York, N.Y.*): I remember in the late 1940s there was a great deal of discussion about the templates for protein synthesis and I recall in this book of Northrop and Summers, or was it Northrop and Kunitz, on the crystallization of a protein, John Northrop had a whole chapter in which he went on and on about the possibility of the protein acting as a template for the synthesis of more protein and at the end he thought, well this is probably not so because you would get the mirror image of the template and we do not get the mirror image.

DAVID BEARMAN (*American Philosophical Society, Philadelphia, Pa.*): You mentioned a rather interesting sort of general phenomenon that has come up in a number of talks here. In 1959 after having been relatively alone in the field, you suddenly found yourself inundated by other approaches and challenged by other people. I just have two questions about that, which might open up some more general discussion as well.

What was it that first changed the perception of the other people who had been there all along and who then changed their research programs so that they begin to address the same kinds of questions? And secondly, what about the commitment that you already had in terms of personnel? And in terms of the material you had chosen? You mentioned that it was significant that you were working in liver; other people were working in other systems. What was it about those commitments that gave other laboratories, which had different commitments and different sets of skills, certain kinds of advantages in going on with the work beyond where you had carried it up to 1959?

ZAMECNIK: I should have mentioned that we were conscious that it would be better for the understanding of protein synthesis if one could establish a cell-free system for *E. coli*, and we tried to do that in 1951 with Novelli who was working in Lipmann's lab. We did observe what might have been cell-free incorporation, but we also had live bacteria present. We were unable to get our preparations absolutely cell free, or at least sufficiently cell free to make us rely on the results we obtained.

In 1953 other investigators studied protein synthesis in supposedly cell-free bacterial systems, which may not have been sufficiently free of intact cells. The perception of the situation became blurred by the conclusion that either of two things might happen: incorporation of amino acids into growing nascent protein chains, or else exchange of individual amino acids into established, completed proteins. This uncertainty caused investigators to shy away from bacterial systems for some time.

A dedicated postdoctoral colleague named Lamborg came to our laboratories in 1957. He started to work on developing a cell-free bacterial system from *E. coli*, a task that took him about three years, by which time he did find one that was active and reproducible. He carefully monitored the number of live bacteria in his cell-free preparations and when they were below 1×10^5 per milliliter, then the cell-free incorporation was reliable; when they were much above that his results were blurred or overshadowed by the much higher synthetic activity of the live bacteria.

Before that paper was published, Tissieres and Watson, our near neighbors, were much interested in bacterial cell-free systems, and Watson said, "Oh, if you can only get a good bacterial system, things will really boom." And so they took the relatively crude system worked out by my colleague Lamborg and me and advanced the level of knowledge by dissections of its component parts, particularly clarifying the role of the ribonucleoprotein particles, and pointing the way to the discovery of polysomes. Within a year or two, in fact, a number of other laboratories, including Nirenbergs, was using this cell-free bacterial system. As the number of skillful investigators entering the field increased, our contributions decreased.

Horowitz: In view of the considerable discussions that have gone on this afternoon about plastein and since Dr. Zamecnik mentioned Borsook and Wasteney in his first slide, I would like to say that when I got to Caltech in the middle 1930s to be a graduate student, Borsook at that time was working with Huffman on measuring the free energies of peptide bond formation by thermal methods. They found the big job is purifying the peptides, and so forth. They found that the free energy formation of a peptide bond is plus 2 to 4 kilocalories per mole. As far as I know that is still the best value. This shows that even out of incorrect hypotheses, if they are carefully followed, useful results may come.

Zamecnik: At that time Borsook and Huffman's data served as a helpful sign post for people that the equilibrium in the reaction was far toward hydrolysis.

Horace Judson (*The New Yorker, New York, N.Y.*): I think that Dr. Bearman's question to Dr. Zamecnik perhaps ought to be rephrased because it would be interesting to know why you did not carry on more with the work that you had begun, particularly since the work that went on in, for example, Nirenberg's lab, was done using a cell-free system that was very close to the one that you and Hoagland had developed and which Tissieres had taken very slightly farther. As Dr. Bearman posed the question he was saying, that the system you had used was the rat liver cell-free system, and therefore there was a change of systems. But in fact, since you had developed the bacteria cell-free system, that part of the supposition in Dr. Bearman's question is gone. At the same time, it is true that this involved people who had been around for a long time, but in some cases, at least in Nirenberg's case, new people were coming into this work. Ochoa of course transferred from other work to get into this. But I wonder whether you can sharpen the answers to why it is that you did not carry on with the work?

I think you once told me in a discussion that the primary reason was that you had been very close to the idea that the ribosome was the thing that carried the specificity. Thus, it was in part a shift of concept over to messenger RNA which changed the outlook. Is this the case?

ZAMECNIK: Yes, that is true. But once they had made this break, I thought they were doing so very well there was no reason to add another laboratory to the chase. We were also interested in purifying transfer RNAs and in the problem of trying to sequence them. And there, Holley had just entered the field and, using a simplified procedure for purifying RNA which our laboratory had just worked out, soon became its front-running competitor. That was 1960 to 65, and we were in that race for a while, but I put my bets at that time on a chemical degradation procedure which had had its origins in the work of Brown, Fried, and Todd, and Whitfeld and Markham, in which they found that periodate oxidation plus addition of an amine resulted in a loss of one nucleotide unit from an RNA chain. We perfected that method a bit and added a phosphomonoesterase, so that we could eliminate one residue at a time. I hoped we could do that repeatedly. Well, we did so but our yields were not good enough. They were of the order of 85% per residue, so that by the time we eliminated three or four residues we were not doing very well. Other people who have pursued that chemical lead in the last 15 years have improved it, but Holley's combination of starting off with a large quantity of material, using counter-current distribution to fractionate it, then use of endonucleases and exonucleases in 7-molar urea columns (which just came on the scene from Tener's laboratory), plus a clever team resulted in his winning that race. I must say that Zachau was not far behind, and in a year or so Raj Bhandary had also sequenced a whole transfer RNA. We shifted our emphasis to transfer RNA about that time because the code-deciphering field had moved too fast; but in the end, so did transfer RNA sequencing.

P. CONDLIFFE (*National Institutes of Health, Bethesda, Md.*): I would like to point out that there were some sort of gross nonscientific factors that may have had a profound effect on some laboratories which decided not to try to rush in with the tide. Between about 1952 and 1962 the number of scientific workers, most of them I guess supported during the initial logarithmic growth phase of grants emanating from Bethesda, approximately quadrupled; and everybody found themselves with a lot of very bright, smart, well-trained competitors. It was very difficult sometimes to keep up. Everybody began to feel the pressure of competition in a way that had not been felt before.

I think it is also true that it was an idea whose time had come. It was obvious that people could jump in, and the field was sufficiently big so that they could all make a contribution, possibly even if they did not win the race themselves; it was possible to make a career.

It is perhaps not a scientific, intellectual, or honorable reason, but nevertheless I think it was a very real thing in the 1950s and 60s.

SHEMIN: I wonder why the emphasis is on winning the race rather than accumulating knowledge?

OLBY: I just wonder whether Nirenberg's greatest contribution possibly was preincubation. Did anybody ever use preincubation to destroy the endogenous messengers before him? And was that maybe how he cleared the way for detecting the good message activity?

ZAMECNIK: Yes, I think so.

FRAENKEL-CONRAT: In regard to your stepwise degradation, by the way, we used it with TMV RNA, up to five nucleotides, but 6,499 remained to go.

OLBY: I wonder if I could take up one point about communication between groups? In your discussion of protein synthesis, I noticed in the diagrams that you showed there were what you referred to as rather vague terms like patternization and so on. These diagrams are after the first attempt to produce some sort of codes, are they not? Caldwell and Henchle, 1950, and Alexander Dounce, 1952. I am not sure whether your attitude was that these early attempts to produce specific codes were too speculative to manage bringing into the diagrams? And in that connection, am I right in saying that the RNA tie club began its existence about 1954–55? That Watson would have been a member of that club? That they circulated a number of documents including Crick's paper on the adaptor hypothesis in 55? Were you excluded from the circulation of that group? And if so, what do you think the reasons were?

ZAMECNIK: I do not know; It is a big world. I was not part of that group, but I mentioned in a publication about 10 years ago that when we found that the ribosome was an important feature in protein synthesis, I talked to Paul Doty, who was the high priest of the RNA field in the Boston area, and said, "How do you think the sequencing step of protein synthesis occurs, and what is the relationship to DNA?" He replied, "I don't know, but there's a young fellow over here named Watson who just happens to be visiting me and I'll ask him if he'll go over and see you. He has a model of the double helix which he and his colleague Crick have recently described." I had not heard about that model, in 1954.

Watson came over and visited me, and I said, "It seems to me that we have evidence going back to Brachet and Caspersson that RNA plays a role in protein synthesis. At least they showed that organs such as the pancreas, in which protein synthesis for export was high, had a high concentration of what they called cytoplasmic nucleic acid." Then Watson showed me his model. Here we have a model of DNA; now where does RNA fit in? If Brachet and Caspersson have a point and the DNA is a generator of the code for protein synthesis, how is its message transmitted? We looked at the model together and I said, "Could RNA possibly fit into a groove here?" Watson just started, shrugged his shoulders, lifted his hands, and went off to look at birds.

KIRSHENBAUM: Dr. Olby has alluded to Nirenberg's preincubations, an important part of his system. But perhaps the most important part was his choice of collaborators. Matthaei, who had been with Katchalsky, had worked on polyphenylalanine (poly Phe) if I remember reading correctly. When they used poly U they made poly Phe and Matthaei, who had known about poly Phe I think, had a leg up in being able to recognize what was being made was indeed poly Phe. So perhaps Nirenberg's most important contribution was in having the correct collaborator. A fortuitous occurrence of unnatural events had occurred to him.

SHEMIN: This is a story I heard, which may be aprocryphal. Since Nirenberg had TMV and other RNAs which pepped up the incorporation of amino acids, Heppel I think suggested that perhaps he use some nonsense RNA so that the real RNA will work and this poly U would not work. Supposedly it was as a negative control that he added the poly U. This is the story I heard and I have no evidence that it is true.

COMMENT: I lived in the same corridor with Gordon Tompkins and Nirenberg, and the idea came I think basically from Gordon Tompkins. It was one of these long arguments; Nirenberg was walking up and down the corridor with him, and Tompkins simply threw the idea out. He said, "For Christ's sake, go out and see Leon Heppel in the 9th or get some poly U, any damn thing he's got."

FRAENKEL-CONRAT: Those are stories that are very widespread and which apparently are apocryphal. I have seen Matthaei's notebooks when those experiments were done and one has to recognize the fact that there were two sets of things which Nirenberg and Matthaei did together. The first was to develop the bacterial cell-free system with preincubation and the other techniques that they were using, and that in fact was where they used poly A as a control on a system that was not designed to have an artificial messenger put in it.

It is apparently the case as far as I was able to determine that Nirenberg and Matthaei both knew that they were deliberately using the poly U because they used two other nonsense RNAs on the same set of experiments. They were deliberately using the poly U as an artificial messenger with the idea that they were going to get incorporation if it worked.

As far as I am able to trace it, the origin of the belief that what Nirenberg and Matthaei did was accidental, the result of a negative control that turned out to be the positive experiment, goes back in fact to Gordon Tompkins who told a number of people in the course of the following summer that that is the way the experiment had been devised. He must have been speaking in fact of the previous work which had been done to perfect the cell-free system with the preincubation before they actually turned around. The way the experiment was carried out, it could not have been designed to use as a negative control. They were using a variety of things to narrow it down to see which one might work.

I go into this at considerable length in a passage in a book I have just been finishing, which will give you a little more of the details.

Seymour S. Cohen received his Ph.D. at Columbia in 1941 and then worked with Wendell Stanley at the Rockefeller Institute. In 1943 studies on the typhus vaccine took him to the Department of Pediatrics of the University of Pennsylvania where he did his work on bacteriophage and virus-induced enzymes. He became an American Cancer Society Professor of Biochemistry in 1957 and Chairman of the Department of Therapeutic Research from 1963 to 1971 and in these positions he initiated studies on D-arabinosyl nucleosides and on polyamines. In 1971 he moved to the University of Colorado Medical Center as Professor of Microbiology and until 1976 also served the American Cancer Society as Chairman of its Council of Analysis and Projection. In 1976, he became Distinguished Professor of Pharmacological Sciences at the State University of New York at Stony Brook. He is working on polyamine biosynthesis in a virus-infection of chloroplasts and has recently described a general approach to the chemotherapy of infectious disease.

Some Contributions of the Princeton Laboratory of the Rockefeller Institute on Proteins, Viruses, Enzymes, and Nucleic Acids

SEYMOUR S. COHEN

Department of Pharmacological Sciences
State University of New York
Stony Brook, New York 11794

THERE HAVE BEEN several routes of investigation, i.e., genetic, virological, and biochemical, to the concept that the nucleic acids determine amino acid sequences in the biosynthesis of proteins. My own work has taken me along several of these converging paths before, during, and after the general acceptance of this fact. Before introducing our main speakers it seems appropriate to the purposes of this symposium to present some personal recollections of one of the less well-known centers, within which some of our present knowledge was generated.

In 1941, 5 years after Bawden and Pirie had demonstrated that tobacco mosaic virus (TMV) contained RNA, I had the good fortune to win a postdoctoral fellowship to work on plant viruses in the group led by Wendell Stanley at the Princeton Laboratory of the Rockefeller Institute. This branch contained the Department of Animal and Plant Pathology of the Institute, and was dissolved in 1950.

As a doctoral student with Erwin Chargaff, my studies on lipoproteins had led me to ask how the RNA of TMV might be attached to the protein of the virus. Stanley agreed that studies of this question and of the structure of virus RNA might be a useful activity. It should be realized that in 1941 RNA was frequently described in the literature as plant nucleic acid and it was not broadly accepted that RNA was a component of all cells, although reports of its presence in animal cells did appear on occasion. RNA, even that of TMV as isolated and characterized by H. Loring, was thought to be no larger than a tetranucleotide, and such a structure was not imagined to possess the complexity essential for a genetic role, that of determining the primary sequences of specific proteins.

Possibly the most significant result of my laboratory work in the years 1941 and 1942 was the demonstration that the RNA isolated after heat denaturation of TMV was a good deal larger than a tetranucleotide. The material I isolated was spontaneously birefringent, highly asymmetric,

0077-8923/79/0325-0303 $01.75/0

and had an average molecular weight of 300,000 daltons. Indeed, I do not doubt today that such a preparation contained infectious RNA. However, I never thought to apply the materials to plants, nor was such an experiment ever suggested in the seminars I presented on this subject at the Princeton or New York branches of the Institute. Furthermore, an experiment to test the infectivity of viral RNA might have been done in many laboratories after 1944 before it was in fact done in Tübingen and Berkeley as late as 1956. It will be of great interest to hear from Dr. Fraenkel Conrat and Dr. Hotchkiss, as well as from Dr. Pirie why they think such experiments were not performed before 1956.

In 1942, I found myself surrounded by great figures of biology and biochemistry. In FIGURE 1, we can see Stanley (1) and the three major figures of his laboratory, Max Lauffer (2), Gail Miller (3) and C. A. Knight (4). Lauffer ran the superb physical chemistry laboratory (with our summarizer H. K. Schachman residing in the dark room). By classical methods of polymer chemistry, Lauffer had determined dimensions of the TMV rod closely approximating those found later in the electron microscope. Miller was producing various protein derivatives of TMV in a search for moieties essential for infectivity. Knight was comparing the amino acid content of various strains of TMV. In retrospect then, the main thrust of Stanley's group at the time was to define the elements of viral structure essential for multiplication and phenotypic expression. I left in 1942 to study the typhus vaccine; the efforts of this entire group shifted to work on influenza virus shortly thereafter.

Other biochemists in the Princeton Laboratory were John Northrop (5) and Moses Kunitz (6), whose great achievements in enzyme isolation had suggested the approach to virus isolation subsequently developed by Stanley. The nucleases isolated and crystallized by Kunitz in the early 1940s helped to provide the most convincing evidence that pneumococcal transforming substance contained DNA as an essential part of the genetic determinant, and that the infectious agent isolated from TMV similarly required intact viral RNA. In 1951 Northrop was the first to state clearly that phage DNA might be the genetic determinant of phage.

The Northrop group also included M. Anson (7) and Roger Herriott (8). Anson's studies on hemoglobin and carboxypeptidase, as well as on protein renaturation, remain as classics of discovery in our knowledge of protein structure. Herriott, whose work on pepsin and pepsinogen is similarly a foundation stone of our discipline, later entered into the excitement of the phage work and he made fundamental contributions to that field. According to Hershey, Herriott had suggested in a personal letter in 1951, that a phage was a "little hypodermic needle full of trans-

FIGURE 1. Staff of the Department of Animal and Plant Pathology of the Rockefeller Institute at Princeton, N.J. 1942. (1) Wendell Stanley, (2) Max Lauffer, (3) Gail Miller, (4) C. A. Knight, (5) John Northrop, (6) Moses Kunitz, (7) M. Anson, (8) Roger Herriott. (From the Rockefeller University Archives.)

forming principles; that the virus as such never enters the cell; that only the tail contacts the host and perhaps enzymatically cuts a small hole through the outer membrane and then the nucleic acid of the virus head flows into the cell."

It is regrettable that I cannot take more time to describe other major figures in this picture. The work I have mentioned, the institution and its unusual biological systems and advanced research tools, the productive individuals, the ambiance, all participated in creating a framework of experience and thought which contributed to the stepwise accretions of knowledge, by which that curious polymer, nucleic acid, and its roles were eventually clarified. By the late 1940s, the exotic DNA of the pneumococcus could also be found as the easily isolable white fibrous DNA of phage, a material comprising half the phage weight. The nucleic acid of these viruses could be labeled easily and analyzed readily. Phage DNA was synthesized specifically in tremendous quantities during virus multiplication, and having come to reasonable conclusions from the work on transformation, we even performed experiments to test if this material might function as part of the phage chromosome. By 1950 the biologists were beginning to get used to the presence of these materials, which were cropping up everywhere. The stage had been set for the Hershey-Chase experiments, and for rigorous work on the structure and functions of viral nucleic acids.

HEINZ FRAENKEL-CONRAT received his medical degree at Breslau and worked for a time thereafter at Edinburgh. In 1937 he came to the Rockefeller Institute and then went to the São Paulo Institute of Brazil where he performed his well-known work on snake venom toxins. In 1942 he went to the Western Regional Laboratory of the U.S. Department of Agriculture, located in Berkeley, where he worked on modification reactions of various proteins including avidin, lysozyme, and insulin. Since 1952 Dr. Fraenkel-Conrat has been at the Virus Laboratory of the University of California at Berkeley, where he has done important work on viruses.

Protein Chemists Encounter Viruses

HEINZ FRAENKEL-CONRAT
Virus Laboratory
University of California
Berkeley, California 94720

THE EARLY 1930s was the era when biochemistry became dominated by great successes in the methodology of protein purification and crystallization. Also this was the era when proteins seemed to be earning the right to their name—the first and foremost components of biological systems. The pituitary hormones, snake toxins, antibodies, and, particularly, all enzymes were found to be proteins. It was thus logical to assume that the mysterious pathogens termed filterable viruses would also be proteins. This was particularly apparent to a young chemist, W. M. Stanley, whose first independent piece of research was to attempt the purification of a virus. Fortuitously he was then located at the Rockefeller Institute at Princeton, where Northrop, Herriott, Kunitz, and Anson were busily developing the methods for the purification and crystallization of pepsin and many pancreatic enzymes and enzyme inhibitors. The title of this talk, "Protein Chemists Encounter Viruses," is not strictly correct for the early years of molecular virology, since neither Stanley nor the other virus researchers prior to 1950 were experienced protein chemists, but rather learned the protein techniques on the job.

Stanley's first investigations at Princeton were actually a series of careful studies of the effects of the newly purified proteolytic enzymes on the infectivity of the not yet pure or crystalline tobacco mosaic virus (TMV). But realization soon followed that the main "protein" in his infectious preparations was actually the virus and that its tendency to form what we now call paracrystalline, i.e., two-dimensionally ordered, aggregates represented a further means of obtaining it pure.[1] With this dramatic discovery, or rather his acceptance of a new concept, furthered by his great gifts as a salesman of science, Stanley had inaugurated the new field of molecular virology, the precursor of molecular biology.

Progress in obtaining purified viruses at Princeton, in England, and elsewhere was then most rapid, hardly inhibited by the controversy whether the tobacco mosaic virus protein, as it was termed by Stanley, was really free of phosphorus and sulfur (0.00%), as at first reported. It took three years before Stanley's laboratory confirmed the analytical

0077-8923/79/0325-0309 $01.75/0

data and accepted the conclusion of Bawden and Pirie that all viruses contained phosphorus and carbohydrate indicative of and corresponding to their nucleic acid content.[2]

Cross-fertilization evidently occurred at Princeton because at that time T. H. Northrop had also purified and characterized a virus, namely, a bacteriophage containing 50% nucleic acid.[3] This was, however, not a first, since the earliest virus purification and characterization, two years before Stanley's isolation of TMV, was also of a bacterial virus. This represented the research done by Schlesinger, at first in Germany and after 1933 in England.[4] That this work did not produce the same impact on the scientific community as the isolation of TMV illustrates the critical importance of the timing and selling of scientific discoveries. Another factor was that Stanley had several close competitors working on the purification of TMV and other plant viruses, who quickly confirmed, extended, and particularly joyously corrected his findings, while Schlesinger was alone ahead of the field and of his time.

Of the many viruses that were then being purified, TMV was the most available for chemical study, and Stanley's laboratory was busily at work in further characterizing it. Since this virus was 95% protein, protein chemistry continued to dominate research during the next two decades, to the point of actual neglect of the possible role and reactivity of the RNA of the virus. The first amino acid analyses, by the crude precipitation methods developed in Max Bergmann's laboratory, used up five precious grams of the virus. As these methods became replaced by the then newly developed microbiological and paper chromatographic techniques, it soon became evident to C. A. Knight that different natural strains of TMV showed differences in their amino acid composition,[5] while the composition of their RNA appeared to be the same. This still seemed to support the protean role of proteins. While the genetic role of DNA had by then been accepted, RNA was still believed to represent only a polyanionic scaffold for the proteins of biological particles. Thus, when this author, a genuine protein chemist in 1951, but naïve in the ways of viruses, proposed to produce mutants of TMV by adding amino acids to the N-terminus of the protein by the carboxyanhydride method, neither Stanley nor other members of the Berkeley Virus Laboratory made discouraging comments regarding this proposal. The analytical consequences of this reaction were ambiguous, and, not surprisingly, no mutants were detected.

Actually it was at that time not known how large this peptide chain was. X-ray diffraction had indicated a regular subunit structure in TMV,[6] but the results of such esoteric procedures were overlooked or not seriously considered by us biochemists. Thus another genuine protein chemist, Ieuan Harris, who has recently died quite tragically, used TMV as a presumptive negative control to test the specificity of carboxypeptidase action, since that viral protein was believed to be about 350,000 amino acids long (mol wt 38×10^6!). The result of Harris's collaboration with C. A. Knight, the finding that the TMV particle was not made up of one enormous protein molecule, but of 2200 chains with C-terminal threonine, represents a beautiful example of scientific serendipity, for it opened up the field of protein subunit chemistry.[7] In my laboratory C.-I. Niu confirmed the genuine nature of the threonine-terminated subunits by hydrazinolysis, and found the reason why only threonine was released by carboxypeptidase action when he was able to isolate the C-terminal hexapeptide which was thr-ser-gly-pro-ala-thr.[8] Where there is a C-terminus there must be an N-terminus and I had just spent a year with F. Sanger, R. R. Porter, and K. Linderstrøm-Lang to learn and/or develop N-terminal techniques when I joined the Virus Laboratory at Berkeley in 1952. Great was my disappointment when no N-terminus was found, particularly when Braunitzer in Schramm's laboratory at Tübingen announced the presence of N-terminal proline in exactly equivalent amount to the C-terminal threonine. Some time was wasted in proving that this was an artifact terminus, resulting from breakage of an asp-pro bond due to the hot TCA treatment of the virus. Then, fortunately, K. Narita, Akabori's most promising pupil, joined us and established for the first time the nature of a blocked N-terminus in a protein, when he isolated acetyl-ser-tyr from TMV protein.[9] He was followed at Berkeley by another superb Akabori pupil, A. Tsugita, and the era of protein sequencing was upon us. TMV protein was the second large protein to be sequenced; in contrast to the first, pancreatic ribonuclease, TMV sequencing was more a competitive than a cooperative effort of two laboratories, and possibly for that reason there were fewer or smaller errors in the first published structures of this then longest sequence of 158 amino acids.[10, 11] The subsequent years saw the sequencing of several strains of TMV by Wittmann and collaborators, and thus the material for the first comparative structural anatomy of a group of proteins, indicative of evolutionary relationships.[12] The probable evolutionary advantages of the blocked N-terminus and the proline near the C-terminus of the TMV protein were also indicated by comparisons of natural strains, and of chemically produced mutants, the latter the work of Tsugita.[13] Tsugita's later work with Streisinger and other coworkers on elucidating the result of a frameshift mutation in T4 phage lysozyme is another exciting contribution of protein chemistry to virology.[14]

An interesting link between the dawn of protein chemistry and our laboratory at Berkeley was the presence there of H. O. L. Fischer, Emil

Fischer's son. His possession of cigarboxes full of samples of Emil Fischer's synthetic peptides yielded useful material for model studies of the chromatographic behavior of peptides derived from viruses.

One line of research that had been going on since Stanley first got interested in viruses in the 1930s was that of protein modification reactions. Alkylating and acylating reagents, formaldehyde, iodine, nitrous acid, and physical agents were used. It was then not yet possible to evaluate the results of these studies in terms of protein chemistry. In terms of their biological effects these modifications are difficult to interpret because the presence and possible reactivity of the RNA in the viruses was not sufficiently considered at that time. In the 1950s, when we knew more, I studied the nature of the single –SH group of the virus protein, and found that its conformational protection in the virus rod was such that it could be selectively iodinated and thus 2200 stable sulfenyl iodide groups could be introduced into a TMV particle—a then new type of bond not usually stable, and not stable even in the isolated TMV protein.[15] The resultant yellow TMV was a novelty to behold. Another modification of the –SH group, substitution with methyl-mercury was of particular usefulness in facilitating the x-ray crystallographic studies of Rosalind Franklin, D. L. D. Caspar, and others, which gave us the fine-structure of the virus rod.[16]

This led to other studies concerned with the conformation of native virus proteins and their quaternary structure in virus particles. There too was a happy hunting ground for protein chemists. Isolation of more or less completely dissociated TMV protein subunits was first achieved in alkali by Schramm and coworkers,[17] and later by me with acetic acid.[18] That the protein was under both of these conditions partially denatured and became spontaneously renatured at neutrality was then not realized. Only after Anderer demonstrated the complete renaturation of urea-denatured TMV protein, and we of guanidine-denatured TMV protein, did it become evident that it was the amino acid sequence of this protein alone that determined its conformation.[19] These results were obtained shortly after Anfinsen and White had demonstrated that reduction plus denaturation of pancreatic ribonuclease was a reversible process, but the result with TMV was in a way more extraordinary and unexpected, because no stabilizing disulfide bonds exist in this protein.

The biological test for successful renaturation of TMV protein was its ability to aggregate at pH 4–6 to rods of identical diameter and electron microscopic appearance as the intact virus. This was first observed by Takahashi with excess TMV protein found in infected plants,[20] and studied in greater detail by Schramm and Zillig in 1954, who tested these rod preparations for infectivity, finding them noninfectious.[21]

At that time we had developed methods to isolate the viral RNA without the damage resulting from the alkaline degradation used in Tübingen. When we allowed the virus protein to aggregate in the presence of such RNA near neutrality (where the protein alone does not form rods), we observed RNA-containing particles that were indistinguishable from the original virus, and were fully infectious.[22] Such reconstituted virus was also as stable over a wide pH and temperature range as the original virus and predominantly of the typical length of 300 nm, in contrast to the rods of protein alone which could be of any length, but were only stable near the isoelectric point of the protein. Thus the RNA was able to stabilize and thereby determine the length of the virus particles.

Since reconstituted virus was very much more infectious than its components, we at first tended to believe that the particulate nature of viruses, and thus the presence of their proteins, were preconditions for their infectivity. However, we, and Gierer and Schramm, soon learned how to handle and test the separated viral RNA, and we were then able to show that the RNA was by itself infectious.[23-25] Even RNA containing no detectable traces of viral protein was effective in eliciting the typical disease symptoms characteristic for the virus strain that it was prepared from; and the protein of the progeny of such RNA infection was identical in composition to that of the strain of origin, even when the RNA had been reconstituted with the required 20-fold of protein from another strain.[26-28] Thus the virus coat protein plays only a protective and infectivity-enhancing role, but the genetic function of the virus —and that is its only essential function—is carried entirely by its RNA. This was the first time that a transmittable disease, or any genetic bit of information, was demonstrated to be carried by a chemically and physically defined single molecular species. Much of this work was done by Bea Singer, another protein chemist to become a nucleic acid chemist at that time. The same kind of results in terms of nucleic acid infectivity and reconstitution were subsequently obtained with many other viral RNAs and DNAs. Thus the main focus of biochemical research turned for two decades to the study of nucleic acids. Fortunately the pendulum has now again turned towards proteins and peptides, which may after all be the more interesting and versatile group of molecules.

REFERENCES

1. Stanley, W. M. 1935. Isolation of a crystalline protein possessing the properties of the tobacco mosaic virus. Science *81*: 644–645.
2. Bawden, F. C. & N. W. Pirie. 1937. The relationships between liquid crystalline

preparations of cucumber viruses 3 and 4 and strains of tobacco mosaic virus. Br. J. Exp. Pathol. *18*: 275–291.

3. Northrop, J. H. 1938. J. Gen. Physiol. *21*: 335.
4. Schlesinger, M. 1933. Biochem. Z. *264*: 6.
5. Knight, C. A. 1947. The nature of some of the chemical differences among strains of tobacco mosaic virus. J. Biol. Chem. *171*(1): 297–308.
6. Bernal, J. D. & I. J. Fankuchen. 1941. J. Gen. Physiol. *25*: 111.
7. Harris, J. I. & C. A. Knight. 1952. Action of carboxypeptidase on TMV. Nature *170*: 613.
8. Niu, C.-I. & H. Fraenkel-Conrat. 1955. C-Terminal amino-acid sequence of tobacco mosaic virus protein. Biochim. Biophys. Acta *16*: 597–598.
9. Narita, K. 1958. Isolation of acetylpeptide from enzymic digests of TMV-protein. Biochim. Biophys. Acta *28*: 184–191.
10. Tsugita, A., D. T. Gish. J. Young, H. Fraenkel-Conrat, C. A. Knight & W. M. Stanley. 1960. The complete amino acid sequence of the protein of tobacco mosaic virus. Proc. Nat. Acad. Sci. U.S.A. *46*: 1463–1469.
11. Anderer, F. A., H. Uhlig, E. Weber & G. Schramm. 1960. Primary structure of the protein of tobacco mosaic virus. Nature *186*: 922–925.
12. Wittman-Liebold, B. & H. G. Wittman. 1967. Coat proteins of strains of two RNA viruses: Comparison of their amino acid sequences. Mol. Gen. Genet. *100*: 358–363.
13. Tsugita, A. & H. Fraenkel-Conrat. 1962. The composition of proteins of chemically evoked mutants of TMV-RNA. J. Mol. Biol. *4*: 73–82.
14. Terzaghi, E., Y. Okada, G. Steisinger, J. Emrich, M. Inouyi & A. Tsugita. 1966. Change of a sequence of amino acids in phage T4 lysozyme by acridine-induced mutations. Proc. Nat. Acad. Sci. U.S.A. *56*: 500–507.
15. Fraenkel-Conrat, H. 1955. The reaction of tobacco mosaic virus with iodine. J. Biol. Chem. *217*: 373–381.
16. Caspar, D. L. D. & R. E. Franklin. 1956. Nature *177*: 928.
17. Schramm, G., G. Schumacher & W. Zillig. 1955. Uber die Struktur des Tabakmosaikvirus. III. Der Zerfall in alkalischer Lösung. Z. Naturforsch. *10b*: 481–492.
18. Fraenkel-Conrat, H. 1957. Degradation of tobacco mosaic virus with acetic acid. Virology *4*: 1–4.
19. Anderer, F. A. 1959. Reversible Denaturierung des Proteins aus Tabakmosaikvirus. Z. Naturforsch. *14b*: 642–647.
20. Takahashi, W. N. & M. Ishii. 1953. A macromolecular protein associated with tobacco mosaic virus infection: its isolation and properties. Am. J. Bot. *40*(2): 85–90.
21. Schramm, G. & W. Zillig. 1955. Uber die Struktur des Tabakmosaikvirus. IV. Die Reaggregation des nucleinsäurefreien Proteins. Z. Naturforsch. *10b*: 493–499.
22. Fraenkel-Conrat, H. & R. C. Williams. 1955. Reconstitution of active tobacco mosaic virus from its inactive protein and nucleic acid components. Proc. Nat. Acad. Sci. U.S.A. *41*: 690–698.
23. Fraenkel-Conrat, H. 1956. The role of the nucleic acid in the reconstitution of active tobacco mosaic virus. J. Am. Chem. Soc. 7*8*: 882.
24. Gierer, A. & G. Schramm. 1956. Die Infektiosität der Ribonukleinsäure des Tabakmosaikvirus. Z. Naturforsch. *11b*: 138–140.
25. Fraenkel-Conrat, H., B. Singer & R. C. Williams. 1957. Infectivity of viral nucleic acid. Biochim. Biophys. Acta *25*: 87–96.

26. Fraenkel-Conrat, H. & B. Singer. 1957. Virus reconstitution. II. Combination of protein and nucleic acid from different strains. Biochim. Biophys. Acta *24*: 540–548.
27. Fraenkel-Conrat, H., B. Singer & A. Tsugita. 1961. Purification of viral RNA by means of bentonite. Virology *14*: 54–58.
28. Singer, B. & H. Fraenkel-Conrat. 1961. Effects of bentonite on infectivity and stability of TMV-RNA. Virology *14*: 59–65.

DISCUSSION OF THE PAPER

Abir-Am: You have spoken about viruses as heavy hunting grounds for protein chemists. Could you tell us a little bit about the relationship of your work to people who were virus geneticists or phage geneticists? What was the difference between the chemical approach adopted by the protein chemists and the more biological approach which geneticists interested in viruses or phages did adopt?

My second question is, how could we explain that for so many years the work on virus done by chemists had very little to do with DNA virus?

Fraenkel-Conrat: The chemical approach is obviously quite different from the genetic approach in that the chemist thinks in terms of structural arrangements of atoms. We heard from Dr. Horowitz earlier how geneticists live in a different world in terms of not thinking of milligrams as seconds, and so forth. At the early stages, when genetics was not yet discussed in molecular terms, there was no relation that I can see.

Horowitz: I think there are two valid approaches to these problems, or maybe more than two: genetics is one and chemistry is another.

Schachman: I would like to comment with regard to that question that there are really two very important laboratories involved in the study of viruses in, say, the late 1940s and 50s, and the approaches were entirely different. One of them dealt with the development of what you might call the chemistry of viruses and that obviously was the one that worked on plant viruses, a natural substance for that kind of research. Animal viruses were almost nonexistent as subjects of study at that time in terms of what we now call molecular virology, if you like.

In contrast to the work on plant viruses was all the magnificent work being done at Caltech started by Delbrück and Luria where the genetics of viruses and the dynamics of viruses were a much more lively activity because the systems were amenable to rapid studies. The turnover time was much shorter. That was the development that you are asking about, and they went entirely different ways. The chemistry was ignored in that laboratory and Dr. Horowitz referred to it in passing. Chemistry was the emphasis in the laboratory at Princeton and of course in Rothamsted and then Berkeley. There really was no relationship. The relationship of genetics to the chemistry of plant viruses was nonexistent.

PIRIE: I have several different points I would like to make. First of all on the history of reconstitution, dismantling and reconstitution of the proteins, I think Robin Hill ought to be borne in mind here. About 1926 he took hemoglobin apart and put it together again and showed at least that it still had the original spectrum and absorbed oxygen and gave off oxygen. This is a piece of work that is very often overlooked now.

Now a piece of mythology has been spread about the tetranucleotide hypothesis. No one who reads Levene's papers could possibly have thought he had demonstrated the tetranucleotide structure. When I used to teach in Cambridge, I think Dr. Gordon will bear this out, I used to ridicule the whole thing because, to anybody who handled nucleic acids and worked with this curious gummy and viscous stuff, it was quite obvious it was a very large molecule. When in 1950 people argued that nucleic acid could not have the complexity needed for specificity, I was amazed because it had never struck me, because of the rather skeptical atmosphere in which I had been brought up, to take the hypothesis seriously. Admittedly, Levene found only four nucleotides in one nucleic acid so he did not have to assume more than four; but he called it only a hypothesis and it should never have been taken seriously.

You asked what became of Schlesinger. Unfortunately he came down with influenza and in the post-flu depression he suicided. He worked in Mill Hill.

You said, perhaps it was a slip of the tongue, that I said all of the viruses were nucleoproteins. We would never say anything like that. I do not generalize in that sort of way. All we said was that the ones we had selected, which were extremely stable and easy to handle, were nucleoproteins, but the further generalization is completely out of character. And one of our reasons for looking at insect-transmitted viruses, on which we did a little work in 1937, was the hope of finding one that was not a nucleoprotein. That was our main objective. We did not attain it.

On the question of the length of TMV: we made an estimate based on Bernal's width measurement coupled with Staudinger's picture of the effect of rods becoming entangled on the physical properties of a solution. One estimate depended on the assumption that the liquid crystal layer formed when a rod needed a sphere with a diameter equal to its length for free rotation, the other on the assumption that it needed only a disk of that diameter and as thick as the rod is wide. We thought the sphere the one that was the most likely, and that gave a length which is very near the now accepted length; that depended of course on Bernal's measurements of width.

On the question of the infectivity of ribonucleic acids someone asked why it took so long to find. I am not saying that we would have found it but under the circumstances we could not have found it because I was in Cambridge, when Bawden was in Rothamsted; while we were taking TMV to pieces the product had to go across country so any infectivity there might have been in partially dismantled TMV would have been lost.

You quoted Knight as having shown amino acid differences. I wonder if you have ever read that paper; if so, you should again. If you add up the amount of amino acid accounted for in the different preparations, it ranges from 92% to 101%. If you have that amount of uncertainty as to how complete the amino acid recovery is, the difference of 1% or 2% in individual amino acids is neither here nor there. In a review article I was unkind enough to do the sums, add up the figures and show that the evidence was negligible. We obviously assumed, or were ready to assume, that there would be differences but not that they had been demonstrated.

And then finally on the curious resistance to proteolytic enzymes of TMV, B. Volcani said at the International Biochemical Congress in Vienna that he had a bacterium that would split native TMV. I wonder if anybody has ever heard any more of that? Volcani is now in San Diego, I think, and he has not pursued the issue. But he said he had an enzyme from the bacterium and that it would split native TMV.

HODGKIN: I should like also to comment on the problem of the length of TMV; this is largely published in the paper by Bernal and Fankuchen in the *Journal of General Physiology* in 1941. There were several varieties of calculation; one involved the shape of the tactoids seen in the liquid crystalline form of the virus preparation which Pirie showed. Another point that was made at the time was that certain of the phenomena, the changes in the distances apart of the particles observed in solution from the lower angle x-ray diffraction effects, suggested continuous, indefinite length. And this is of course correct. It can be seen in the electron microscope. There is a unit rod but the rods aggregate end to end, as recognized by insight from the x-ray diffraction effects before the phenomenon was observed in the electron microscope, a year or two later.

SMITH: This is a very minor comment about the reversible denaturation of proteins. First, on Pirie's remark about Robin Hill's regeneration of native hemoglobin, that was very beautiful work that Robin Hill did in 1925–26; he separated the heme from the globin and in fact used reconstituted hemes containing different metals to add back to the protein and regenerated, in the case of the iron heme, protein of native properties, but Hill did not recognize that he had denatured the globin in the process of separating the two. Credit for that really belongs to Anson and Mirsky, who pointed out that the globin was completely denatured when heme was separated and that the reconstituted protein did represent a regeneration of the native state and all of its properties. Similar is the case of serum albumin, which was first done by Mona, Spiegel, and Adolph, and then by Hsien Wu and by Anson and Mirsky, where they got the native crystalline protein back again.

Once again the most spectacular work of the 30s was in the Northrop laboratory; there was Kunitz's work, for example, on the denaturation of trypsin in which later Anson and Mirsky even measured the equilibrium properties and all the thermodynamic properties of the equilibrium between native and denatured trypsin and calculated all the conventional parameters, and of course got active trypsin back again. And similarly, Herriot's work on pepsinogen, which he completely unfolded and denatured and regenerated back to the native pepsinogen that was fully convertible to active pepsin and so on. All of this very classical work on the regeneration of denatured proteins was done in the late 1920s and 30s, long before what we might call the modern era.

FRAENKEL-CONRAT: Yes, I think we are in fact inclined to forget this era, partly because protein chemistry acquired a new aspect with the determination of new sequences. Then you could write the amino acid sequence of something that folds and unfolds. Until then it was all probably mysterious, but that does not mean it was not just as valid retrospectively.

ROLLIN D. HOTCHKISS received his graduate education in chemistry at Yale University. In 1935 he then began working at the Rockefeller Institute where he has since remained. Dr. Hotchkiss participated in, among many other projects, the important DNA work performed by O.T. Avery's group.

The Identification of Nucleic Acids as Genetic Determinants

ROLLIN D. HOTCHKISS
Rockefeller University
New York, New York 10021

THE WHOLE DEVELOPMENT of modern genetics is constituted of a sequence of great steps linking the formal concepts of classical genetics with the sciences of matter. There was, first, the identification of the formal concept of *linkage group* with the cytological entity, *chromosome*, by a series of experiments lasting well into the 1920s. Another great connection, between the unit gene product and the enzyme, had been propounded by Garrod, furthered by Wright and Scott-Moncrieff, and made experimental by Beadle and Tatum in the early 1940s. A good 20 years of refinement were required before the one gene, one enzyme idea was adequately restated as one cistron makes one RNA and (usually) therefrom one polypeptide. Even today we learn of new steps in the processing and assembly of these primary gene products.

The DNA Revolution, which I will discuss, concerned the operational identification of the DNA molecule as the gene material itself. It took place in the quarter century of 1930 to 1956. It was made operational and experimental by the new field of transfer genetics, introduction of heritable traits by artificial means, which began with the phenomenon of type transformation described by the British microbiologist Griffith in 1929.[1] He showed that heat-killed encapsulated pneumococci somehow induced specific capsule production in unencapsulated bacteria —all within the subcutaneous tissues of infected mice.

During the first half of the quarter century I mentioned, i.e., 1930–1943, the Griffith experiment was systematized in the laboratory of O. T. Avery at Rockefeller Institute with co-workers Dawson, Sia, Alloway, and finally, Maclyn McCarty and Colin MacLeod. The 1944 paper of Avery with the last two workers[2] definitely indicated that the active "transforming principle" for encapsulation is DNA and that DNA preparations from different specific encapsulated strains of pneumococci can confer specifically their own types heritably upon the recipient bacteria and lead to production of more DNA of the specific donor type.

I shall lean upon your knowledge of this much-quoted work and prin-

0077-8923/79/0325-0321 $01.75/0

cipally review the following 10 to 12 years, 1944 to 1956, for the reason that two other disparate ideas about those years have been made public and have achieved a certain following among those who came along later. On the one hand, we often hear it implied that the Avery work had made it so abundantly clear that genes were made of DNA that this generalization should have been forthrightly made then and there, and all its implications brought out. Several authors (e.g., H. V. Wyatt[3]) have hinted that Avery, MacLeod, and McCarty must have been excessively conservative, or rather remiss, or even uncomprehending, not to emphasize such broad implications. Somewhat oppositely, we have heard it stressed that the Avery findings were not immediately incorporated (by other scientists) into the general working knowledge of the time (G. Stent uses the term "premature" for the discovery[4,5]). Connected with this latter view is the stated or implied one that it was not until the Hershey-Chase 1952 experiments[6] on bacteriophage DNA transfer and the Watson-Crick DNA model[7] of 1953 that necessary convincing steps were made to initiate the DNA era.

I want to assure you that Avery and colleagues, and those of us who pursued the subject in that next decade, were well aware of genetic implications, but simply had the professional responsibility that other interpreters did not, to test out those implications in at least some concrete particulars. And to carry forward the new field of genetic chemistry, we were faced with the difficult task of devising several new tools and types of experiments to make such demonstrations—tools and experiments that have since come to be familiar or even "obvious."

The significant Hershey-Chase experiments of 1952 on the DNA-phage relation will be taken up later on. The Watson-Crick hypothesis of binary DNA structure eventually contributed vastly to underlying molecular theory, but when enunciated in 1953,[7] it was an inspiration and not yet a demonstration of how genes are copied. Like any physical model, it had the great advantage that it portrayed explicitly within itself exactly the process that would be required to prove it experimentally. That proof required another four to five years before the beautiful work of Taylor and of Meselson and Stahl with semiconservative replication.

Avery's transformation result of ten years before was not a theory but something different—an elegant demonstration of an unpredicted relationship—a discovery in search of its proper generalization.

In one of McCarty's last papers on transformation, in 1946,[8] he and Avery superbly put the genetic expectations in the following words: "It is possible that the nucleic acid of the R pneumococcus is concerned with innumerable other functions of the bacterial cell, in a way similar to that in which capsular development is controlled by the transforming substance. The deoxyribonucleic acid from Type III pneumococci would then necessarily comprise not only molecules endowed with transforming activity, but in addition a variety of others which determine the structure and metabolic activities possessed in common by both the encapsulated (S) and unencapsulated (R) forms. If these considerations have any foundation in fact, the task of discovering the chemical basis of biological specificity of deoxyribonucleic acids becomes extremely complex, since a given preparation will represent a mixture of a large number of entities of diverse specificity."

Notice that while shrewdly assigning DNA a variety of roles, McCarty and Avery reserve the term transformation for change of capsule type, and in fact, do not go so far as to propose that the various other activities predicated for the DNA would in fact be induced by a DNA transfer *in vitro*. These are very clear genetic implications, even if they do not predict in detail how other markers should behave in experiment. I have not seen these remarks quoted, even by Wyatt,[3] who deplores the absence of genetic generalization in the classic Avery paper of two years before.

I should like to tell you how a participant looked at the developing ideas of that period; to do so I shall run as rapidly as I can through that decade with several questions and their relevant experiments. I commenced to work with Avery on transformation in 1946, after asking for the privilege eight years earlier but being put off by World War II and other duties. At the time, I began to frame the questions listed in TABLE 1. During the two years until Avery retired from Rockefeller, his thoughts continued to be concerned principally with the first three chemically oriented questions. Before embarking on the story of how a few of us tackled those and the whole list of questions, I want to quote for you the responses of various known members of the scientific community through that decade as one or the other of these questions came to their minds. I believe these and my table will give you a reliable picture of what it meant to face then the intellectual task of assimilating the new insights.

Some scholars accepted in essence the chemical findings but might be concerned about the biological meaning of capsule transformation in such unlikely creatures as bacteria. Others sensed a genetic meaning of the phenomenon but were inclined to doubt the adequacy of the chemical data, based as they were in part on the enzyme, deoxyribonuclease, not then well known.

Let us see what various people said: for brevity I shall have to paraphrase, but I have tried to do so without distortion. On the chemical nature of the transforming agent: Sir Alexander Haddow, in 1944,[9]

TABLE 1

QUESTIONS CONCERNING THE NATURE AND FUNCTION OF BACTERIAL TRANSFORMING AGENTS*

(1) Is the active transforming agent exclusively made of DNA, does it contain a different active component, or is it the combination (say a nucleoprotein) which is active?

(2) What is the nature of the serum helper factor?

(3) How is DNA chemically differentiated in different mutants?

(4) How does DNA induce genetic change in bacteria?

(5) Are traits other than surface antigens controlled by DNA in pneumococci? Can they be demonstrated *in vitro*?

(6) Are all traits in *pneumococci* controlled by DNA?

(7) Are traits in other bacteria transformed by DNA? In other cells?

(8) If traits are transferable on some fairly general basis, does DNA show other properties of genes, e.g., autonomy and independence? Are mutations reflected in changed DNA? Are purely adaptive traits not transferred? Are linked sets of traits carried by DNA? Can they recombine? Are all of the traits carried by a DNA inserted at one time? Is every gene a separate DNA particle? Does transforming DNA become inserted into cell DNA? Does inserted DNA produce enzymes as products?

*As formulated ~ 1946.

wrote that the product, polysaccharide, and the agent, apparently DNA, were clearly distinct. Gulland, Barker, and Jordan, 1945, in an annual review[10] reported strong evidence that the agent is a sodium salt of DNA. Kalckar, in 1945,[11] also accepted the identification and wrote of its widespread significance. D. D. Woods in 1947[12] considered the chemical nature rigorously proved, and concluded that nucleic acid controls polysaccharide formation. Chargaff in 1947[13] reported that he was led by the Avery paper to look for evidence of specificity in DNA. J. Greenstein at the same symposium expressed cautious acceptance.[14] Alfred Mirsky (in 1947 and following years) frequently proposed that there must be chromosomal protein in the transforming agent, overlooked in the tests and analyses used.[15, 16] Protein, of course, would have been highly eligible as capable of a range of compositions and configurations. Although he did not seek contact with those of us who were investigating that very question, Mirsky by 1950 had come round to analyzing the DNA content of cells,[17] and the base compositions of different DNAs.[18] The carbohydrate chemists Haworth and Stacey in 1948[19] wrote of the importance of an agent "essentially DNA" but still recalled for one more time Stacey's earlier stand that the agent must be a carbohydrate priming its own biosynthesis.

As to the biological meaning of transformation: biologists and geneticists had to overcome the obstacle that it was by no means agreed in the early 1940s that pneumococci or other bacteria had the equivalent of genes. Accustomed to teaching the historically important and impressive parallels between the segregation and distribution of the gene set and of the chromosomes, geneticists tended to take the laws of diploid chromosome behavior to be the very laws of genetics. Organisms that are haploid and replicate mitotically do not lend themselves to segregation, gamete formation, and backcrossing, nor dominance tests. I well recall the time in 1951 when I was asked by a classical geneticist how I could presume to be dealing with genes so long as I "couldn't show a hybrid cross." (He might have asked me to show that our bacteria had compound eyes, or two sets of wings.) I replied that for me the transformations themselves were the genetic transfer equivalent to a cross. Norton Zinder found himself three years later similarly asked at an Oak Ridge Symposium what his transduction work "had to do with genetics." He leaped to his feet and responded with a delightful compound of simplicity and grandeur, "it *is* genetics!"

But there were other serious doubts; with every cell a member of the germ line, bacteria might have no need for genes to concentrate the total inheritance into a transmissible totipotent gamete. Every part, from wall to wall and back, might have derived from a precursor identical to itself. At just about this time Luria and Delbruck,[20] followed by Demerec,[21] showed that stable bacterial variants arose as the result of discrete rare events, much resembling the mutations of higher forms. So at least one could begin to speak of genetic determinants in bacteria.

What were biologists saying about transformation? Even before the Avery paper several had commented on the Griffith work, interestingly enough, mostly in terms directly related to their own principal working concepts. Avery himself was at that time drawing a parallel between the "complete antigen complex" of pneumococci and the transforming "principle." James Murphy[22] of Rockefeller pointed out a resemblance to a tumor agent; Wendell Stanley of the Plant Pathology Department in 1938 remarked that the transforming agent seemed like a virus.[23] His colleague, C. A. Knight in 1947[24] repeated this comparison and at around that time began to shift his attention from the protein of their plant viruses to the nucleic acid components.

Stanley has variously reported that he found the Avery work significant,[25] and that its significance escaped him and most biochemists.[26] Dobzhansky in 1941 described Griffith's transformation as a specific genetic mutation and advised attention to its possible generality.[27] Sonneborn in 1943[28] proposed that it was a specific activation of a resident gene (like those he found affecting the antigens of *Paramecium*). R. A.

Gortner in a textbook[29] in 1938 described it as the first specifically directed genetic change. The bacteriologist Langvad-Nielson[30] of Sweden and the virologists Berry and Dedrick[31] attempted their own transformation experiments following the Griffith procedure.

After the 1944 report from the Avery laboratory there were several immediate reactions. Several workers not technically concerned in genetics indicated rather instant acceptance. An editor of the *Journal of the American Medical Association* in 1944 suggested far-reaching implications in biology, growth, differentiation, and cancer.[32] G. E. Hutchinson, scientist-commentator, in 1945 designated DNA as at least a fragment of a gene, extremely fundamental.[33] Macfarlane Burnet in 1943 reported in a letter[34] that he learned that "Avery has isolated the gene." Marshak and Walker in 1945[35] wrote that the nucleic acid evidently builds in an altered structure; the gene can be a nucleoprotein. Sir Henry Dale in 1946 likened the agent to a gene in solution, "appearing to be DNA."[36] J. A. Harrison in a review in 1947 emphasized the self-reproducibility and mentioned the gene analogy.[37] Other dutiful comments were made through the years.[38, 39] The subject was certainly not being ignored!

Geneticists themselves, ranged from optimistic to rather skeptical: Sewall Wright in a 1945 review[40] wrote that transformation by apparently pure DNA has great possible significance, for the chromosome and other self-duplicating entities. Andre Boivin, who in 1945,[41] following Avery's lead, reported a possible but not confirmable transformation in *Escherichia coli*, wrote enthusiastically then and until 1948[42] about controlled mutations, newly opened horizons for the biochemistry of heredity for bacteria and other cell types, dissolving the bacterial chromosome without loss of function, and so on. His optimistic suggestions have been well enough borne out in later work, but they were largely ignored or even scorned, and constitute all the evidence we need to state that, in contrast with Stent's conclusion, it was early generalizations and not the discoveries themselves that were premature. It was not in the ethos of that time for an explorer to announce the "path to the summit" before he had at least tested it, and even if he spoke glowingly of his hopes, no one paid much attention. And, contrary to Wyatt's wistful reappraisals, even had Avery said much more about possible genetic implications, no one seemed able to imagine just how to design experiments to test the genetic implications.

The role played by the transforming agent (for reasons indicated in TABLE 1) was not so easily perceived. I. C. Gunsalus in 1948 described it as exerting an environmental influence on metabolism[43] and C. Lindegren in the same year called it a selective influence.[44] Hermann Muller, with his strong mechanistic sense, conjectured about the nucleoprotein nature of gene substance,[45] and before and after 1944 simply considered the transforming agent to be a gene-like entity, conscientiously considering the possibilities for purity and complexity in its chemical nature.

Perhaps we can take George Beadle's growing acceptance of genetic transformation as a reading of informed professional consensus: within 1948 in two articles he spoke of the agent as a transmuting agent,[46] and then *either* a transmuting agent *or* a part of the genetic system[47]; by 1951 he states that type-specific DNA is a component of the gene, and tremendously important[48]; by 1956 he simply called it the primary genetic substance.[49] Tatum and Perkins in 1950 gave a similar choice: either a specific gene mutator or a self-replicating particle.[50] The idea of a specific mutator was of course not correct; formal genetics did not yet have a concept, or a word, for a unit transfer process.

Lederberg in 1948 called attention to the literature on transformation-like effects of various culture filtrates,[51] but felt that the characterization well begun by Avery should be urgently continued because of the gene-like activity. In 1949 he felt that the identity of the agent was still not clear,[52] and in 1953 and 1954 he wrote that the meaning of transformation was not yet well understood.[53] Nevertheless much earlier, in Ryan's and Tatum's laboratories he had made tests of possibilities of transforming his own systems. In 1951, Norton Zinder, then with Lederberg, wrote me asking for a reliable sample of DNase for testing its effect on his newly discovered transducing preparations.

Fano, Caspari, and Demerec in 1950 suggested, in a medical handbook article on genetics,[54] that transformation is an evidence for the predominant nucleic acid nature of genes. Demerec himself in 1951 first reacted to my own transformation of drug resistance as an environmentally produced mutation, but Evelyn Witkin who was in the room quickly agreed with me that it must signify the *transfer* of a mutated property. N. Horowitz and H. Mitchell in a 1951 review suggested that DNA recombination may have to be looked for.[55] Jack Schultz, interested since 1935 in the nucleic acid of chromosomes, viewed it as the prosthetic group of a nucleoprotein, but in 1952 was ready to grant that DNA performs the function of the gene, or "perhaps of the whole chromosome."[56] B. Ephrussi, married to Harriett Taylor, one of the earliest of Avery's transformation coworkers, stated in 1953 that, although introduction of pure DNA was equivalent to a gene transfer, he had no doubt that transformation was only the signal representing the complete autoreproducing system containing other autonomous components.[57]

By the time of the 1956 Symposium on the Chemical Basis of Heredity at Johns Hopkins,[49] any remaining concerns were in the nature of requests for more detail, and at the meeting some of us were supplying such details. Such authorities as Beadle and Benzer simply asserted that DNA is the primary substance of heredity, and the symposium volume itself displays the DNA model as its symbol.

What had been happening in the years intervening to bring this relative conviction to so many by about 1956? I should like to suggest that a part of the answer can be found by returning to what was happening in a few laboratories where some of us were busy working on the genetic transfers by DNA, between 1944 and 1956.

My own first effort, in 1946, was to show that any of a variety of normal serum albumins, purified away from some natural masking lipid material, would supply the protein factor needed for transformation.[58] Until then the system required such mysterious and exotic supplements as chest fluid or ascitic fluid from an acutely infected human patient to supply this factor.[2, 8]

By 1947, as Avery, now in his last year at Rockefeller, watched with considerable interest and encouragement, I had begun a series of chemical studies. By simple good fortune we were at this time in a position to make the actual first demonstration that crystalline pancreatic DNase did, as we all expected, inactivate the transforming agent.[59] The activity of the best recrystallized DNase, made by the hand of Moses Kunitz

TABLE 2

CHEMICAL ANALYSES OF TRANSFORMING AGENT*

Analysis	Successive Stages of Purification† 1	2	3	4	Calf Thymus DNA
Ratio N : P	1.97	1.66	1.88	1.72	1.72
α-amino N: total N	0.020	0.015	0.011	0.005	—
P : A_{260}	3.6	6.8	4.5	4.4	4.2
deoxypentose: A_{260}	0.013	0.074	0.081	0.086	0.093
Half maximal activity, μg/ml	0.30	0.08	0.075	0.070	—
Activity recovered, %	(100)	87	66	46	—

	Substance Exposed to Strong Acid Hydrolysis: Clupeine	Bovine Serum Albumin	Transforming Agent Step 4	Adenine
α-amino acid as % of total N	27.6	69.6	0.69	1.75
Glycine N as % of total N	0.24	1.59	0.695	1.78
	0.24	1.75	0.71	1.80

*Data taken from References 59 (upper part) and 60 (glycine etc. in lower part).

†Preparations: 1, crude precipitate; 2, RNase-treated; 3, Ca precipitated; 4, multiply precipitated.

himself, was barely three times that of the purified preparations McCarty had been making by that time.

Then using progressively purer samples of pneumococcal transforming DNA, I found that although protein itself could not be detected, the sensitive Van Slyke ninhydrin method revealed a moderate release of α-amino acid during strong acid hydrolysis. As the DNA was purified, the amount so released diminished as traces of protein were being removed, although the insensitive nitrogen : phosphorus ratio could not show it (TABLE 2). From the most highly purified product the rate of amino acid release, however, was much slower than that from any protein under the same conditions. I was able to show that the normal constituent base, adenine, is decomposed by strong acid at this same slow rate (FIGURE 1) and the transforming agent gave so little rapidly released amino acid that it could not possibly have contained more than 0.2% its weight of protein. These 1948 results were presented in Paris at a symposium[59] and have only appeared in French; they were not abstracted and were little referred to in this country. They were soon supplemented by specific analyses which showed that all of the amino acid released was in fact glycine, the normal product of slow acid decomposition of adenine (TABLE 2). I could now estimate that no more than 0.02% of protein, if indeed any, could be present in purified transform-

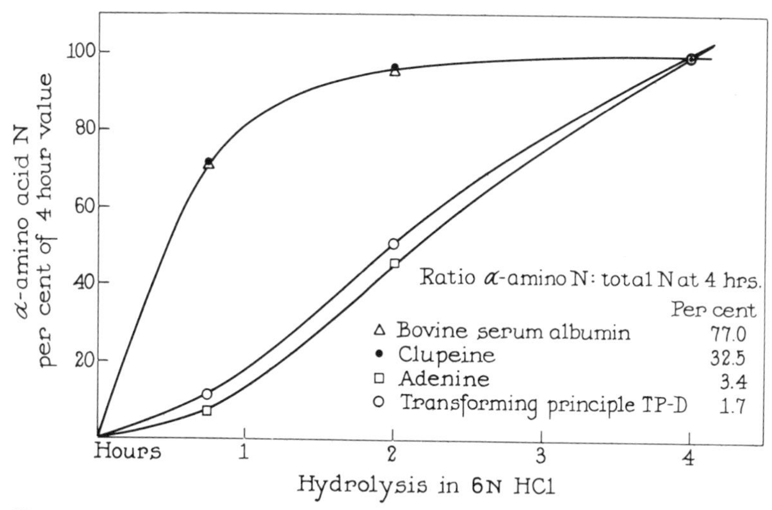

FIGURE 1. Measurement of α-amino acid liberated by strong acid hydrolysis at 120°C, showing early complete liberation from two proteins, serum albumin and clupeine, and small, slow linear release from transforming agent and pure adenine. Determination by ninhydrin gasometric method. (From Hotchkiss.[59])

ing agent.[60] A few champions of protein still spoke up at rare intervals for a while after that, as they thrilled to discover that even this low number, multiplied by Avogadro's large one, gave a finite number for possible active virus particles, e.g., which might be present. At the same 1952 conference, Stephen Zamenhof presented evidences extending the physical correlations Avery and McCarty had made between the DNA physical integrity and transforming activity.[61] His data lacked quantitative precision (being based only upon a logarithmic titer of an end-point dilution, plotted on an arbitrary scale) but they did show that enzymatic, acid, base, and UV inactivation of transforming activity occurred at thresholds of physical change in the DNA. Later years of course brought demonstrations of heat depurination and DNA melting, which Muriel Roger and I explored, and the denaturation studies of Marmur, Lane, and Doty.

The *Escherichia coli* bacteriophages had been described as "deoxyribonucleoproteins" for a number of years. I believe that our emphasis on the question of whether nucleic acid or protein was the transforming agent stimulated the framing of the questions that Hershey and Chase asked in their significant 1952 work on the genetic material of T 2 phage.[6] The idea of labeling separately the protein and the DNA with sulfur and phosphorus surely benefited from the milieu in which we were asking our questions. In a letter written to me by Alfred Hershey on April 18, 1949, he asks for my data: "I hear from Adams and Delbruck that you have convinced yourself that Avery's stuff is really DNA. I would like to be able to say something more definite about this at the SAB meeting. If you are willing, I would appreciate hearing something about what you are doing. Best regards. . . ." I sent him a manuscript version of the 1948 symposium paper, with an informal account of my further analyses of base and glycine liberation from acid hydrolysis. A few weeks later I received a letter including these remarks: "Thanks very much . . . the experiments are very beautiful. . . . My own feeling is that you have cleared up most of the doubts. Some people may cling to the virus theory a little longer, perhaps. . . ." I find that letter all the more interesting in that its writer was within three years to show that the virus activity would be explained by the DNA theory, rather than the opposite!

It is worth noting that the isotope method could not have been useful to me, for my lower limit, 0.02% for protein in transforming agent, was perhaps 100 times lower than even the best possible resolution of the isotope measurements in the phage work. Hershey could at first remove no more than 80% of the protein label as unessential for infection, and find no more than 80% of the DNA label in infected cells.[6] (Hershey never seemed to me to overstate his case, by the way.) Later refinements could reduce these margins to a few percent, but nevertheless, a sulfur-free phage phosphoprotein such as anyone was free to postulate, would have cast grave doubts on the simple picture. It was the elegance of the concept that made an immediate response; and it was in an atmosphere that I think had ripened. What was most impressive for us was that the sulfur was not conserved in the phage progeny, while a good part of the phosphorus label was, and we had to assume that a total virus genome was being transferred.

Salvador Luria had some difficulties on accepting the DNA as the phage genetic core; his example shows how the conscientious expert faces problems not felt by others less thoughtful or experienced in a subject. In a review in 1947,[62] Luria had clearly accepted the DNA nature of the transforming agent, and later he tells us that he also tried some sort of transformation experiments himself. Yet, in 1950 and 1952, seriously concerned by the rather low percentage of phage atoms of any kind to reappear in progeny, and impressed by the DNA-free phage heads seen in infected cells, he had made the postulate that the DNA-free "donuts" were already genetically nucleated, and DNA was later added as a kind of final "baking" (as he put it) of the phage.[63] Just as he was learning of Hershey and Chase's work, he was writing that no stable genetic core seemed to pass from infecting phage to progeny, and that both protein and DNA seemed to be involved.[64] In Olby's fascinating book on the history of the DNA model,[65] one can find quoted a letter from Luria to Hershey showing the interested but doubtful response he felt at the first news. This kind of conscientious and expert reaction is so easy for the outsider or latecomer to misjudge when he is loaded with "convincing" hindsight!

The most interesting chemical problem was to define the possible basis of chemical differentiation and specificity in DNAs. Since the identification of the four common bases, adenine, guanine, cytosine, and thymine, around 1910, there had been a tendency to relate structural and even analytical data to an equimolar combination of these four bases: the unit tetradeoxynucleotide. Such a speculative calculation is common and natural for organic chemists to make with low-molecular-weight compounds, but also natural for others to question for many reasons. The tetranucleotide hypothesis was proposed around 1910 and by 1930 was implanted as one of the very few statements about DNA which could be found in textbooks of biochemistry. It doubtless accounted for skepticism by many nonspecialists about the possibility for specific DNAs. The source universally quoted was the 1931 monograph by Levene and Bass.[66] The book makes clear that there were no quantitative methods for isolating or estimating individual bases or nucleotides in the state of the art even in the 1930s. But nevertheless: "on the basis of isolation of the four bases in equimolecular proportions . . . it is warranted to attribute . . . a tetranucleotide structure."

Levene's whole presentation of the tetranucleotide hypothesis in the monograph is confined to a table, reproduced here in TABLE 3. Here we see base contents calculated for a uniform tetranucleotide, and observed values somewhat lower but approximately in proportion. One can calculate unambiguously that the analysis of Levene and Mandel given here (adjusted for deoxypentose, phosphate and esterification, sodium salt formation, etc.) leaves some 17% of the mass unaccounted for. This should not be surprising, considering the approximate recoveries (as distinct from analyses) dependent on relative insolubility of one or another derivative, and necessity of correction for mineral content, as Levene notes. But it should tell us that we need not take the values too seriously. This is what I meant and I suppose what Gulland, Chargaff, and Pirie meant when we all stated that we didn't feel obliged to take the tetranucleotide calculations seriously. But others did, persons not perhaps acquainted with these perilous calculations, and the assumptions they rest upon.

You will see that Levene and Bass lean also on other more closely agreeing values from Steudel (third column). But I must now point out a remarkable thing: these "Found" values attributed to Steudel[67] can be "found" in Steudel's paper all right (lower part, TABLE 3), but they are strictly *calculated* values, also derived for an arbitrary tetranucleotide. Thus what seems like double support for the tetranucleotide hypothesis collapses into one set of analyses (Levene's) and a redundant quotation from another worker who made the same idealized drastic assumption. Also curiously enough, Steudel's calculation was made for a copper salt, so his values do not show a tell-tale identity with Levene's for a sodium salt. It is of course clear that Steudel's found value is not in good agreement with any calculated expectation. His final paragraph (TABLE 3) nevertheless indicates his satisfaction that "for such complicated investigations, the figures speak clearly" for equivalent molecular ratios of the nitrogenous bases.

The inadequacy of the tetranucleotide idea does not end with a lack of primary data. By 1940 many workers, including Levene (who died in that year) had realized that DNA from all sources had a relatively high molecular weight. This should have meant that all bets were off as to molar ratios, but unfortunately there was a tendency to cling to the idealized tetranucleotide unit and simply polymerize it. Gulland pointed

TABLE 3

THE TETRANUCLEOTIDE ASSUMPTION IN THE LITERATURE

[From *Nucleic Acids* by P.A. Levene & L.W. Bass, 1931]

TABLE XVIII.—*Theoretical and Experimental Percentages of Purines and Pyrimidines in Thymonucleic Acid.*

	Theory		Found	
	$C_{43}H_{50}N_{15}P_4O_{30}$ (Hexosenucleic acid)	$C_{39}H_{51}N_{15}P_4O_{25}$ (Desoxypentosenucleic acid)	Steudel*	Levene and Mandel**
Adenine	9.40	10.00	9.56	8.17
Guanine	10.50	11.30	10.72	9.15
Cytosine	7.70	8.29	7.86	7.0
Thymine	8.77	9.24	8.93	8.0

* *Z. physiol. Chem.*, 49, 406 (1906).
** *Biochem. Z.*, 10, 215 (1908).

[From H. Steudel, loc. cit., 1906]

Die Zusammensetzung der Nucleinsäuren.

Sollen aus der Nucleinsäure nun je ein Molekül Guanin, Adenin, Cytosin und Thymin hervorgehen, so würde sich für $C_{40}H_{53}Cu_2N_{15}O_{26}P_4$ berechnen:

Verlangt:	Guanin	10,72 %
	Adenin	9,58 %
	Cytosin	7,86 %
	Thymin	8,93 %

Experimentell ist dazu von mir folgendes gefunden worden: Die 100 g lufttrockenes nucleinsaures Kupfer mit einem Stickstoffgehalt von 8,5 % N entsprechen 58,82 g vom trockenen Salze, und für dieses ergeben sich an Ausbeuten:

Gefunden:	Guanin	9,01 %
	Adenin	10,68 %
	Cytosin	4,26 % [1])
	Thymin	8,33 % [1])

Die Übereinstimmung ist besser, wie man sie eigentlich bei solchen komplizierten Untersuchungen erwartet, und die Zahlen sprechen eine deutliche Sprache: **Die 4 stickstoffhaltigen Komponenten der Nucleinsäure und nur diese vier kommen im molekularen Verhältnis in der Säure vor.**

Note that Levene and Bass[66] quote calculated values of Steudel[67] as "found". The figures allow a calculation of the molecular weight unit and mass unaccounted for:

	Mol. Wt.	Missing (%)
Levene-Bass, theory	1347	0.4*
Levene-Mandel, found	1616	17*
Steudel, theory	1410	0
found	1765	20

*If Na salt; otherwise even greater deficit.

out this weakness, for one.[68] If we simply take the molecular weight of 10^5, Steudel's data show around 250 base residues varying from 39 to 81 for individual bases, and Levene's about the same total, varying from 60 to 64 of each. (Chargaff's later data show them paired, at 68 and 87 residues, about 309 in all.[49])

As you know, analyses eventually showed a finely graded compositional differentiation of DNA base distribution. What I believe must have been the first base chromatogram for DNA is represented in FIGURE 2. Looking for traces of amino acid in DNA hydrolysates on paper chromatograms, I found none that way, but in 1947 started looking on the blank paper for the purine and pyrimidine bases. Elution and spectroscopic assay showed high purity of some of the bases but guanine and adenine were considerably degraded. This same chromatogram shows another scoop, the first recovery of a modified base from DNA. The fast moving cytosine shoulder was, as suggested, 5-methylcytosine, confirmed by Wyatt later from this same source, thymus DNA. I could present at the same Paris symposium[59] in 1948 approximate base composition data, the first actual demonstration of the four bases in transforming agent, with uracil disappearing and thymine being retained during purification. When I discovered that Chargaff's laboratory was developing their qualitative[69] chromatography also into quantitative[70] analysis, I retired from that field to do more biological things, after describing my method for analyzing base mixtures.[79]

By the 1950's there were a good thirty or more transformations reported with pneumococci, or in *Hemophilus* (work of Alexander and Leidy). Most of these were transformations of various surface antigens, and that aspect has been continued by Ravin and Austrian and Bernheimer with increasing chemical definition. About 1949 I attempted to measure the rate of production of surface antigen in transforming cells (experiment represented in Cairns *et al.*[71]) and found that the cells became coated with antigen between 30 and 90 minutes after exposure to DNA.

Drug resistance transformation, which we began to study in our laboratory in 1949 offered new, selective traits and the possibility of quantitation. By the time of a Cold Spring Harbor Symposium in 1951 we could describe[72] a quantatitive scoring of the actual number of transformant cells produced from a corresponding known number of cells exposed to DNA.[72] Our yields were something like 2% to 5%; later we could obtain up to more than 12% (although other bacterial species respond less efficiently).

Somewhat unexpectedly, my evidence demonstrated essentially independent transfer of the traits of penicillin- or streptomycin-resistance

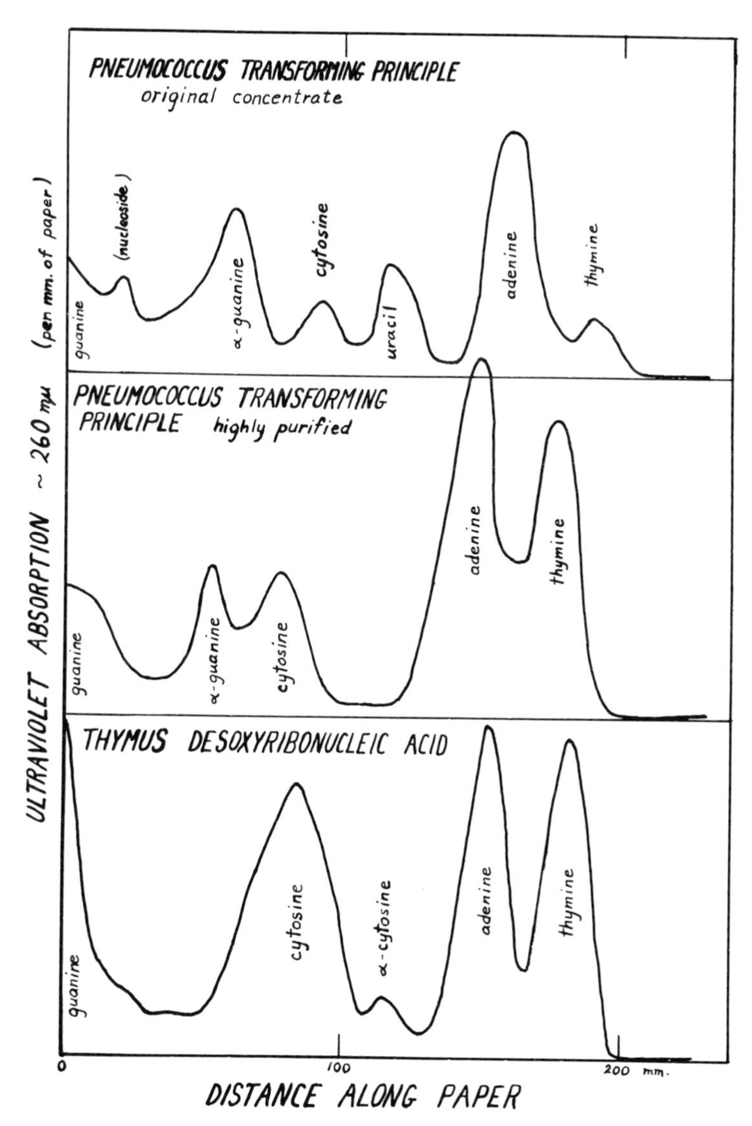

FIGURE 2. Early chromatograms of DNA hydrolysates, made in 1947 and January 1948, with butanol-ammonia paper chromatography. There is some destruction of adenine and guanine, but these early experiments showed the removal of uracil (and RNA) during purification of transforming agent, some apparent differences between pneumococcal and calf thymus DNAs, and also the first recovery from a defined DNA of a rare base, a fast-migrating form of cytosine (seen as a shoulder-peak on the cytosine peak in thymus DNA), correctly postulated to be 5-methyl-cytosine. (From Hotchkiss.[59])

and the capsule trait into different recipient cells.[72] This fragmentation of the microbial genome was destined to become the rule for transformation, transduction, even conjugation, and therefore the basis for fine-structure genetic mapping.

The quantitatively assayable trait made possible several other findings for that 1951 paper: resistance to penicillin, which occurred in a sequence of cumulative mutation steps, was also transferred by DNA in a recapitulation of the same steps. By contrast, streptomycin resistance arrived by mutation, and was transformed commonly, in a single step.

A pleasant adaptation of these principles permitted a very efficient transfer of traits without isolation of DNA. In what we called a lysate transformation,[72] drug-sensitive *donors* exposed to either of the drugs, would die and release active DNA (it was DNase-sensitive) and transform a drug-resistant recipient to some other trait carried by the donor. This kind of process we proposed, might go on in mixed populations in nature.

Quantitation also gave us a more adequate measure of the fact that full development of the transferred drug resistance required more than 90 min. This was the forerunner of many quantitative studies of gene transfer and expression. Before long the concentration of DNA could be related to the yield of actual transformation events—the true biology of molecules.[73] From such dose-response curves we estimate the number of active transforming units of (normal or treated) DNA in the linear part, and from the slope, whether one or two particles are required (nor-

TABLE 4
INTERACTIONS OF TYPE III TRANSFORMING AGENTS*

Strain	Form Induced by Action of Transforming Agent Derived from					
Treated	SIII-1	SIII-1a	SIII-1b	SIII-1c	SIII-2	SIII-N
R36A	SIII-1	SIII-1a	SIII-1b	SIII-1c	SIII-2	SIII-N
SIII-1	0	SIII-2	SIII-2	SIII-N	SIII-2 & SIII-N	SIII-N
SIII-1a	SIII-2	0	0	SIII-N	SIII-2 & SIII-N	SIII-N
SIII-1b	SIII-2	0	0	SIII-N	SIII-2 & SIII-N	SIII-N
SIII-1c	SIII-N	SIII-N	SIII-N	0	SIII-2 & SIII-N	SIII-N

*Designations used: SIII indicates strains producing Type III pneumococcal capsule; substrains SIII-1, -1a, -1b, -1c, and -2 show low to moderate production of capsular polysaccharide; SIII-N is normal fully encapsulated Type III; R36A not encapsulated. (From Ephrussi-Taylor.[74])

mally it is one). We can judge the affinity of DNA for cells by the competition in the plateau region.

The 1951 Symposium also brought a stimulating report by Harriett Ephrussi-Taylor on transformations of cells with an intermediate degree of encapsulation.[74] She was able to show that two strains with defective capsule formation could interact in one direction to make either fully encapsulated cells or the donor type (TABLE 4). Had there been a reproducible way to derive those defective strains from the fully encapsulated one, there could have been no doubt that it represented genetic recombination, as we now are sure it did.

For a short time after 1951 we rejoiced in the simple view that different genes were carried on different particles of DNA. This would have paralleled the "beads-on-a-string" concept of chromosome structure, and perhaps suggested protein as an eligible linking agent. But by 1954, Julius Marmur and I had discovered the linked transfer of an enzymic trait, mannitolphosphate dehydrogenase, with streptomycin resistance.[75] Quantitative work allowed us to show[76] that the nonselective as well as selective alleles could be transferred (FIGURE 3).[76] Thus, we seemed to have linkage, with partial recombination, for all forms of these unrelated traits, and drug resistance could be driven out of cells by linked cotransfers.

The fact that genes carried by DNA were connected to each other by nothing more or less than DNA, provided an adequate end of the beads-

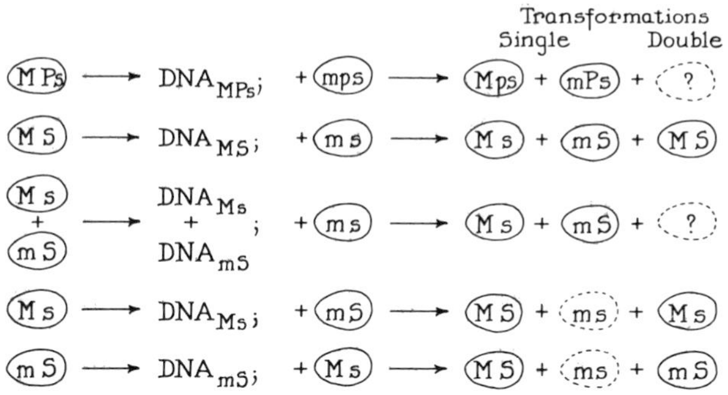

Transformations
Single Double

(MPs) ⟶ DNA_{MPs}; + (mps) ⟶ (Mps) + (mPs) + (?)

(MS) ⟶ DNA_{MS}; + (ms) ⟶ (Ms) + (mS) + (MS)

(Ms) + (mS) ⟶ DNA_{Ms} + DNA_{mS}; + (ms) ⟶ (Ms) + (mS) + (?)

(Ms) ⟶ DNA_{Ms}; + (mS) ⟶ (MS) + (ms) + (Ms)

(mS) ⟶ DNA_{mS}; + (Ms) ⟶ (MS) + (ms) + (mS)

FIGURE 3. Double transformations to two heritable properties. Schematic diagram showing linkage (cotransfer) of mannitol-phosphate dehydrogenase and streptomycin markers in their selective forms, M and S, and in combination with their nonselective forms, m and s, representing mannitol nonutilization and streptomycin sensitivity. Penicillin resistance, P, does not show linkage to mannitol. Ellipsoidal outlines represent cells; dotted outlines indicate possible doubly nonselective configurations that would not be detected. (From Hotchkiss.[80] Related data appear in References 75 to 77.)

on-a-string model. I tried only one more way to convince myself that the mannitol and streptomycin genes were biologically, rather than physically, brought together. Assembling by mutation or transformation a large series of strains with various sequences of insertion of markers (TABLE 5), Dorothy Lane and I found that they all showed the characteristic ~20% cotransfer of the marker combination in their DNA's.[77] Therefore each marker had a fixed site in the DNA to which it found its way, and not random adventitious sites created by the transformation process.

The fact that the mannitol utilization trait was inducible brought us in 1955 one more correlation of DNA with gene. While induced and noninduced cells were very different in content of enzyme and activator, they should not be different in gene content. In agreement with this postulate their DNAs proved to be equally effective in introducing the trait.[78]

Thus, by 1956, gene-like activity of DNA represented many traits, transferred in units resembling the mutations that produced them, often

TABLE 5

MULTIPLE MARKER STRAINS OF PNEUMOCOCCI WHOSE DNAS ALL SHOW LINKAGE BETWEEN MANNITOL AND STREPTOMYCIN MARKERS*

Transformant Strains	Derived Mutants or Back-Mutants
MS	
Ms	MS
mS	
SM	sM
sM	
sPM	
PFMS	PfMS, PFmS
PMFS	
SMFP	
MPSF	mPSF
FSMP	
FSPM	
sMPF	
sPMF	
mFSP	
fPMS	

*Marker symbols: M, mannitol utilization; P, S, F, penicillin, streptomycin, and sulfonamide resistance, respectively. Order of symbols is the order of single marker introduction into strain. Lower case symbols indicate drug sensitivity, wild-type or back-mutant. Linkage, about 20%, found only for the M, S, m, and s pairs. (From Hotchkiss.[77])

moving independently, but also occupying linked sites which could carry various alleles of a gene, yet allow them to recombine into other combinations, retaining the same activity level regardless of the state of expression in the cell. These results had been presented in numerous seminars and symposia, and published, somewhat belatedly perhaps, as various energetic chairmen insisted on seeing that the papers did eventually see daylight.

I hope I have also removed for you some of the mystery of that decade. It was a time of maturation of a new field. But I see this maturation as the growth of an infant science and not the delicate nurturing of a "premature" one as Stent has suggested and others have tried to explain. As a midwife and nurse in the upbringing of the infant, I want to report that it was a normal healthy one. The fact that the infant by the age of twenty or so had composed brilliant concertos, sonatas, and symphonies need not raise in anyone's mind the question, why didn't he compose a rondo or cadenza before the age of ten?

As I have mentioned, but not stressed, the same broad pattern of discovery, consolidation, and reorientation is woven in and out of the whole connected history, not only in the DNA-transformation part. It required some 15 years to go from Griffith to Avery; some 5 to 8 years for full validation of the Watson-Crick replication model; 17 years to deal with plant viruses as nucleoproteins, 20 years to treat nucleoproteins as nucleic acid and protein separately, 30 years from Garrod to Beadle and Tatum for enzyme-gene relations, to mention only a few. While some may choose to infer that the slow process concerned integration and circulation of the ideas, I have a lively sense that in the DNA revolution at least, the slow process consisted in imagining and designing the new kinds of experiments that could give force and generality to those ideas.

Further concern with this supposed paradox seems fruitless to me. Let us by all means keep a healthy impatience to see work and its interpretation go efficiently—but not expend that impatience as time-warped tourists in an incompletely perceived past. As this symposium has shown, our past has been very good indeed to us; it is up to us all to help to keep the future as productive.

REFERENCES

1. Griffith, F. 1928. J. Hyg. *27*: 113–159.
2. Avery, O. T., C. M. MacLeod & M. McCarty. 1944. J. Exp. Med. *79*: 137–158.
3. Wyatt, H. V. 1972. Nature *235*: 86–89.
4. Stent, G. S. 1972. Sci. Am. 227: 84–93; 1972. Adv. Biosci. *8*: 433–449.

5. Stent, G. S. 1969. The Coming of the Golden Age: A View of the End of Progress. The Natural History Press. New York, N.Y.
6. Hershey, A. D. & M. Chase. 1952. J. Gen. Physiol. *36*: 39–56.
7. Watson, J. D. & F. H. C. Crick. 1953. Nature *171*: 737–738; 964–967.
8. McCarty, M. & O. T. Avery. 1946. J. Exp. Med. *83*: 89–96.
9. Haddow, A. 1944. Nature *154*: 194–199.
10. Gulland, J. M., G. R. Barker & D. O. Jordan. 1945. Ann. Rev. Biochem. *14*: 175–206.
11. Kalckar, H. M. 1945. Ann. Rev. Biochem. *14*: 283–308.
12. Woods, D. D. 1947. Ann. Rev. Microbiol. *1*: 115–140.
13. Chargaff, E. 1947. Cold Spring Harbor Symp. Quant. Biol. *12*: 28–34.
14. Greenstein, J. P., C. E. Carter & H. W. Chalkley. 1947. Cold Spring Harbor Symp. Quant. Biol. *12*: 64–94.
15. Mirsky, A. E. 1946. J. Gen. Physiol. *30*: 117–148.
16. Mirsky, A. E. 1947. Cold Spr. Harbor Symp. Quant. Biol. *12*: 15–16.
17. Mirsky, A. E. & H. Ris. 1949. Nature *163*: 666–667.
18. Daly, M. M., V. G. Allfrey & A. E. Mirsky. 1950. J. Gen. Physiol. *33*: 497–508.
19. Haworth, N. & M. Stacey. 1948. Ann. Rev. Biochem. *17*: 97–114.
20. Luria, S. E. & M. Delbruck. 1943. Genetics *28*: 491–511.
21. Demerec, M. 1945. Proc. Nat. Acad. Sci. U.S.A. *31*: 16–24.
22. Murphy, J. B. 1935. Bull. Johns Hopkins Hosp. *56*: 1–31.
23. Stanley, W. M. 1938. Handb. Virusforsch. Vol. 1: 491. Springer. Vienna.
24. Knight, C. A. 1947. Cold Spring Harbor Symp. Quant. Biol. *12*: 115–121.
25. Stanley, W. M. 1963. Viruses, Nucleic Acids, and Cancer. Williams and Wilkins. Baltimore, Md.
26. Stanley, W. M. 1970. Arch. Environm. Health *21*: 256–262.
27. Dobzhansky, T. 1941. Genetics and the Origin of the Species. :47. Columbia Univ. Press. New York, N.Y.
28. Sonneborn, T. M. 1943. Proc. Nat. Acad. Sci. U.S.A. *29*: 329–343.
29. Gortner, R. A. 1938. Outlines of Biochemistry. 2nd Ed. :547. Wiley. New York, N.Y.
30. Langvad-Nielsen, A. 1944. Acta Path. Microbiol. Scand. *21*: 362–369.
31. Berry, G. P. & H. M. Dedrick. 1936. J. Bacterial. *31*: 50; *32*: 356.
32. Editorial. 1944. J. Am. Med. Assoc. *126*: 964.
33. Hutchinson, G. E. 1945. Amer. Sci. *33*: 56–57.
34. Burnet, F. M. 1943. Letter quoted in Changing Patterns: An Atypical Biography (1969) Elsevier. New York, N.Y. (From Olby.[65])
35. Marshak, A. & A. C. Walker. 1945. Science *101*: 94–95.
36. Dale, H. 1946. Proc. R. Soc. *185A*: 127–143.
37. Harrison, J. A. 1947. Ann. Rev. Microbiol. *1*: 19–42.
38. Shimkin, M. B. 1950. Ann. Rev. Med. *1*: 179–198.
39. Several reviewers, 1949–53 in Ann. Rev. Microbiol. as follows: McClung, L. S. *3*: 395–422; Watson, D. W. & C. A. Brandly. *3*: 195–220; Shrigley, E. W. *5*: 241–264; Kaplan, R. W. *6*: 49–76; Wyss, O. & F. L. Haas 7: 47–82.
40. Wright, S. 1945. Ann. Rev. Physiol. 7: 75–106.
41. Boivin, A., A. Delaunay, R. Vendrely & Y. Lehoult. 1945. Experientia *1*: 334–335.
42. Boivin, A. R., Vendrely & C. Vendrely. 1948. Colloques Internat. C.N.R.S. *8*: 67–78.
43. Gunsalus, I. C. 1948. Ann. Rev. Microbiol. *2*: 71–100.

44. LINDEGREN, C. C. 1948. Ann. Rev. Microbiol. *2*: 47–70.
45. MULLER, H. J. 1941. Cold Spring Harbor Symp. Quant. Biol. *9*: 290–308; also MULLER, H. J., C. C. LITTLE & L. H. SYNDER. 1947. Genetics, Medicine and Man. :1–34. New York, N.Y.
46. BEADLE, G. W. 1948. Amer. Sci. *36*: 71–74.
47. BEADLE, G. W. 1948. Ann. Rev. Physiol. *10*: 17–24; 1948. Ann. Rev. Biochem. *17*: 727–752.
48. BEADLE, G. W. 1952. Dyer Lecture, U.S. P.H.S. Publ. No. 142, U.S. Gov. Printing Office. Washington, D.C.
49. Symposium on The Chemical Basis of Heredity, 1956. (Publ. 1957). W. D. McElroy & B. Glass, Eds. Johns Hopkins Press. Baltimore, Md. (Articles by G. W. BEADLE, S. BENZER, E. CHARGAFF, H. EPHRUSSI-TAYLOR, S. H. GOODGAL & R. M. HERRIOTT, R. D. HOTCHKISS, S. ZAMENHOF.)
50. TATUM, E. L. & D. D. PERKINS. 1950. Ann. Rev. Microbiol. *4*: 129–150.
51. LEDERBERG, J. 1948. J. Heredity *2*: 145–198.
52. LEDERBERG, J. 1949. Ann. Rev. Microbiol. *3*: 1–22.
53. LEDERBERG, J. & E. L. TATUM. 1954. Sex in Microorganisms. AAAS. Washington, D.C.; But see LEDERBERG, J. 1955. Perspectives and Horizons in Microbiology. Rutgers Univ. Symposium.
54. FANO, U., E. CASPARI & M. DEMEREC. 1950. Chapter on genetics. *In* Medical Physics. O. Glasser, Ed. Vol. 2: 365–385. Year Book Publishers. Chicago.
55. HOROWITZ, N. H. & H. K. MITCHELL. 1951. Ann. Rev. Biochem. *20*: 465–486.
56. SCHULTZ, J. 1952. Exp. Cell Res. *3* (Suppl. 2) :17–43.
57. EPHRUSSI, B. 1953. Nucleo-Cytoplasmic Relations in Microorganisms. Clarendon Press. Oxford.
58. HOTCHKISS, R. D. & H. EPHRUSSI-TAYLOR. 1951. Fed. Proc. *10*(1): 200.
59. HOTCHKISS, R. D. 1949. Colloq. Internat. C.N.R.S. *8*: 57–65.
60. HOTCHKISS, R. D. 1952. *In* Phosphorus Metabolism. W. D. McElroy & B. Glass, Eds. Vol. 2: 426–436. Johns Hopkins Press, Baltimore, Md.
61. ZAMENHOF, S. 1952. *In* Phosphorus Metabolism. W. D. McElroy & B. Glass, Eds. Vol. 2: 301–328. Johns Hopkins Press, Baltimore, Md.
62. LURIA, S. E. 1947. Bact. Rev. *11*: 1–40.
63. LURIA, S. E. 1950. Science *111*: 507–511.
64. LURIA, S. E. 1952. Sympos. Soc. Gen. Microbiol. *2*: 99–113.
65. OLBY, R. 1974. The Path to the Double Helix. Univ. of Washington Press. Seattle.
66. LEVENE, P. A. & L. W. BASS. 1931. The Nucleic Acids. Chemical Catalog Co. New York, N.Y.
67. STEUDEL, H. 1906. Z. Physiol. Chem. *49*: 406–409.
68. GULLAND, J. M. 1944. J. Chem. Soc. :208–217.
69. VISCHER, E. & E. CHARGAFF. 1947. J. Biol. Chem. *168*: 781–782.
70. VISCHER, E., S. ZAMENHOF & E. CHARGAFF. 1949. J. Biol. Chem. *177*: 429–438.
71. HOTCHKISS, R. D. 1966. *In* Phage and the Origins of Molecular Biology, J. Cairns, G. S. Stent & J. D. Watson, Eds. :180–200. Cold Spring Harbor Laboratory, Cold Spring Harbor, N.Y.
72. HOTCHKISS, R. D. 1951. Cold Spring Harbor Symp. Quant. Biol. *16*: 457–461.
73. HOTCHKISS, R. D. 1957. *In* Chemical Basis of Heredity :321–335. Johns Hopkins Press. Baltimore, Md.
74. EPHRUSSI-TAYLOR, H. 1951. Cold Spring Harbor Symp. Quant. Biol. *16*: 445–456.

75. HOTCHKISS, R. D. & J. MARMUR. 1954. Proc. Nat. Acad. Sci. U.S.A. *40*: 55–60.
76. HOTCHKISS, R. D. 1955. J. Cell. Comp. Physiol. *45* (Suppl. 2): 1–22.
77. HOTCHKISS, R. D. 1956. *In* Enzymes: Units of Biological Structure and Function. : 119–130. Academic Press. New York, N.Y.
78. MARMUR, J. & R. D. HOTCHKISS. 1955. J. Biol. Chem. *214*: 383–396.
79. HOTCHKISS, R. D. 1948. J. Biol. Chem. *175*: 315–332.
80. HOTCHKISS, R. D. 1954 (Publ. 1955). Harvey Lectures. Academic Press. New York, N.Y.

DISCUSSION OF THE PAPER

ABIR-AM: You say that it is the accumulation of the work between 1944 and 56 that you just described which led to acceptance of the view that DNA is the genetic material? Would you like to tell which decisive factor had a particular influence?

HOTCHKISS: There is a kind of fallacy in the assumption that there must always be some highly decisive factor and others play little part. I think that it was an accumulation, and I think most of all that a good deal of it did not shown in the final printed record because all of us were talking to each other a great deal. All the material that I presented and wanted to show you here was given in many lectures all over and is not necessarily in the final record. I think we historians must take into account the exchange between individuals in the form of informal communication. And there was very much of that going on at that time.

Erwin Chargaff received his graduate degree in chemistry in Vienna in 1928. From there he went to Yale, the Institute of Hygiene in Berlin, the Pasteur Institute in Paris, and finally in 1935 the College of Physicians & Surgeons at Columbia, where he later became chairman of the Biochemistry Department.

How Genetics Got a Chemical Education

ERWIN CHARGAFF
350 Central Park West
New York, New York 10025

NOWADAYS WE ARE, of course, familiar with such terms as "molecular biology," "molecular genetics" or "molecular pharmacology." The curious umbrella of molecularity under which the various biological disciplines practice a previously unimaginable form of togetherness testifies to the extent to which chemistry—the very science of molecules and, therefore, one of the few not tolerating the all-embracing adjective—has acted as a cement holding together the several branches of biology. In this respect biochemistry naturally shares the role of chemistry, and nobody has yet come forward with such a thing as molecular biochemistry. If we had to give up molecules, what would be left of us? Well, perhaps molecular biology would be left.

There were, however, times when the biological sciences were not yet domesticated and did not march nicely in pairs on the leash of chemistry. They were robust fellows, each with his own code of honor and jealous of his independence; and they left each other more or less alone. I remember these days quite vividly, and how astonished we students were when at the chemistry colloquium at the University of Vienna a stray botanist or pathologist put in an appearance. In an even earlier generation the great Karl Landsteiner was one of the exceptions. When I met him in 1935 in Siasconset on Nantucket Island and he told me, walking on the beach or sitting in his somber house, of his early days in Emil Fischer's laboratory in Berlin, I was surprised about the wide range of his scientific interests. Now, when I am much older than he was at that time, I realize that this form of openness is no longer possible, and the sciences, as they have grown together, have become more hermetic than they ever were. There existed, of course, and there still exist, a few people able to break through the boundaries.

Of the exact sciences, physics and somewhat later chemistry were the first to develop greatly. Before they reached the stage at which they could support the biological sciences, it is not surprising that biologists had little use for chemistry and physics. For this reason, biochemistry, not to speak of biophysics, represents a relatively late development. As was to be expected, among the biological disciplines, physiology was perhaps the first to experience a need for chemical assistance; and here

0077-8923/79/0325-0345 $01.75/0 1979, NYAS

lies, in fact, the origin of biochemistry. It began as a branch of physiology and was, for a long time, designated physiological chemistry.

The "molecular revolution"—10% advance and 90% verbiage—came about, however, in a surprisingly different fashion. Not so much physiology as two other branches of biology, microbiology to a great extent and immunology to a lesser, were involved; and the real beneficiary was a fourth branch, namely, genetics. Of immunology I do not want to speak here. I already have mentioned one of the great men who brought about the association of immunology and chemistry, Karl Landsteiner; and it is moving and surprising to me that many of us know the appealing figure of Michael Heidelberger, the true founder of what is essentially a science in itself, immunochemistry.

I should, however, like to say a few words about genetics, doing justice to the title of my talk.

If we are to believe the Oxford English Dictionary, the word *genetics* was first used in 1905 or 1906. The man usually credited with the introduction of this term is the English biologist William Bateson. It is one of the youngest sciences, being dated from the year 1900, when Gregor Mendel's observations were rediscovered. To the extent that its first tools consisted in breeding experiments, genetics also is a very old science indeed: animal and plant breeders practiced an early form of applied genetics. For some reason popular ideas of heredity have always involved some kind of "blood and soil" mythology. Therefore, I could read only recently in a book on the Bach family that some of Johann Sebastian Bach's blood still rolls through the vessels of a Mr. Colson. The author would clearly have been unable to tell me, more scientifically, what percentage of J .S. Bach's DNA still was around. Even if he could have done so, I daresay it would have been of no interest whatever. DNA does not compose heavenly cantatas, nor even musical trash, although a lot of other trash has been produced through its help. If the ballyhoo had taken place a hundred years earlier, Johann Strauss would undoubtedly have composed a "Double Helix Polka." Our times would at best be capable of producing a dance that might perhaps be called the "dobble-wobble."

Even the early workers who experimented with living cells or tissues or with blood must have had a perception of the ghost-like presence of chemistry in all they were doing. They knew, of course, some chemistry and they knew that chemistry was the science of substances, of compounds. They must have realized that protoplasm was composed of compounds that, at least on one level of their existence, obeyed the laws of chemistry. But despite the early appearance of such men as Friedrich

Miescher or Hoppe-Seyler I do not think that many bridges existed between biology and chemistry. The reasons why the early biologists kept their distance from the exact sciences do not all speak against the profundity of their perceptions. Reductionism had not yet entirely taken over their ways of thinking, as it has done now, and what one could call the technicalization of biology—the wheels and the gears and the pulleys, the fuel, the lubricants, the templates, and so on—had not yet won its shallow victory. There was still some reverence left, an awe before the everlasting mystery of life. Those that were good among these old men stepped softly.

When I turn to the early stage of genetics I get the impression, perhaps mistakenly, that the initial exponents of this science were particularly unable or unwilling to think in terms of chemistry. Once the gene, as the unit of heredity, was defined and its localization in the chromosomes made probable, there was, of course, enough reason to assume that this unit was a substance or a conglomerate of substances and, thereby, subject to the scrutiny of the chemist. I am not able to determine how much speculation on the chemical nature of the gene did take place in the early days of genetics; but I am sure that a geneticist, had he been pressed to reveal his thoughts, would have guessed that the gene may be a protein, for proteins were at that time the receptacles of all that was mysterious and refractory. Of course, the term "biological information" could not yet exist; those were the times of the log table, if not the abacus, and not of the computer.

I am not sure if I am right in saying that if one wants to decipher an as yet unread language, simple, primitive texts may be more useful than complex, poetic documents. In the case of the chemical basis of heredity this is, however, unquestionably true. Without the use of phages and of microorganisms little could have been achieved in this regard. This stage was reached in the early forties, i.e., at the time when Oswald T. Avery and his collaborators began to work on the transformation of pneumococci, and Delbrück and Luria on the phages of *E. coli*, although Avery's laboratory was much more receptive to the application of chemistry than was the other group. It is really with Avery that there began what I have called, in my title, the chemical education of genetics. It is, perhaps, characteristic of the way in which science operates that the educator was neither a geneticist nor a chemist. The learning process was slow and weary: the card-carrying members of the guild and their assorted acolytes refused as long as possible, and even beyond that point, to take cognizance.

The path of careful, conscientious, and responsible research that led

Avery and collaborators[1] to the recognition that the hereditary units, the genes, were composed of DNA has been described excellently in the Dubos book.[2] The amazing difficulties that this truly epochal observation experienced before being accepted have been narrated comprehensively in Olby's book.[3] I should also like to refer to my reviews of the books of Dubos[4] and Olby.[5]

How profound the impression was that Avery's discovery made on me I have attempted to relate more than once.[6, 7] I shall return to it presently. One should have thought that if I, a simple chemist only distantly interested in the mechanisms of heredity, was so deeply moved by the sudden appearance of a giant bridge between chemistry and genetics, the practitioners of the latter science would have been alerted even more forcefully. I had, however, at that time the impression that this was far from being the case. In preparing the present essay I wanted to confirm the accuracy of my recollection. I went, therefore, back to the library and looked through a few genetics texts that were current in the period following the 1944 paper by the Avery group.

The oldest book I consulted was the fourth edition, published in 1950, of a widely read introduction by Sinnot, Dunn, and Dobzhansky.[8] The last two of the three authors, both Columbia professors, I had known very well during their lifetimes. The names Beadle, Delbrück, Lederberg, Luria, and Tatum appear in the index, and so does, more modestly, DNA as a component of chromosomes and some viruses. The name of Avery, however, does not appear; and so far as I can see—I did not again read the entire 500 pages—his discovery is not mentioned.

In a textbook published three years later,[9] Avery is listed in the index. My joyful exclamation was, however, stifled when I discovered that it was, alas, the wrong Avery. The "tetranucleotide theory of Levene" is discussed at length in ancient terms; but, again, no ripple of the wave of the future.

The next candidate is a small monograph on the biochemistry of genetics, published in 1954.[10] As was to be expected of so intelligent an author as J. B. S. Haldane was, the discovery of pneumococcal transformation by DNA is mentioned (p. 49); but there is little evidence of an awareness of what Avery's discovery meant in terms of the chemical structure of DNA and its role in the chromosomes. The nucleoproteins of the nucleus are regarded as catalysts (p. 117), perhaps in fixing ATP (p. 126). If the first two books rate an F, this one would rate C+.

My last witness is a book published in 1958.[11] Most of the standard names can be found in the index, but neither Avery nor Crick and Watson. One should have thought that enough time had then elapsed to digest the digestible, not to speak of the precooked, such as the double helix. Mention is made of transformation being induced by a "nucleic acid, of a specific type"; but that is about all. Fourteen years after his discovery, and three years after his death, Avery did not even rate honorable mention. One gets the impression that the tenaciously engrained conception of the classical gene acted as a vanishing cap for its real unraveler.

The title I have chosen is, therefore, possibly wrong, and it should read: "How Genetics Refused to Get a Chemical Education." But "genetics" is, of course, as vague an entity as "the people"; and the collective is made up of all sorts of individuals, each one doing his own thing. Besides, it is a fortunate fact that amateurs often are better in advancing science than are the professionals. Nothing more deadening than being a specialist, an expert. You lecture before a perpetually somnolent audience—the people change, but they are equally bored or obtuse—or, if you are lucky, you teach in a workshop on a beautiful island, and you teach them to become as you are; whereas what a scientist ought to do is to teach others to become as different from himself as possible. *Vive la différence!* should be the battle cry. Instead, it is "like begets alike," until at the end dismal sociobiology takes over to tell us that you must be programmed in your genes to attend Asilomar. Scientific life nowadays would be funny if it were not sad.

So let me think of better times. I have been trying to recollect when I first heard of nucleic acids: probably during my university time, but I cannot have learned more than what I learned about insect pigments or anthocyanins. As a post-doc at Yale I saw, however, T. B. Johnson every day, and there the purines and pyrimidines made themselves known plentifully. As a young *Assistent* at the Bacteriology Department of the University of Berlin I earned a little extra money writing abstracts for the *Centralblatt*, and one day I got the newly published book by P. A. Levene[12] for review. This was late in 1931 or at the beginning of 1932. I read it dutifully, but I do not remember any more what I said about it. The book certainly was not particularly pleasant to read, although I have kept my review copy to this day.

In my own work I encountered DNA when Seymour Cohen studied the composition of rickettsiae,[13] and together with him and Aaron Bendich we came across RNA in our work on the thromboplastic protein.[14] What a job it was to do a spectrum on the Hilger spectrograph! Besides, the wet blanket of the tetranucleotide hypothesis extinguished all enthusiasm for these unpleasant laboratory curiosities. But at about the time when I wrote those two papers, deoxyribonucleic acid captivated my attention in a much more compelling manner.

Was it in the at-that-time still pleasant dining room? Was it in one of the cheerless corridors of the College of Physicians and Surgeons? Anyway, somebody came and told me to read a paper by Avery in the *Journal of Experimental Medicine*. This was the article I mentioned before.[1] Associations of thought normally cannot be reconstructed after the event, for they have the logic of dreams; but it was obvious to me that I must work on the chemistry of the nucleic acids. The road to take was, in fact, clearly delineated before my eyes; what I did not know was how to get to the beginning of the road. I knew that we had to find methods for the complete and accurate analysis of the nitrogenous components and the sugars of several DNA specimens widely separated as to their origins. Since most specimens would be difficult to come by, the methods, moreover, had to be applicable to minute amounts.

The immediate problems, then, were (1) to develop procedures for the quantitative analysis of each of the purines and the pyrimidines present in a DNA; (2) to establish satisfactory balances in terms of total N and P; (3) to identify the sugar, or the sugars, present in a given nucleic acid; (4) to secure a variety of intact DNA specimens.

Memory, unless it is committed to writing (and even then), is the most evanescent of gifts. How many are there still left who remember what it meant to determine the composition, let alone the exact composition, of a nucleic acid in, say, the year 1945? If you consider that at that time the quantitative analysis even of a protein could not be achieved—and proteins had been studied much longer and more intensively than the nucleic acids—you will conclude, and rightly so, that nothing could be done for the nucleic acids. Levene's book[12] epitomized the situation. For instance, on page 113 of this treatise the following statement will be found: "No methods exist for the quantitative determination of the individual purines when present in mixtures." For the qualitative isolation of the constituent purines, procedures requiring 50 g of nucleic acid are described, and the same is true of the pyrimidines.

It is, therefore, perhaps not surprising that in the absence of any means of ascertaining the truth about the composition of the nucleic acids a form of mock democracy was observed by the investigator who proclaimed: "All nucleic acid bases are equal." That some could be more equal than others did not even dawn on the archreductionist. This led to the baseless tetranucleotide theory.* I only regret that P. A. Levene

* Much later some people came and told me that they never had believed in the tetranucleotide structure. This may be so; but by sitting solidly on one's haunches, while having hunches, one does not advance science.

did not call it the "Central Tetranucleotide Dogma," as sillier times would undoubtedly have done. To topple a dogma is more fun than to disprove a theory, for topple or disprove we certainly did.

When in 1946, together with Ernst Vischer and Charlotte Green, I set out to develop a quantitative micromethod for the complete analysis first of DNA, and a little later also of RNA, we were favored by an unusual conjunction of lucky circumstances. First of all, the introduction of paper chromatography two years earlier, for the qualitative separation of amino acids,[15] had shown the possibility of separating, and in most cases identifying unambiguously, minute quantities of organic substances. Secondly, at the end of the war, in the beginning of 1946, excellent photoelectric quartz spectrophotometers became available commercially, and we were among the early acquirers of a Beckman instrument. And thirdly, and most importantly, the possession by the purines and pyrimidines of specific and characteristic absorption spectra in the ultraviolet, together with the arrangements mentioned just now, rendered feasible for the first time the development of an exact procedure for quantitative microestimation by paper chromatography. Our first preliminary note was submitted in April 1947[16] and a detailed paper in June 1948.[17]

One month later, we sent in two papers on the complete quantitative analysis of several DNA preparations: one dealt with the purines and pyrimidines of the DNA of calf thymus and beef spleen,[18] the other with those of the DNA of tubercle bacilli and of yeast.[19]

One curious circumstance attending the publication of these papers deserves mention because it illustrates the ignorance about nucleic acids that then prevailed among the scientific elite. I had, at that time, already published something like 75 articles in the *Journal of Biological Chemistry* without ever having one sent back by the editor for clarification or revision. The papers about DNA composition, however, were returned to me with a particularly silly objection. How could I, the editor asked, express the composition of a DNA as moles of adenine or guanine, cytosine or thymine, per gram-atom of phosphorus, since the purines and pyrimidines did not contain any phosphorus? After I had repeated, in my answer to the editor, part of the introductory lecture on the nucleic acids, which at that time I was already giving to the first-year medical students at Columbia, we achieved grudging reconciliation.

The emphasis that I placed on molar relationships underlines the fact that I approached the problem as a chemist. The phosphorus and nitrogen contents of a nucleic acid preparation can be ascertained by elementary analysis, and we did not rest satisfied until our analytical methods permitted recoveries of the total bases in the range of 96% to 98%. This is no longer done in experimental studies on DNA. Sometimes complete nucleotide sequences are now published without any proof that these correspond to the total P, N, and purine and pyrimidine contents of the specimen. In this way, half-truths are piled on quarter-truths until one day the entire structures will collapse. In my opinion, molecular biology is disregarding chemistry at its own peril.

Another utopian attempt of ours was the investigation of the nature of the deoxy sugar in every DNA specimen prepared by us. It is true, nothing but deoxyribose was found; but there was no reason to assume beforehand this to be the case. In other words, someone had to do what modern scientists would consider as "dirty work," and we were not loath to do it.

I have recently described the path that led us to our present view of the chemical nature of DNA,[7] and I should not wish to repeat myself here. When the time had come for me to attempt a first summary,[20] this is what I wrote:

> We started in our work from the assumption that the nucleic acids were complicated and intricate high-polymers, comparable in this respect to the proteins, and that the determination of their structures and their structural differences would require the development of methods suitable for the precise analysis of all constituents of nucleic acids prepared from a large number of different cell types. These methods had to permit the study of minute amounts, since it was clear that much of the material would not be readily available. . . . The desoxypentose nucleic acids from animal and microbial cells contain varying proportions of the same four nitrogenous constituents, namely, adenine, guanine, cytosine, thymine. Their composition appears to be characteristic of the species, but not of the tissue, from which they are derived. The presumption, therefore, is that there exists an enormous number of structurally different nucleic acids; a number certainly much larger than the analytical methods available to us at present can reveal. . . . Desoxypentose nucleic acids from different species differ in their chemical composition, as I have shown before; and I think there will be no objection to the statement that, as far as chemical possibilities go, they could very well serve as one of the agents, or possibly as the agent, concerned with the transmission of inherited properties.

I am not a historian of science—if there is such a thing—and I am, therefore, not sure that I am correct in saying that this is among the early statements concerning chemically encoded biological information. My claim is perhaps strengthened by another passage from the same review, affirming the biological importance of nucleotide sequence:

> We must realize that minute changes in the nucleic acid, e.g., the disap-

> pearance of one guanine molecule out of a hundred, could produce far-reaching changes . . . ; and it is not impossible that rearrangements of this type are among the causes of the occurrence of mutations.

One other short paragraph of the same article[20] carried the seeds of the future. It reads:

> The results serve to disprove the tetranucleotide hypothesis. It is, however, noteworthy—whether this is more than accidental, cannot yet be said—that in all desoxypentose nucleic acids examined thus far the molar ratios of total purines to total pyrimidines, and also of adenine to thymine and of guanine to cytosine, were not far from 1.

How this statement came to be inserted into the galley proofs of the review article,[20] I have recounted in my forthcoming book.[7] Our first results, marred by the initial necessity of determining the purines and pyrimidines separately and by an indirect procedure of demonstrating the separated spots on the filter paper, did not lead obviously to such a conclusion. Without the complete balance of recoveries in terms of nucleotide phosphorus, which we established in all our analyses, we should never have come to the recognition of the remarkable pairing rules. The first two observations we made on the basis of these balances were (1) the recoveries of purines were invariably much higher than those of pyrimidines: a difference attributable to the different hydrolysis procedures then employed for the liberation of these two classes of compounds; (2) even in the first studies on DNA of beef tissue,[18] and despite the higher yield of purines that I have mentioned, the molar ratios of adenine to guanine were very similar to those of thymine to cytosine: the average for A/G was 1.3 and that for T/C 1.4.

When I gave a series of lectures in 1949, I mentioned these and related observations, but when it came to rewriting them in the form of the *Experientia* review,[20] I hesitated first to emphasize any compositional regularities, owing perhaps to my skeptical and antidogmatic character. But in the meantime we had begun to improve our initial methods considerably by introducing formic acid hydrolysis for the simultaneous liberation of all nitrogenous constituents and by using a suitable UV lamp for the demonstration of the separated adsorption zones on the filter strip. The rapidly accumulating new results encouraged me to insert the well-known paragraph.

The relationships in DNA which probably contributed a great deal to the chemical education of biologists, are as follows. (1) $A + G = T + C$; (2) $A = T$; (3) $G = C$; and as a logical consequence of these three equations: (4) $A + C = G + T$, *i.e.*, the sum of the 6-amino compounds equals that of the 6-oxo derivatives. The last-mentioned regularity, the equality of 6-amino and 6-oxo compounds, also applies, in the absence of the other regularities, to the *total* RNA of a cell.[21] Not unrelated to this as yet unexplained finding may be later observations from my laboratory, namely, that in microbial DNA the separated heavy and light strands, although complementary to each other with respect to base composition, both exhibit the same equivalence of 6-amino and 6-oxo bases.[22-24]

To my knowledge, there have been no follow-up studies of the last-mentioned observations in other laboratories. This is, perhaps, not surprising, since in the present turmoil prevailing in molecular biology all chemical concepts appear to have been displaced. Our original findings led, of course, to the double helix proposal of Crick and Watson.

Is the title that I have chosen for this brief account justified? Did genetics, did biology in general, receive a chemical education during the period through which I have lived? To what extent did my own laboratory participate in this effort? The answer to the last question I shall have to leave to others. But as concerns chemistry supplying the foundation of the life sciences in our days, not just an underpinning, the answer is Yes. Even the most recalcitrant geneticist, even the most nebulous of immunologists, can no longer disregard the very science of substances that is chemistry. The victory may have been, however, a Pyrrhic one. In teaching them the nomenclature, we may not have taught them the skill. Compounds that had to be prepared in the laboratory, in a painful struggle for purity, now are supplied at great cost by sloppy merchants, painlessly, dirtily. For this reason, large areas of molecular biology appear to operate in a cloud of ambiguity that may turn out to be lethal.

In any event, it looks as if the smattering of chemistry acquired by biology has killed biochemistry. Does anyone care? Has the history of the decline and fall of a science ever been written? Can it be, except at a distance of many centuries?

Daily we are told about the great success of one science or the advances made by another. What is success in science, what is advance? Is there a final goal, is there such a thing as a better truth? Like everything else, the sciences are governed by the principle of change; but I should hesitate to replace the letter G in that word by the letter C, as was done by the late Jacques Monod. In fact, *Change and Inertia* would be a better title than *Chance and Necessity*. The waves roll on and break, and then there come other waves. Each is different, although they all are waves. We try to adapt ourselves to whatever force carries us at any given time. Confident of leaving our imprint on them, we are actually shaped by the wave of the moment.

When genetics was ripe for a chemical education, the educators arose in unexpected quarters. The foremost was a biologist with a deep regard for chemistry, among the others was a chemist with a great reverence for life. As is true of all teachers in our time, they reaped little honor. There will be other teachers and other ingratitudes. What they will teach, I cannot predict, for I am convinced that the wave of extreme reductionism that has been carrying us is about to spend itself.

REFERENCES

1. Avery, O. T., C. M. MacLeod & M. McCarty. 1944. Studies on the chemical nature of the substance inducing transformation of pneumococcal types. Induction of transformation by a desoxyribonucleic acid fraction isolated from pneumococcus Type III. J. Exp. Med. *79*: 137–158.
2. Dubos, R. J. 1976. The Professor, the Institute, and DNA. Rockefeller University Press. New York, N.Y.
3. Olby, R. 1974. The Path to the Double Helix. University of Washington Press. Seattle.
4. Chargaff, E. 1977. Experimenta lucifera. Nature 266: 780–781.
5. Chargaff, E. 1976. Review of Olby (Reference 3). Perspect. Biol. Med. *19*: 289–290.
6. Chargaff, E. 1971. Preface to a Grammar of Biology. Science *172*: 637–642.
7. Chargaff, E. 1978. Heraclitean Fire. Rockefeller University Press. New York, N.Y.
8. Sinnott, E. W., L. C. Dunn & T. Dobzhansky. 1950. Principles of Genetics. 4th edit. McGraw-Hill Book Company. New York, N.Y.
9. Hovanitz, W. 1953. Textbook of Genetics. Elsevier. New York, N.Y.
10. Haldane, J. B. S. 1954. The Biochemistry of Genetics. Allen & Unwin. London.
11. Srb, A. M. & R. D. Owen. 1958. General Genetics. Freeman. San Francisco.
12. Levene, P. A., & L. W. Bass. 1931. Nucleic Acids. Chemical Catalog Co., New York, N.Y.
13. Cohen, S. S. & E. Chargaff. 1944. Studies on the composition of rickettsia prowazeki. J. Biol. Chem. *154*: 691–704.
14. Chargaff, E., A. Bendich & S. S. Cohen. 1944. The thromboplastic protein: Structure, properties, disintegration. J. Biol. Chem. *156*: 161–178.
15. Consden, R., A. H. Gordon & A. J. P. Martin. 1944. Qualitative analysis of proteins: A partition chromatographic method using paper. Biochem. J. *38*: 224–232.
16. Vischer, E. & E. Chargaff. 1947. The separation and characterization of purines in minute amounts of nucleic acid hydrolysates. J. Biol. Chem. *168*: 781–782.
17. Vischer, E. & E. Chargaff. 1948. The separation and quantitative estimation of purines and pyrimidines in minute amounts. J. Biol. Chem. *176*: 703–714.
18. Chargaff, E., E. Vischer, R. Doniger, C. Green & F. Misani. 1949. The composition of the desoxypentose nucleic acids of thymus and spleen. J. Biol. Chem. *177*: 405–416.

19. Vischer, E., S. Zamenhof & E. Chargaff. 1949. Microbial nucleic acids: The desoxypentose nucleic acids of avian tubercle bacilli and yeast. J. Biol. Chem. *177*: 429–438.
20. Chargaff, E. 1950. Chemical specificity of nucleic acids and mechanism of their enzymatic degradation. Experientia *6*: 201–209.
21. Elson, D. & E. Chargaff. 1955. Evidence of common regularities in the composition of pentose nucleic acids. Biochim. Biophys. Acta *17*: 367–376.
22. Karkas, J. D., R. Rudner & E. Chargaff. 1968. Separation of *B. subtilis* DNA into complementary strands, II. Template functions and composition as determined by transcription with RNA polymerase. Proc. Nat. Acad. Sci. U.S.A. *60*: 915–920.
23. Rudner, R., J. D. Karkas & E. Chargaff. 1968. Separation of *B. subtilis* DNA into complementary strands, III. Direct analysis. Proc. Nat. Acad. Sci. U.S.A. *60*: 921–922.
24. Karkas, J. D., R. Rudner & E. Chargaff. 1970. Template properties of complementary fractions of denatured microbial deoxyribonucleic acids. Proc. Nat. Acad. Sci. U.S.A. *65*: 1049–1056.

DISCUSSION OF THE PAPER

Edsall: I would supplement Dr. Chargaff's remarks about Avery by mentioning another item. I happened to note in looking at the list of recipients of the Passano Award for medical research that Avery received that award in 1949. The citation described his earlier work on the various specific carbohydrates of various types of pneumococci, which was certainly work that well deserved such an award. However, the citation made no mention of his work on DNA; yet this was in 1949, about six years after the DNA paper was published.

Olby: I cannot really stay sitting down in the face of the very provocative remarks Prof. Chargaff has made about recent history. Perhaps I should not try to say that one attempts to write living history, but I would like to take issue with you over the question of recent history on the following grounds.

First, I think we have behind us now a large number of examples of historical interpretations of older science which, in the light of what is now revealed, display a deplorable misinterpretation of the way in which major discoveries were made, for instance, not only in the case of Lavoisier, but certainly in the case of Darwin, where we are only just beginning to come to grips with the sort of intellectual challenge and the state of information in which he made his developments; we are only beginning to come to grips with what it was like in the 1830s because of the availability now of so many of the notebooks which hitherto had not been brought together in Cambridge. What I would like to suggest therefore is that confining ourselves to much earlier periods does not necessarily guarantee that we will write better history of science.

The second point I would make is that if you are only presented with the documents from a period some way back there may be certainly in some cases a greater tendency to adopt a presentists standpoint, that is to say, to interpret the discoveries of earlier periods in the light of the successes that have subsequently been recognized and have led to important developments. And this whig interpretation of history, as it is referred to, is something which one ought to be able to guard against if one speaks with people who actually made those discoveries. I do agree that this is by no means a guarantee, but I do think one has a better chance if one actually can speak with the people.

The last point is that Prof. Chargaff has suffered from a number of interviewers, including myself no doubt, and I must put the selfish point of view on this, namely, that to get into the publications of field that you have not been trained in it is a tremendous advantage to have discussion with the person that did the work. I must say what a tremendous help this has been to me and to many others I know who would be tempted to do the same type of thing. And again I feel that this is an advantage for writing recent history.

I would have thought that Prof. Chargaff would agree on this point. I would like to know that facts in science of course are very dependent upon the circumstances in which they are viewed and interpreted, and that although the scientist is seeking the truth of course, the truth is not necessarily as easily attained. The hard so-called facts of science are not so easily to be assured of as perhaps many empiricist scientists have in the past assumed. One only has to take examples from the facts about the genetic determination resulting from individual nucleotides changes. How were those facts seen in the 1950s and how are they seen now? What seem to have been hard, good solid facts are not considered so at the present time.

In a way we are both in glass houses on this. Your position I am sure is a stronger one than the historian's, but we are not the only people who suffer from the relativism of the approach to truth.

FRUTON: To sum up what has just been said, there is nothing more permanent than a theory, and there is nothing more temporary than a fact.

HOTCHKISS: I would like to just comment on a couple of facts. Dr. Edsall has quoted one of Avery's medals, a late one in which his work on transformation was not mentioned, but the actual citation I made to Henry Dale was at the time of the awarding of the Copley medal in England in 1946 and that is when he said that Avery had done something like having discovered the gene in solution.

Another point refers to Dr. Chargaff's experience with the *Journal of Biological Chemistry*. My own paper on paper chromatography was also rejected with a comment I considered invidious. It said, "Yours is the most strange way of reporting optical data I have ever seen."

There was a curious overlap in the work of Dr. Chargaff and myself in that his first paper utilized the mercury salts of the purines and the solvents were picoline and quinoline, having themselves a high ultraviolet absorption. It did not seem likely that he planned doing quantitative analyses by ultraviolet. So it was until I knew he was going forward very thoroughly that I continued developing my own method. I was working somewhat single-handedly; I would have needed a steamroller and hard hat to go on after he started.

But this question of selective sources that he brought up himself seems to me to have weakened one of his own points. He looked in classic genetic textbooks for the evidence of the effect of the Avery work. Now this would be something like the last place to look for a reduction to practice. The transistor, which had the most explosive history in our world today, took 10 to 12 years to become reduced to practice. Geneticists you see were teaching the lovely (and I submit it is a lovely development) formal science of genetics, which is the only formal biology we had for years and years. They were repeating the lovely development of Mendel and Morgan and were content with this. So it was a long time before those textbooks, which essentially are the lecture series of the various school teachers, showed any need for the new material.

But if you looked in the bacterial genetics compilations, the bacterial genetics volumes that began to appear right in the 1940s, you would find all of this work reported.

HAUROWITZ: Dr. Chargaff mentioned that the application of chemistry, or if you want, the molecular aspects in biology, came late. It is strange that in the Austrian-Hungarian monarchy all the medical faculties had departments of biochemistry, whereas in Germany only two of the 20 universities had departments of biochemistry: Tübingen and Leipzig. I was wondering why this happened?

It is interesting because of the large number of biochemists among the Nobel laureates in the Austrian-Hungarian Universities. I mention here only Otto Loewi and Fritz Pregl in Graz, Hans Fischer in Wien (later in Munich), and Albert Szent Györgyi in Hungary (Szeged).

HOTCHKISS: I think it is really very astonishing that the idea that the nucleic acids are the determinant in genetics encountered great resistance apparently in interesting the chemists and biochemists. I am a survivor of the old-time biochemistry and I remember the end of the 1920s everybody was convinced that nucleic acids play a decisive role in genetics because this was a massive phenomenon, so to speak, from a purely morphological point of view, and nobody doubted that this substance, which is present in such a well-defined form within the living cells and which plays such a tremendous role in cell division, has something to do with the transmission of the inherited material.

I believe that this shows only that to get through with a new idea in science you must show the people that you can do something about it. It took such a great time to get through to everyone the importance of Avery's discovery. For quite a few years, so to speak, people did not know very well what to do about it. This is a purely pragmatic attitude, which is a very dominant feature in science. And, therefore, it is not very surprising that it took some time.

I am somewhat interested in the question of discovery, which seems to come in here a bit. I often think of this little allegory: Suppose my uncle in New Hampshire, who, let us say, is a chicken farmer, says, "I think there may be life on Mars." And then suppose that in later years it turned out there was life on Mars. Now, was he a discoverer because he was the first person to have issued that suggestion? I guess we would say no. But suppose the life on Mars turned out to be chickens. Do we now say that he, being an expert, was the kind of person who had the right to make that prediction? I think we again know something about the answer.

Now I have a case like this in mind. I followed Fraenkel-Conrat on a platform supported by The New York Academy of Sciences, published in 1957, under the sponsorship of the National Science Foundation. The topic was viruses and I was asked to discuss Fraenkel-Conrat's paper. I went into careful thought about whether his RNA demonstrations were adequate demonstrations of the transfer of a viral property. At the end I suggested that part of the dilemma we were facing was that RNA and DNA viruses were always made in cells that contained both RNA and DNA:

"Perhaps the confusing relations between RNA and DNA may be eliminated by the speculation that, as a genetic determinant, RNA was replaced during biochemical evolution by the more metabolically stable DNA. Cell lines have preserved the RNA entities which evolutionwise were primary to DNA and may have allowed them to store their information in DNA and thereby become subservient to it metabolically.

"This secondary position, which in a sense has been forced on RNA because of its lower stability and perhaps because of its failure to become organized in the double helix . . . ."

Viruses, as products of retrograde evolution by loss of function, may have had the choice of either RNA or DNA to get themselves made in the ample environment of the host cell. What is hidden in there is a toppling of the central dogma, that RNA makes DNA. And we now can say that we know about reverse transcriptase. So I was predicting. Did I make a discovery? No, of course not. First of all I did not sway the world; I did not influence anybody with this thought. Second, it was a kind of friendly speculation; it did not carry any experimental data. But it bears upon the very topic of what does constitute something useful because of course you cannot ever be sure that it did not slightly influence somebody. As a final last thought I would suggest that what we have when we have all of the necessary data we call a demonstration; and when we do not have all the necessary data we call it a theory.

Howard K. Schachman was trained as a chemical engineer at M.I.T. and then turned to physical chemistry at Princeton University where he received his Ph.D. in 1948. At that time he joined the Department of Biochemistry and the Virus Laboratory at the University of California, Berkeley; for the period from 1969 to 1976 he was Chairman of the Department of Molecular Biology and Director of the Virus Laboratory. His research for many years was focused on the development of ultracentrifugal methods for the study of biological macromolecules and their interactions. More recently his laboratory has been engaged in the study of allostery with the regulatory enzyme, aspartate transcarbamoylase.

Summary Remarks: A Retrospect on Proteins

HOWARD K. SCHACHMAN*

Department of Molecular Biology and Virus Laboratory
Wendell M. Stanley Hall
University of California
Berkeley, California 94720

I AM PLEASED THAT Doctors Srinivasan and Fruton invited me to participate in this "Retrospect on Proteins," thereby giving me an opportunity to hear the historical accounts from scientists who were directly engaged in many of the discoveries. It is evident from some of the discussion, as well as the talks, that it is difficult to understand the process of discovery. Often we do not know how discoveries were made or why researchers did certain experiments. In some instances we have difficulty identifying individuals responsible for discoveries. Perceptions of the discovery process vary so much in terms of psychological and sociological factors that understanding how discoveries were made may be more difficult than making the discoveries themselves.

What were our views of protein structure, synthesis and function 50 years ago and what do we know about these subjects today? How many of the questions and controversies of the past have been resolved? How was our knowledge acquired? What factors led to rapid progress in some areas and what were the impediments in others? Before confronting these difficult questions let us consider a simpler one. Has the scope of research on proteins changed over the past 50 years? A brief look at the funds available for the support of research provides an answer, although it may suffer from inaccuracies. About 10 years ago (without any data) I contrasted the average yearly grants for research on proteins and nucleic acids and my summary is shown in FIGURE 1. In the 1930s and 1940s protein chemists received annual grants of about $30,000, and nucleic acid biochemists received what they deserved, namely nothing. By 1953 this situation had changed radically as the support for research on nucleic acids rose dramatically and there was a concomitant decrease in the number of dollars available to protein chemists. A perturbation occurred

* This paper was prepared during the author's tenure as a Scholar-in-Residence at the Fogarty International Center, National Institutes of Health, Bethesda, Maryland 20014

0077-8923/79/0325-0363 $01.75/0 © 1979, NYAS

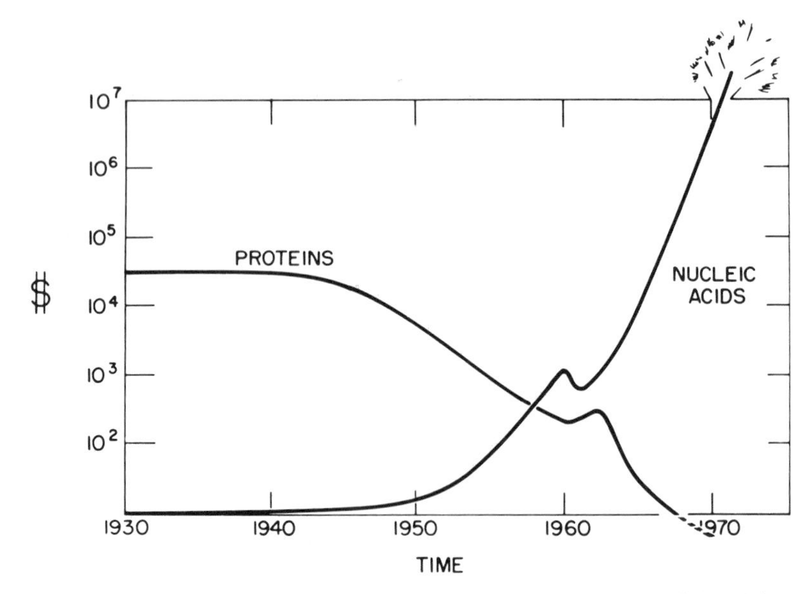

FIGURE 1. Financial support for research on proteins and nucleic acids expressed in terms of average yearly grant in dollars as a function of time.

about 1960 which gave the protein researcher renewed (if short-lived) hope. By 1962 the trend of the late 1950s was restored with ever increasing amounts of money being devoted to research on nucleic acids.

The historians may wish to investigate the anomalous funding pattern in 1960 but my own advice would be to ignore it. It may be attributed to the surge of research on protein biosynthesis and the prompt recognition that such investigations were the domain of the nucleic acid biochemist (there are other theories which I can propound but they depend on whom I wish to insult). The issue of importance in 1967 when I prepared this figure (without the constraint of data) was simply: Would the funds for research on proteins asymptotically approach zero or would the curve actually go through zero? As you can see from the curve, our "computer-based" extrapolation for the late 1960s indicated that all of the funds would be given to the nucleic acid workers so that their activity could be maintained at a level to which they had become accustomed.

Although this apocryphal portrayal of the history of protein research from 1930 to the present proved unduly pessimistic I though the trend would give some clues about the progress and the dynamics of the field. Many young biochemists, who as yet show little interest in the history of their subject and are noticeably absent, may feel that the "action" today is not in protein research. Perhaps that is why there has been a symposium here on the history of protein research.

Let us now try to be serious (at least for a short time) and examine some of the answers to basic question that have troubled protein biochemists during the past 50 years.

Yes, proteins do have definite molecular weights. To be sure, values for a given protein fluctuated widely for a period of time as theories were sharpened, pitfalls were overcome, experimental techniques became more accurate, and preparations of proteins were improved. Indeed, during this period of changing values we had to contend with skepticism (or perhaps cynicism), which can be paraphrased as "the molecular (or particle) weight of tobacco mosaic virus must be somewhere between zero and infinity." Though that opinion has stood the test of time without requiring revision, it has not proved very enlightening. Much of the progress in this field stems from the pioneering efforts of Svedberg and his collaborators through their invention, development, and application of the ultracentrifuge. I need not dwell further on this subject since this work and the demonstration of the remarkable homogeneity of many proteins have been discussed here by Doctors Williams and Edsall. This work in Svedberg's laboratory and that conducted by Standinger and his group led to the acceptance of the concept of macromolecules, which belatedly replaced that advocated by the colloid chemists. To be sure, Svedberg's proposal that proteins were composed of a basic unit with a molecular weight of 17,000 proved to be an oversimplification. It is striking, nonetheless, that our present view of the subunit structure of larger proteins was already evident in the 1930s as a result of research with the ultracentrifuge.†

Proteins also have definite shapes. For many years beginning in the 1930s physical chemists utilized hydrodynamic theories and experimental data from ultracentrifuge, diffusion, and viscosity studies to calculate the shapes and sizes of proteins in terms of axial ratios for rigid ellipsoids of revolution. Although the estimates varied with time, it was a major triumph when the predictions of the size and shape of tobacco mosaic virus were largely confirmed when the particles were observed

† During the glorious period from 1927 to the present the ultracentrifuge was an indispensable tool for studying proteins (and nucleic acids). When different molecular weights for a given protein were obtained by various authors, or even the same author at different times, I used to dub this marvelous instrument, the "ultrasubterfuge." Now as I look at the young investigators at a typical university seminar, I realize that many of them have not even seen an ultracentrifuge, to say nothing about using it. Hence today I describe the ultracentrifuge as a gadget into which you pour Sephadex at one end and watch polyacrylamide and molecular weights come out the other end. That's history!

directly by electron microscopy. Indeed, frictional coefficients and intrinsic viscosities were used very early to distinguish between native proteins as compact, globular macromolecules and denatured proteins as swollen, flexible, unfolded, and perhaps elongated particles. So frequent were our calculations of the axial ratios of protein molecules that I summarized the various conformations some years ago as shown in FIGURE 2. The compact globular protein with a defined secondary and tertiary structure is shown there along with the denatured unfolded polypeptide chains with various degrees of swelling caused, for example, by electrostatic repulsion. Included in my exhibition are two of the prize-winning models, the α helix and the myoglobin molecule.

While pointing to these landmarks in the history of protein chemistry I can't resist emphasizing the different ways in which important knowledge was obtained from x-ray crystallography. On the one hand there was the direct effort of Kendrew, Perutz, and their colleagues to deduce the structure of two proteins. In contrast we can summarize the approach of Pauling and Corey that the way to study proteins was not to examine proteins. Instead they advocated and pursued a searching investigation of the small molecules that are the building blocks of all proteins in order to discover the general rules or laws that govern the folding of polypeptide chains. Both views have triumphed and we are richer today because both routes were followed. The historians may find it of interest

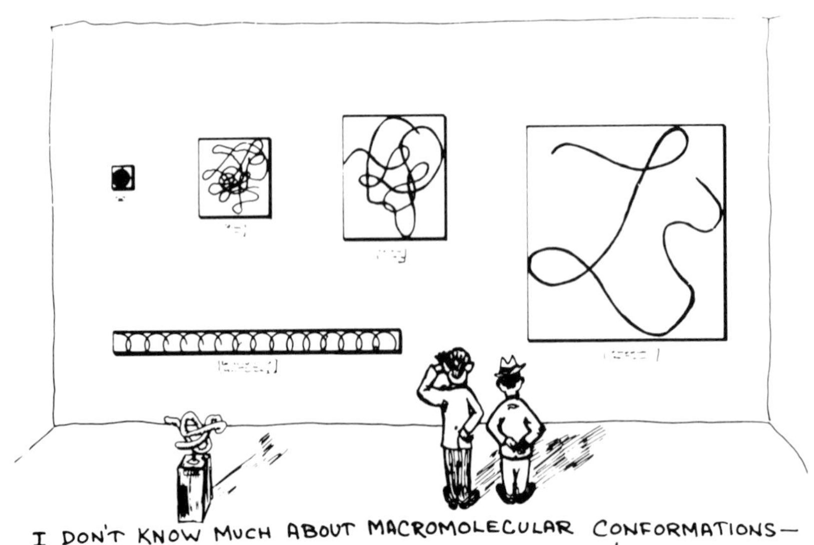

I DON'T KNOW MUCH ABOUT MACROMOLECULAR CONFORMATIONS—
BUT I KNOW WHAT I LIKE!

FIGURE 2. Macromolecular conformations of native and denatured proteins.

to ponder what would have happened to protein crystallography had the original investigations not been conducted on myoglobin and hemoglobin, which have such high contents of α helix. If progress had been slower because rod-like segments weren't detected at a relatively early stage and if the investigators had not derived considerable satisfaction and encouragement in finding these structural features, might they have given up and would the history Doctor Hodgkin presented have been very different?

Although most of us take on faith the exquisite knowledge of protein structure deduced from crystallographic studies, we should recognize that x-ray diffraction has had its skeptics in the past. This is illustrated by the following picturesque criticism[1] of Bragg's interpretation of the x-ray diffraction pattern of crystals of sodium chloride:

> Professor W. L. Bragg asserts that "In sodium chloride there appear to be no molecules represented by NaCl. The equality in number of sodium and chlorine atoms is arrived at by a chess-board patterns of these atoms; it is a result of geometry and not of a pairing-off of the atoms."
>
> This statement is more than "repugnant to common sense." It is absurd to the *n* . . . th degree, not chemical cricket. Chemistry is neither chess nor geometry, whatever x-ray physics may be. Such unjustified aspersion of the molecular character of our most necessary condiment must not be allowed any longer to pass unchallenged. A little study of the Apostle Paul may be recommended to Professor Bragg, as a necessary preliminary even to x-ray work, especially as the doctrine has been insistently advocated at the recent Flat Races at Leeds, that science is the pursuit of truth. It were time that chemists took charge of chemistry once more and protected neophytes against the worship of false gods: at least taught them to ask for something more than chess-board evidence.

As Doctor Hodgkin suggested, "it is fascinating to look back at the earlier observations and speculations. . . ."

Just as we observed the accumulation of evidence that proteins have a definite size and shape so also have we witnessed numerous demonstrations that they have well-defined primary structures. Investigators today doubtless find it difficult to understand why earlier workers questioned whether proteins had a unique amino acid sequence. Before this question was answered clearly by Sanger we were subjected to a theory based on inadequate data, a theory based on inaccurate data, and even a theory which neglected or disregarded all existing data. The various theories have been discussed by Doctors Fruton and Smith and we have heard from Doctor Gordon about the important roles chromatography and electrophoresis played in the determination of the primary structures of proteins. As Doctor Haurowitz pointed out, protein chemists 25 years ago were troubled about the microheterogeneity of proteins. Now we no longer face evidence of inhomogeneity with such foreboding. Instead when we encounter microheterogeneity we think more in physiological terms and consider proteolysis or posttranslational modification such as glycosylation, nucleotidylation, and phosphorylation.

On several occasions during this symposium we heard of the early work of Anson, Mirsky, Pauling, and Wu dealing with the denaturation of proteins and its reversal. There was a substantial amount of research in the 1930s showing that unfolded polypeptide chains produced by the denaturation of proteins could refold to give structures having the properties of the native proteins. But despite the beautiful experiments, we were not prepared to accept the view, now so widely adopted, that the primary structure dictates the secondary, tertiary, and even quaternary structures of biologically active proteins. According to the physical chemists the entropy change in the conversion of randomly coiled polypeptide chains into a unique three-dimensional architecture was too unfavorable. Indeed, workers at that time who claimed success in the renaturation of proteins may have been dismissed with the comment, "the protein wasn't completely unfolded and there were still regions of folded structure about which the remainder of the chain could fold correctly." Now we have come full circle. As a result of the research on ribonuclease by Anfinsen and his colleagues and the related studies on other proteins we have adopted the opposite position: "If you can't succeed in regenerating active proteins from the unfolded chains, you are a poor experimentalist."

Although we are prepared to accept the view that proteins have well-defined secondary, tertiary, and quaternary structures maintained by many noncovalent interactions, few of us can furnish a satisfactory theoretical explanation of the forces responsible for their architectures. There has been little discussion here of "the rise and fall" of the hydrogen bond as the *sine qua non* of protein structure. For many years it was fashionable to attribute the stability of native proteins to that bond. But the widely accepted stabilization energy was shown to be illusory when the effect of H_2O was recognized. Accordingly we resorted to the hydrophobic bond and now that fashion is having its difficulties. Perhaps 50 years from now the participants in a symposium on the history of proteins will express astonishment that their predecessors had so much trouble in the 1970s understanding how and why protein molecules achieve their three-dimensional structures.

Largely as a result of the research of Linderstrøm-Lang we now think of proteins as relatively flexible, dynamic molecules which breathe and have motility. Indeed, the phrase "conformational change" so dominates our thinking that whenever we do not understand complex biological interactions involving proteins we postulate that a conformational change has occurred. Some day we will understand (as we are beginning to in the case of hemoglobin, thanks to Perutz) whether the change is direct and local, i.e., restricted to the site of interaction, or whether it is indirect and gross, i.e., propagated throughout the entire protein molecule. In the interim we might profit from looking back at a very early criticism[2] of a proposal dealing with conformations‡:

> Not long ago I expressed the view that the lack of general education and of thorough training in chemistry of quite a few professors of chemistry was one of the causes of the deterioration of chemical research in Germany. A consequence of this lamentable state of affairs is the spread of the weed, the seemingly learned and profound but in fact trivial and superficial speculative philosophy. Fifty years ago this kind of philosophy was eradicated by the advance of precise science, but now pseudo-scientists fetch it back from the limbo of human errors. Like an old whore, it is given a new dress and a lot of makeup and smuggled into polite society, where it does not belong. Will anyone to whom my worries may seem exaggerated please read–if he can–a recent memoir by Herr van't Hoff on "The Arrangements of Atoms in Space," a document crammed to the hilt with the outpourings of a childish fantasy. This Dr. J. H. van't Hoff, employed by the Veterinary College at Utrecht, has, so it seems, no taste for accurate chemical research. He finds it more convenient to mount his Pegasus (evidently taken from the stables of the Veterinary College) and to announce how, on his daring flight to Mount Parnassus, he saw the atoms arranged in space.
>
> It is typical of the present times–uncritical and even anticritical–that a virtually unknown chemist from a Veterinary College arrogates to himself making pronouncements on one of the ultimate problems of chemistry, that of the arrangement of the atoms in space, which may never be solved and to supply a solution to this problem with a sureness and audacity, nay impudence, which can but amaze the genuine scientist.

If van't Hoff had so much trouble with the tetrahedral carbon atom, is it any wonder that we today have difficulty in being precise about the meaning of "conformational change"?

Much of the progress that I have sketched is attributable to the development of elaborate, sensitive, accurate techniques which permitted workers to entertain questions that earlier seemed beyond attack. It was of great interest, therefore, to hear Doctor Holmes describe the origins and setting for the investigations of protein metabolism in the 1890s.

‡ The English translation by Krebs[3] was characterized by him as "a somewhat pale reflection of the far more vitriolic German text."

These studies and the later ones by Rose, which were described by Doctor Carter, illustrate how much progress could be achieved without the sophisticated tools available today. The recognition of the role of amino acids in nutrition and the detection of those that were essential constitute a brilliant chapter in the history of protein metabolism. This remarkable research based on feeding experiments was extended in much greater detail when it became possible to determine the fate of dietary amino acids through the use of isotopic tracers. Doctor Ratner, in her summary of the elegant studies of the Schoenheimer-Rittenberg group (to which she and Doctor Shemin contributed greatly) described how much advanced, detailed planning was required for each experiment in the early days of research with amino acids labeled with deuterium and ^{15}N. How these isotopes were exploited successfully in detailed metabolic investigations was illustrated by Doctor Shemin who took us on a tour with glycine. I found it striking in the historical accounts of Doctors Carter and Ratner how the researches of Rose and Schoenheimer and their respective colleagues complemented one another. Clearly, the concept that cellular proteins undergo continued turnover by synthesis and degradation represents one of the landmark ideas in biochemistry.

And how are proteins synthesized? This fascinating story was summarized by Doctor Zamecnik. He took us through the original ideas based on the reversal of proteolysis through the explosive period of crucial discoveries of amino acid activation, transfer RNA, and the ribosome as the site of peptide chain initiation and elongation, to the demonstration of messenger RNA as the vehicle for dictating the sequence of amino acids in the newly synthesized protein. How complicated it seemed for a while. But then how deceivingly simple it appeared as our knowledge increased. As Doctors Zamecnik and Horowitz pointed out, this history is far from complete since discoveries of the past few years make it clear that we have much to learn about gene overlaps and intervening DNA sequences. Doctor Horowitz's historical account of the gene-protein relationship leading to the "one gene–one enzyme" theory began in 1902 with the findings of Garrod, which were ignored for so long until the classical studies of Beadle and Tatum. The belated recognition of the role of nucleic acids as genetic determinants was described in detail by Doctors Hotchkiss and Fraenkel-Conrat in their presentations of the research on transformation and tobacco mosaic virus.

Looking back at the evolution of our knowledge of proteins over the past 100 years has been worthwhile even though we have not discovered any general or unique formula which describes how we learned what we know. On the contrary, it appears that our understanding of protein structure, functions, biosynthesis, and metabolism was acquired in diverse ways.

In some instances newly developed instruments and techniques permitted workers to do experiments that provided interesting results leading to other fields of research activity. To illustrate this point I cite the impact of isotopes in the demonstration of turnover in proteins, of chromatography in the determination of the structure of proteins, and of the ultracentrifuge in studies of the homogeneity, size, shape, and subunit composition of proteins. Often the tool that proved so powerful was developed for reasons having little to do with curiosity about proteins.

In other instances the investigators were motivated by specific concepts and the research followed from the ideas. Even when the hypotheses or proposals were found later to be incorrect, significant progress resulted. It is relatively unimportant, for example, that the energy of stabilization of proteins due to hydrogen bonds was overestimated by a large amount. What is important is that the inflated value coupled with other factors was used brilliantly in fashioning the concept that the folding of the polypeptide backbone in proteins was a crucial factor in determining the architecture of proteins. From this idea came the α helix. Perhaps the historians would like to speculate how our understanding of protein structure would have progressed if the role of water in decreasing the stabilizing effect of hydrogen bonds had been appreciated earlier. Might it not also be of interest to consider where we would be if the relative simplicity of the folding of the backbone had been neglected and instead all intellectual efforts had been focused on the more complicated problem of the effect of the side chains.

During the discussion of the history of protein synthesis we saw how difficult it is to obtain a precise and accurate picture of the process whereby knowledge was acquired. Why did certain workers do specific experiments? Why did not they or other workers recognize the impact of certain findings and follow leads that a little later seemed so obvious? The history of science is full of examples of valuable observations that were unexploited for relatively long periods. Frequently we can cite good reasons for the delay in accepting or appreciating a new concept. Other times, as we look back, we have difficulty in rationalizing the delay. We haven't spent much time here examining retrospectively the converse situation wherein a proposal was accepted too rapidly. Are there examples where progress in protein biochemistry was seriously impeded because of too rapid an acceptance of claims or theories? In the process of discovering do we suffer more from too rapid an adoption of new ideas than from their belated acceptance?

It is important to recognize the impact of individuals, groups, schools, and even countries in influencing the pattern of scientific studies.§ For many years great contributions in the fields of electrophoresis and chromatography stemmed from relatively small research groups in Uppsala. We can also cite the magnificent research on proteins in the Carlsberg Laboratory under the leadership of Kjeldahl, Sörensen and Linderstrøm-Lang. To a remarkable extent, x-ray diffraction studies on proteins were concentrated in England, and investigations of the dynamic state of proteins were the result of the group at Columbia University.

Although at this symposium we were asked to look back, to summarize what we have learned, to attempt to understand how we learned it, and to clarify the factors that contributed to (or even interfered with) the learning process, I would like to spend these closing remarks in looking forward. What can be expected in a symposium on the history of proteins to be held 50 years from now? Will there be talks on amino acid analyses, sequence determinations, structures from x-ray diffraction, dynamics of chains from nuclear magnetic resonance spectroscopy, folding of polypeptide chains to give proper tertiary structures, specific association of subunits to form regulatory enzymes, interactions of ligands with receptors, and control mechanisms for the synthesis and degradation of proteins? Or will the symposium neglect all these topics because they ceased being of interest? Already some advocate determining amino acid sequences in proteins from the analysis of the nucleotide sequences in the appropriate DNAs. Others are claiming that the secondary (if not the tertiary) structure of proteins can be predicted from the primary structure. Thus, to continue with the fantasy, all we need do is synthesize the appropriate DNA with allowances for insertions in the nucleotide sequence and proper reading frames in order to produce a stable enzyme needed to remedy some metabolic defect.

And who will be at the symposium? Will there be more historians than scientists? Will the scientists be younger and the historians older? How many scientists will be working on proteins? How will their research be supported? The answer to this last question is likely to be of paramount importance since the history of proteins in the next 50 years will depend on the research performed. Hence it is appropriate that my closing remarks (like my opening comments) be directed toward funding patterns. Most investigators concede that the probability of receiving financial support is related to the originality of the grant request. But some consider the relationship to be perverse or bizarre. According to this latter view, summarized in FIGURE 3, grant requests showing no origi-

§ The sociologists might wish to examine ethnic factors which may have influenced some young talented individuals to become biochemists in the 1930s and 40s.

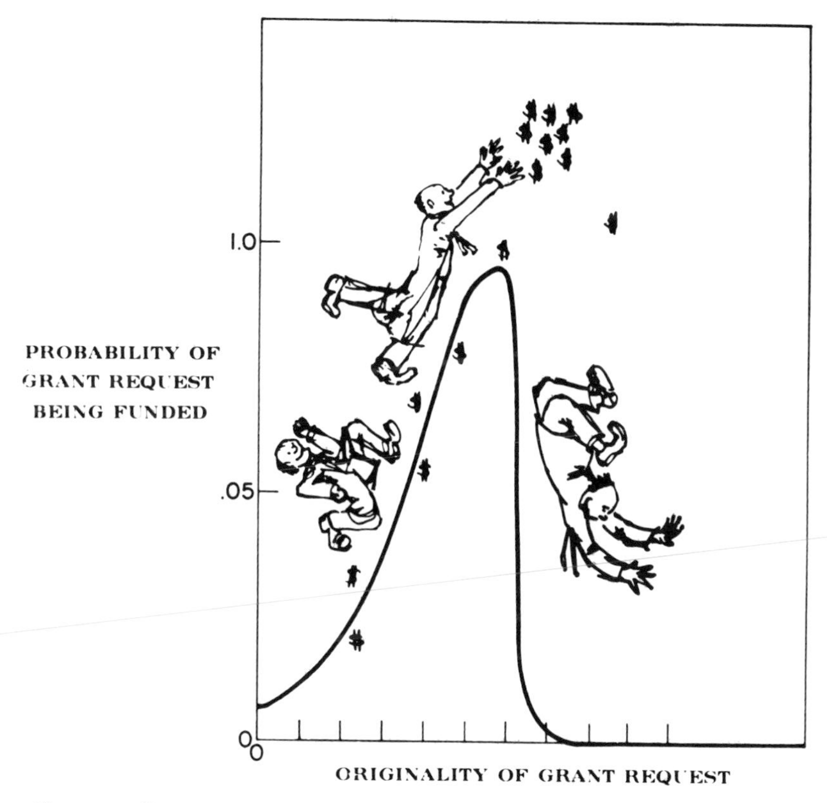

FIGURE 3. Prospect of grant request being funded in relation to the originality of the research proposal.

nality have a small, but finite, probability of being funded. These cynics will concede that a little originality is good. Therefore, as seen by the gradual increase in the curve, the prospects of financial support are enhanced as originality increases. But only to a certain level. As these pessimistic (and perhaps immodest) critics say: "Our proposal was so original that the reviewers could not understand it; no wonder we didn't receive any funds."

Clearly the history we learn depends on who is the historian.

REFERENCES

1. ARMSTRONG, H. E. 1927. Nature *120*: 478.
2. KOLBE, H. 1877. J. Prakt. Chem. *15*: 473–477.
3. KREBS, H. A. 1966. *In* Current Aspects of Biochemical Energetics. N. O. Kaplan & E. P. Kennedy, Eds. :83–95. Academic Press. New York, N.Y.

Author Index

(Italicized page numbers refer to comments in Discussions)